MICROWAVE-ASSISTED ORGANIC SYNTHESIS

A Green Chemical Approach

MICROWAVE-ASSISTED ORGANIC SYNTHESIS

A Green Chemical Approach

Edited by

**Suresh C. Ameta, PhD, Pinki B. Punjabi, PhD,
Rakshit Ameta, PhD, and Chetna Ameta, PhD**

Apple Academic Press

TORONTO   NEW JERSEY

Apple Academic Press Inc.	Apple Academic Press Inc.
3333 Mistwell Crescent	9 Spinnaker Way
Oakville, ON L6L 0A2	Waretown, NJ 08758
Canada	USA

©2015 by Apple Academic Press, Inc.

First issued in paperback 2021

Exclusive worldwide distribution by CRC Press, a member of Taylor & Francis Group
No claim to original U.S. Government works

ISBN 13: 978-1-77463-355-7 (pbk)
ISBN 13: 978-1-77188-039-8 (hbk)

Library of Congress Control Number: 2014954051

Library and Archives Canada Cataloguing in Publication

Microwave-assisted organic synthesis: a green chemical approach/edited by Suresh C. Ameta, PhD, Pinki B. Punjabi, PhD, Rakshit Ameta, PhD, and Chetna Ameta, PhD.

Includes bibliographical references and index.
ISBN 978-1-77188-039-8 (bound)
1. Organic compounds--Synthesis. 2. Microwave heating. 3. Green chemistry. I. Ameta, Suresh C., author, editor II. Punjabi, Pinki B., author, editor III. Ameta, Rakshit, author, editor IV. Ameta, Chetna, author, editor

---

| QD262.M52 2014 | 547'.2 | C2014-907154-X |

Apple Academic Press also publishes its books in a variety of electronic formats. Some content that appears in print may not be available in electronic format. For information about Apple Academic Press products, visit our website at **www.appleacademicpress.com** and the CRC Press website at **www.crc-press.com**

ABOUT THE EDITORS

Suresh C. Ameta, PhD

Prof. Suresh C. Ameta is serving as Director of the Pacific College of Basic & Applied Sciences, and Dean of Faculty of Science at PAHER University in Udaipur, India. He has served as Professor and Head, Department of Chemistry, North Gujarat University, Patan (1994) and M. L. Sukhadia University, Udaipur (2002–2005), and Head, Department of Polymer Science (2005–2008). He also served as Dean, Postgraduate Studies for a period of four years (2004–2008). Prof. Ameta has occupied the coveted position of President of the Indian Chemical Society, Kolkata, and is now life-long Vice President (2002–continue) of the society. He has been awarded a number of prizes during his career, such as the national prize (twice) for writing chemistry books in Hindi, the Prof. M. N. Desai Award, the Prof. W. U. Malik Award, the National Teacher Award, the Prof. G. V. Bakore Award, and the Life Time Achievement Award by the Indian Chemical Society, among others. He has successfully guided 71 students to PhD degrees.

With more than 300 research publications to his credit in journals of national and international repute, he is also the author of around 40 undergraduate- and postgraduate-level books. He has also written chapters in books published by Nova Publishers, USA; Taylor & Francis, UK; and Trans-Tech Publications, Switzerland. He is co-editor of the book *Green Chemistry: Fundamentals and Applications*, published by Apple Academic Press, USA. He has completed five major research projects from different funding agencies, such as DST, UGC, CSIR, and the Ministry of Energy, Government of India. With experience of about 43 years of teaching and research experience, Prof. Ameta has delivered lectures and chaired sessions in national conferences throughout India and has acted as a reviewer for a number of international journals.

Dr. Ameta obtained his masters degree from the University of Udaipur, India, and was awarded a gold medal in 1970. He also secured first position in MPhil in the year 1978 at Vikram University, Ujjain (MP). He obtained this PhD degree from Vikram University in 1980.

Pinki B. Punjabi, PhD

At present Dr. Pinki B. Punjabi is serving as Associate Professor, Department of Chemistry, Mohanlal Sukhadia University, Udaipur, India, for the last 27 years. Her research interests are photocatalysis and photo-Fenton reactions. She has also contributed to microwave-assisted organic synthesis and polymer synthesis. Twenty-two research students have been awarded PhD degrees under her supervision, and six students are currently working on their degrees, and one is postdoctoral fellow. She has more than 100 research papers in journals of national and international repute. She has attended various conferences and delivered invited lectures. She has contributed to undergraduate- and postgraduate-level books. She has also written chapters in books published by Nova Publishers, USA; Taylor and Francis, UK; Trans-Tech Publications, Switzerland; and Apple Academic Press, USA.

Dr. Pinki B. Punjabi obtained her MSc, MPhil, and PhD degrees from Vikram University, Ujjain, securing gold medals in both postgraduate degrees.

Rakshit Ameta, PhD

Rakshit Ameta, PhD, is Associate Professor of Chemistry, Pacific College of Basic and Applied Sciences, PAHER University, Udaipur, India. He has 8 years of experience in teaching and research. He has published more than 50 research papers and at present is guiding seven research students for their PhD theses, and two students have already obtained their PhDs under his supervision.

Dr. Rakshit Ameta has received various awards and recognition in his career, including being awarded first position and the gold medal for his MSc and receiving the Fateh Singh Award from the Maharana Mewar Foundation, Udaipur, for his meritorious performance. He has served at M. L. Sukhadia University, Udaipur; the University of Kota, Kota; and PAHER University, Udaipur. He has over 50 research publications to his credit in journals of national and international repute. He holds one patent, and two more are under way. Dr. Rakshit has organized three national conferences as Organizing Secretary at the University of Kota in 2011 and PAHER University in 2012 and 2013. He is also planning to organize the fourth conference at PAHER University in 2014. He has delivered invited lectures and has chaired sessions in conferences held by the Indian Chemical Society and the Indian Council of Chemists. Dr. Rakshit was elected as council member of the Indian Chemical Society, Kolkata (2011-2013), and Indian Council of Chemists, Agra (2012-2014). He has written five degree-level books and has contributed chapters to books published by several international publishers.

Chetna Ameta, PhD

Dr. Chetna Ameta is presently working as Assistant Professor of Chemistry at Sukhadia University, Udaipur, India, where she was awarded her PhD degree. She was a recipient of a Junior Research Fellowship (JRF) and a Senior Research Fellowship (SRF) from the Indian University Grants Commission (UGC). With 16 research papers to her credit, she has also contributed a chapter to the book *Green Chemistry: Fundamentals and Applications*, published by Apple Academic Press, Inc. Her research interests are in microwave-assisted organic synthesis and green chemistry.

CONTENTS

LIST OF CONTRIBUTORS

Ameta Chetna
Department of Chemistry, M. L. Sukhadia University, Udaipur, India E-mail: Chetna.ameta@yahoo.com

Ameta K. L.
Department of Chemistry, Modi Institute of Technology and Science, Lakshmangarh, India, E-mail: klameta77@hotmail.com

Ameta Rajat
Amoli Organics Pvt. Ltd., Vadodara, India, E-mail : ameta_rajat@rediffmail.com

Ameta Rakshit
Department of Chemistry, PAHER University, Udaipur, India. E-mail: rakshit_ameta@yahoo.in

Ameta Suresh C.
Department of Chemistry, PAHER University, Udaipur, India, E-mail: ameta_sc@yahoo.com

Benjamin Surbhi
Department of Chemistry, PAHER University, Udaipur, India, E-mail: surbhi.singh1@yahoo.com

Bhanat Kumudini
Department of Chemistry, M. L. Sukhadia University, Udaipur, India, E-mail: kumudini23@gmail.com

Dashora Purnima
Department of Chemistry, PAHER University, Udaipur, India, E-mail: purnimadashora@gmail.com

Jain Abhilasha
Department of Chemistry, St. Xavier's College, Mumbai, India, E-mail: jainabhilasha5@gmail.com

Jangid Nirmala
Department of Chemistry, M. L. Sukhadia University, Udaipur, India, E-mail: nirmalajangid.111@gmail.com

Joshi Meenakshi
Department of Chemistry, PAHER University Udaipur, India, E-mail: meenakshijoshi909@gmail.com

Kalal Sangeeta
Department of Chemistry, M. L. Sukhadia University, Udaipur, India, E-mail: sangeeta.vardar@yahoo.in

Kothari Seema
Department of Chemistry, Pacific College of Engineering, Udaipur, India, E-mail: seemavkothari@rediffmail.com

Kumawat Poonam
Department of Chemistry, PAHER University, Udaipur, India, E-mail: poonam.kumawat41@gmail.com

Kunwar Neelam
Department of Chemistry, PAHER University, Udaipur, India, E-mail: neelamkunwar13@yahoo.com

Meghwal Kiran
Department of Chemistry, M. L. Sukhadia University, Udaipur, India, E-mail: meghwal.kiran1506@gmail.com

Panchal Shikha
Department of Chemistry, PAHER University, Udaipur, India, E-mail: shikha_dpr99@yahoo.co.in

Parsoya Priya
Department of Chemistry, PAHER University, Udaipur, India, E-mail: priyapp1987@gmail.com

Pathak Arpit
Department of Chemistry, M. L. Sukhadia University, Udaipur, India, E-mail: arpitpathak2009@gmail.com

Punjabi P. B.
Department of Chemistry, M. L. Sukhadia University, Udaipur, India, E-mail: pb_punjabi@yahoo.com

Sharma Sanyogita
Department of Chemistry, Pacific Institute of Technology, Udaipur, India, E-mail: sanyogitasharma22@gmail.com

Sharma Shewta
Department of Chemistry, PAHER University, Udaipur, India, E-mail: rakshit_ameta@yahoo.in

Solanki Meenakshi Singh
Department of Chemistry, PAHER University, Udaipur, India, E-mail : meenakshisingh001989@gmail.com

Soni Dipti
Department of Chemistry, PAHER University, Udaipur, India, E-mail: soni_mbm@rediffmail.com

Tak Paras
Department of Chemistry, PAHER University, Udaipur, India, E-mail: parastak2011@gmail.com

Tripathi Abhilasha
Department of Chemistry, PAHER University, Udaipur, India, E-mail: abhilashatripathi001@gmail.com

Trivedi Monika
Department of Chemistry, PAHER University, Udaipur, India, E-mail: monachem.01@gmail.com

Vardia Jitendra
Amoli Organics Pvt. Ltd., Vadodara, India, E-mail : jitendravardia@yahoo.com

Vyas Ritu
Department of Chemistry, Pacific Institute of Technology, Udaipur, India, E-mail: ritu24vyas@gmail.com

LIST OF ABBREVIATIONS

APIs	Active Pharmaceutical Ingredients
ATRP	Atom-Transfer Radical Polymerization
BAT	Bromoalkylthiophenes
BIDC	Benzimidazolium Dichromate
BSA	Bovine Serum Albumin
CF	Continuous Flow
CMSS	Carboxymethylated Sago Starch
CMWS	Continuous Microwave Sterilization
CNF	Carbon Nanofiber
CNTs	Carbon Nanotubes
CP	Conducting Polymers
CRP	Controlled Radical Polymerization
CTMABC	Cetyltrimethylammonium Bromochromate
CTMACC	Cetyltrimethylammonium Chlorochromate
DEAD	Diethyl Acetylenedicarboxylate
DMAD	Dimethyl Acetylenedicarboxylate
DMC	Dimethyl Carbonate
DMS	Dimethylsulphate
DYN	Dynamic
EPS	Expanded Polystyrene Waste
FAME	Fatty Acids Methyl Esters
FFA	Free Fatty Acids
F-MCRs	Fluorous Multicomponent Reactions
FNC	Functionalized Nanocarbon
FRP	Fiber-Reinforced Plastic
F-SPEs	Fluorous Solid-Phase Extractions
GQDs	Graphene Quantum Dots
HMS	Hexagonal Mesoporous Silicas
HMTA	Hexamethylentetramine
IR	Infra-Red
MAOS	Microwave-Assisted Organic Synthesis
MMF	Monomethylformamide
MMT	Montmorillonite
MPV	Meerwin-Ponndorf-Verley
MS	Molecular Sieves
MWCNTs	Multi-Walled Carbon Nanotubes

NIS	Near Infrared Spectroscopy
NMP	nitroxide-Mediated Radical Polymerization
OSHA	Occupational Safety And Health Administration
PAMAM	Polyamidoamine
PAN	Polyacrylonitrile
PANI	Polyaniline
PCNDs	Photoluminescent Carbon Nanodots
PEG	Polyethyleneglycol
PEMFCs	Proton Exchange Membrane Fuel Cells
PET	Polyethylene Terephthalate
PLSNs	Polymer Layered Silicate Nanocomposites
PMDA	Pyromellitic Dianhydride
PNC	Polymer Nanocomposites
PTFE	Polytetrafluoroethylene
RAFT	Reversible Addition Fragmentation Chain-Transfer
RCM	Ring-Closing Metathesis
RGO	Reduced Graphene Oxide
ROMP	Ring-Opening Metathesis Polymerization
SARs	Structure–Activity Relationships
SI-FRP	Surface-Initiated Free Radical Polymerization
SPS	Pulsed Power Mode
SWNTs	Single-Walled Carbon Nanotubes
TBAB	Tetrabutylammonium Bromide
TBADT	Tetrabutylammonium Decatungstate
TBAHS	Tetrabutylammonium Hydrogen Sulfate
TBATB	Tetrabutylammonioum Tribromide
TBCA	Tribromoisocyanuric Acid
TBDMSCL	Tert-Butyldimethylsilyl Chloride
TEA	Triethanolamine
TFA	Trifluoroacetic Acid
TMAFC	Tetramethylammonium Fluorochromate
TMS	Tetramethylenesulfone
TMSCL	Trimethylsilyl Chloride
UHP	Urea-Hydrogen Peroxide

PREFACE

Microwave-assisted organic synthesis has been considered better than conventional or traditional heating because this technique provides a newer approach to synthetic chemistry, which is ecofriendly. Some important advantages of this technique are a higher rate of reaction resulting in reduced reaction time, improved yield, and better quality of product. Because of these advantages, it is considered a sustainable process—fulfilling the majority of green chemistry measures. Nowadays, microwave-assisted organic synthesis is promoted due to solvent-free reactions, which avoid polluting solvents and reduction in the generation of toxic byproducts and emission of harmful gases.

Most of the reactants are overheated by the traditional way of heating, and this may result in the decomposition of the reactants, reagents and/or products, but this is not the case with microwave heating as these radiations pass through the walls of reaction vessels.

In this era of environmental consciousness, the role of chemistry and chemist is to incorporate processes and design products, which can completely eliminate or minimize the generation of pollutants in the form of waste. Consequently, any ideal synthesis or chemical approach must include this element of environmental awareness. Nowaday, a new kind of chemical revolution is brewing, that is, green chemistry.

The fundamental idea of green chemistry is that the manufacturer of a chemical needs to consider what will happen to human life after that particular chemical is generated and introduced into the society. The fate of wastes generated during the production of that chemical or after its use is also a question of major concern. These principles address several concerns, such as the use of various solvents, the amount of chemical wastes produced, the use of catalyst and reagents from quantity and reusability point of view, atom economy (maximum incorporation of atoms of reactants), the number of chemical steps (energy efficiency), the use of safer chemicals and reaction conditions, etc. Practically, it is very difficult for a new synthetic protocol to satisfy all of the 12 principles of green chemistry. The more principles a protocol satisfies, more the developed process will be considered relatively greener.

Therefore, the demands for green and sustainable synthetic methods in the fields of health care and fine chemicals, combined with the pressure to produce these substances expeditiously and the more in an environmentally benign way, pose significant challenges to the synthetic chemical community. These objectives can be conveniently achieved through the development of aqueous synthetic protocols using microwave heating. It should be noted that rapid development of green organic chemistry is due to the recognition that environment friendly processes and products will prove to be economical and efficient in the long-term as they circumvent the need for treating pollutants and byproducts generated by conventional synthesis.

Microwave-assisted heating under controlled conditions is an invaluable technology because it not only often dramatically reduces reaction time, typically from days or hours to minutes or even seconds, but it also enhances the rate of reaction and fulfills the major aims of green chemistry by reducing side reactions resulting in increased yield and improved reproducibility.

Microwave chemistry has become a central tool in this fast paced, time sensitive field, and it has blossomed into a useful technique for a variety of applications in organic synthesis where high yielding protocols and facility of purifications are highly desirable. Faced with the increasing demands of novel drug targets, there is considerable current interest to accelerate this technology associated with combinatorial and computational chemistry.

Microwave radiations have proved to be highly effective as a heating source for chemical reactions. They provide rapid and homogeneous heating, which has certain advantages, such as reaction rate acceleration, milder reaction condition, and higher chemical yield. In short, microwave enhanced chemical reactions are safer, faster, cleaner, and more economical than conventional reactions. They help in developing cleaner and greener synthetic routes. However, at present, they are also associated with some limitations like scalability and applicability, which will be overcome in years to come.

—**Suresh C. Ameta, PhD, Pinki B. Punjabi, PhD,**
Rakshit Ameta, PhD, and Chetna Ameta, PhD

INTRODUCTION

Production of different chemicals on large scale based on the varied needs of the society has created environmental pollution. The world is facing a serious outcome of this action because of byproducts and improper disposal of wastes. Green chemistry has come to our rescue at this stage. Most of the organic syntheses require toxic solvents, more reaction time, drastic conditions of temperature, etc., and, therefore, conventional methods of organic synthesis are less preferred these days. Microwave-assisted organic synthesis enters the scene here. This technique is considered a green chemical approach as it reduces reaction time from days or hours to minutes or, some time, to even seconds along with reduction of side reactions, increased yields, using less solvents or almost solvent-free, solid supported reactions, improved purity, etc. The main focus of this book is on various reactions like cycloaddition, rearrangement, elimination, substitution, oxidation, reduction, condensation, coupling, polymerization, nanomaterials, synthesis of heterocycles, industrial applications, etc., under microwave irradiation. The time is not far off when this methodology will almost replace the existing and cumbersome methods of organic synthesis.

CHAPTER 1

INTRODUCTION

SURESH C. AMETA

CONTENTS

1.1 HISTORY

The history of microwave heating is quite interesting. In 1946 during Second World War, Dr. Percy L. Spencer was experimenting on Magnetron in radar research. He was surprised to observe that a candy bar in his pocket melted while working near magnetron. Out of curiosity, he tried another experiments and placed some popcorn kernels near the vacuum tube (Magnetron). He could see the popcorn sputtering, cracking and popping in the laboratory. Then he carried out further experiments with an egg, when egg began to tremor and quake. One of his colleagues had a closer look, suddenly egg exploded and hot yolk splashed on his face. These all observations indicated towards generation of high temperature during exposure to microwaves. Immediately, an idea flashed into his mind that when candy bar, popcorn and egg can be heated to this extent, why not some other materials? This can be considered the real beginning of field of applications of microwaves in kitchens and later on in chemical synthesis. The first microwave oven was produced by Raytheon Corporation in 1954 and was named as 1161 Radarange. The use of microwave oven in kitchen for cooking any food and other materials has its advantages like rapid, homogeneous and inner core heating along with its own disadvantages.

Long back, it was observed that water is heated dramatically on exposure to microwaves. Then this interesting properties of heating materials by microwaves was used for cooking food materials and the domestic microwave oven came into existence (in late forties). For some years, these microwave ovens were used for heating and cooking foods and were limited to kitchens only. First time, microwaves were used to analyze moisture in solids and a few years later, it was used for drying different organic materials. Then microwave radiations were used for different processes of chemical analysis like extraction, digestion, ashing, etc.

It was Gedye et al. (1986) and Giguere et al. (1986), who made an attempt to use microwaves for chemical synthesis. This was the beginning of microwave-assisted organic synthesis (MAOS). Microwave radiations have been used for inorganic reaction since late seventies, even earlier than their use in organic synthesis. However, the potential of microwave heating for organic synthesis attracted interests of synthetic chemists soon after the first report appeared in late eighties.

Thereafter, various modifications were made in the microwave instrument and number of publications in these reactions went on increasing day by day. In almost last three decades, a wide variety of compounds have been synthesized using microwaves and it has enriched the fields of pharmaceutical drugs, nanoparticles, biomaterials, polymers, etc.

Microwave-assisted organic synthesis has been considered better than conventional or traditional heating. This technique provides a new approach in synthetic chemistry, which is ecofriendly. Some important advantages of this technique are higher rate of reaction resulting in reduced reaction time, improved yield and better quality of product. Because of these advantages, it is considered a sustainable process-fulfilling majority of green chemistry measures. Nowadays, this method of organic synthesis is promoted due to solvent-free reactions, which avoids polluting solvents and reduction of generation of toxic by-products and emission of harmful gases.

Most of the organic reactions are carried out by using heating equipments like water bath, oil baths, sand bath and heating mantles. This heating is relatively slow because the heat is convicted from outer surface of the vessels to the inner core of the solution/liquids resulting into a temperature gradient. Over heating of the reactants may also result into the decomposition of the reactants, reagents and/or products, but in the case of microwave heating, the microwave radiation passes through the walls of reaction vessels. Here, only the reactants and solvents are heated and vessel remains unaffected by heating. Secondly, the rise in temperature is uniform throughout the sample.

It is a nonconventional technique of organic synthesis, which shows higher selectivity in the formation of products and quite easier work-up. In these conscious days of deteriorating environment, it is a welcome addition to the list of green chemical synthetic methods. This method is also energy saving in this era of energy crisis as nobody will like to use traditional method of heating for driving a chemical reaction in a desired direction. Various other nonconventional synthetic methods are also available like sonochemical, electrochemical, enzymatic, etc. but every method has its own disadvantages and/or limitations. Microwave-assisted organic synthesis has some limitations also like use of nonpolar solvents and scalability, but the second one has overcome by fabrication and modification of microwave instruments.

The microwave region lies between 1 cm and 1 m in the electromagnetic spectrum. Most of the domestic and commercial microwave instruments are designed to operate at 2.45 GHz (corresponding to wavelength 12.2 cm) just to avoid any interference with radar and telecommunication activities.

Microwave-assisted organic synthesis has expanded reaction diversity and it also makes some difficult reactions possible, which is otherwise very cumbersome. It requires very less reagents and avoids many toxic solvents. Even some reactions, which are carried out conventionally at 0 °C or so for hours and require inert atmosphere as well as low boiling solvent like diethyl ether, can be carried out in presence of microwave oven at 60 °C and that too in few minutes in presence of high boiling solvents like tetrahydrofuran and at ambient conditions. Sidewise purging nitrogen gas to create inert atmosphere can also be avoided. The best example is the synthesis of Grignard reagent by alkyl halide.

MAOS is considered a part of green chemistry as it provides solid support and/or solvent-free reactions (Varma, 1999). The growth of this technology and its use in organic reactions is slow as compared to combinatorial chemistry and computational chemistry. This is all because of the fact that the fundamentals of microwave dielectric heating were not very clear at that moment. This is further supported by lack of reproducibility and controllability as well as its unsafe nature. With the development of the solvent-free reactions, the safety aspects were improved making

Microwave Induced Organic Reaction Enhancement (MORE) chemistry an attractive field of research for organic chemists, but the increased rate and reduction of time has been the main force behind this progress.

Many organic reaction have been successfully carried out on microwave exposure like Oxidation (Chakraborty and Bordoloi, 1999; Galica et al., 2013; Hashemi et al., 2005; Reddy et al., 1999), Reduction (Desai, 2005; Gadhwal et al., 1999; Feng et al., 2001; Kanth et al., 2002; Varma and Saini, 1997), Alkylation (Davis et al., 2013; Kumar and Gupta, 1996; Rajabi and Saidi, 2004; Runhua et al., 1994), Acylation (Gadhwal et al., 1998; Lai et al., 2013; Paul et al., 2003) Arylation (Gonzalez-Arellano et al., 2009; Verma et al., 2013), Cyclization (Alajarin et al., 1992; Ericsson and Engman, 2004; Ganto et al., 2011; Rama Rao et al., 1992; Puciova et al., 1994; Singh et al., 2006), Cycloaddition (Garrigues et al., 1996; Kaddar et al. 1999; Wu et al., 2006; Patrick et al., 2007; Chakraborty et al., 2012), Addition (Amore et al., 2006; Rao and Meshram, 2013), Condensation (Villemin et al., 1993; Gupta et al., 1995; Kumar et al., 1998; Agrawal and Joshipura, 2005; Pasha and Nizam, 2012), Substitution (Kad et al., 1996; Hu et al., 1999; Sagar et al., 2000; Ju et al., 2006; Štefane et al., 2012), Elimination (Navratilova et al., 2004; Sarma et al., 2012), Protection (Kad et al., 1998; Hajipour et al., 1999; Corsaro et al., 2004; Walia et al., 2013), Deprotection (Meshram et al., 1999; Chakraborty and Bordoloi, 1999; Hosseinzadeh et al., 2002; Dandepally and Williams, 2009), Rearrangement (Bosch et al., 1995; Yu et al., 1999; Patil et al., 2002; Kotha et al., 2004; Deodhar et al., 2010), Esterification (Kim et al., 1997; Lami et al., 1999; Yang et al., 2008; Didem, 2013), Organometallic reactions (Shaabani, 1998; Van Atta et al., 2000; Albrecht et al., 2009; Kimura et al., 2012), Synthesis of heterocyclics (Filip et al., 1996; Ben-Alloum et al., 1997; Feng et al., 1998; Goncalo et al., 1999; Jnaneshwara et al., 1999; Soukri et al., 2000; Bentiss et al., 2000; Dallinger and Kappe, 2007; Kaur et al., 2012), nanomaterials (Liu et al., 2007; Qiu et al., 2012; Mahdi et al., 2012;), etc., but this is not an end to the possible applications of microwave-assisted organic synthesis. Many more chapters will be added to this field in years to come and the future trends of this field will decide a golden pathway for it. Some excellent reviews are available in the field of microwave-assisted organic synthesis. (Li, 1993; Caddick, 1995; Galema et al., 1998; Lidstrom et al., 2001; Loupy, 2002; Hayes, 2004; Kappe and Stadler, 2005; Bogdal 2005; Kappe and Dallinger, 2006; Polshettiwar and Varma, 2008; Ameta et al., 2011; Jacob, 2012; Ameta et al., 2014).

1.2 ADVANTAGES

1.2.1 INCREASED RATE OF REACTION

Microwave heating enhances the rate of certain chemical reactions by 10 to 1000 folds as compared to conventional heating. This is due to its ability to substantially increase the temperature of a reaction. Many organic reactions are completed within

8–10 h or even more under conventional heating while these reactions could be completed within 2–5 min or even less under microwave radiations.

At present, there are two main theories that explain the rate acceleration caused by microwaves. These theories are based on experiments conducted on the following set of reactions:

- Liquid phase reactions: The rate acceleration in liquid phase reactions by microwave radiations can be attributed to the superheating of solvents. This superheating of solvents enables the reaction to be performed at higher temperatures and results in an increase in the rate of the reaction.
- Catalytic reactions: The rate acceleration in solid-state catalytic reactions by microwave radiations is attributed to high temperatures on the surface of the catalyst. The increase in the local surface temperature of the catalyst results in an enhancement of the catalytic action leading to an increase in the rate of reaction.

1.2.2 EFFICIENT SOURCE OF HEATING

Heating by means of microwave radiations is a highly efficient process and results in a significant energy saving. This is primarily due to the fact that microwaves just heat up the sample and not the apparatus (reaction vessel) and therefore, energy consumption is much reduced. A typical example is the use of microwave radiation in the ashing process.

1.2.3 HIGHER YIELDS

In majority of chemical reactions, microwave radiations produce higher yields as compared to conventional heating methods. The conventional heating provides the yields of the products in the range of 40–50% or even lower, which can be increased to 80–95% or even higher under microwave exposure.

1.2.4 UNIFORM HEATING

Microwave radiations, unlike conventional heating methods, provide uniform heating throughout a reaction mixture. In conventional heating, the walls of the bath (oil/water/sand) and reaction vessel get heated first and then the solvent. As a result of this, the heat is distributed in the bath and there is always a temperature difference between the walls of reaction vessel and the solvent. In the case of microwave heating, only the solvent and the solute particles are heated, which results in uniform heating of the solvent. This feature allows the chemists to place reaction vessels at any location in the cavity of a microwave oven. It also proves vital in processing multiple reactions simultaneously, or in scaling up reactions that require identical heating conditions.

1.2.5 SELECTIVE HEATING

Selective heating is based on the principle that different materials respond differently to microwaves. Some materials are transparent to microwaves where as others absorb these radiations. Therefore, microwaves can be used to heat a combination of such materials, for example, the production of metal sulfide with conventional heating requires weeks because of the volatility of sulfur vapors while rapid heating of sulfur in a closed tube results in the generation of sulfur fumes, which can cause an explosion. However, in microwave heating, since sulfur is transparent to microwaves, only the metal gets heated. Therefore, reactions can be carried out at a much faster rate with rapid heating, without the threat of any explosion.

The ability of metal powders to couple to microwave radiation has been used to accelerated a range of solid-state reactions. Although metal objects cause extensive arcing within a microwave cavity, metal powders couple in a more conventional manner with the microwave radiations and experience high heating rates. This property has been used for the synthesis of metal chalcogenides and metal cluster compounds of the Group 5 and 6 elements.

1.2.6 ECOFRIENDLY CHEMISTRY

Reactions conducted through microwaves are cleaner and more environmental friendly than conventional heating methods. Microwaves heat the compounds directly and therefore, use of solvents in the chemical reaction can be reduced or in some cases even eliminated. A method was developed to carry out a solvent-free chemical reaction on sponge like material (alumina) with the help of microwave heating. The reactants are absorbed on alumina and exposed to microwaves. It was observed that they react at a faster rate than conventional heating. The use of microwaves has also reduced the extent of purification required for the end products of chemical reactions.

1.2.7 GREATER REPRODUCIBILITY

Reactions with microwave heating are more reproducible as compared to conventional heating because of uniform heating and better control process parameters. The temperature of chemical reactions can also be easily monitored. This is of particular relevance in the lead optimization phase of the drug development process in pharmaceutical companies.

1.3 LIMITATIONS

The limitations of microwave chemistry are mainly, its scalability, limited applications and health hazards, which can be solved in years to come.

1.3.1 LACK OF SCALABILITY

The yield obtained by using commercial microwave apparatus is limited to a few grams. Although there have been developments in the recent past relating to the scalability of microwave equipment but still there is a gap that needs to be filled by making this technology scalable. This is particularly true for reactions at the industrial production level and for solid-state reactions.

1.3.2 LIMITED APPLICABILITY

The use of microwaves as a source of heating has limited applicability for materials as only polar molecules absorb them. Microwaves cannot heat materials such as sulfur (nonpolar), which are transparent to these radiations. Although microwave heating increases the rate of reaction and yield in certain reactions, but it also results in reduction in yield as compared to conventional heating methods in some reactions.

1.3.3 SAFETY HAZARDS

Although manufacturers of microwave heating apparatus have made microwaves a safe source of heating, even then some times, uncontrolled reaction conditions may result in undesirable results, for example, chemical reactions involving volatile reactants under superheated conditions may result in explosive conditions. Moreover, improper use of microwave heating for rate enhancement of chemical reactions involving radioisotopes may also result in uncontrolled radioactive decay.

1.3.4 HEALTH HAZARDS

Health hazards related to microwaves may be caused by the penetration of microwaves. While microwaves operating at a low frequency range are only able to penetrate the human skin, higher frequency range microwaves can reach other organs of the body also. Research has proved that prolonged exposure to microwaves may result in the complete degeneration of body tissues and cells. It has also been established that constant exposure of DNA to high frequency microwaves during a biochemical reaction may result in complete degeneration of the DNA strand.

1.4   GREEN CHEMICAL APPROACH

'Preventing pollution and minimizing waste generation will gradually clean up sins of the past' (Clark, 1995). In this era of environmental consciousness, the role of chemistry and chemist is to incorporate processes and design products, which can completely eliminate or minimize the generation of pollutants in the form of waste. Consequently, any ideal synthesis or chemical approach must include this element of environmental awareness. After almost for more than one and half centuries of the first chemical revolution, a new kind of chemical revolution is brewing, that is, Green Chemistry.

The fundamental idea of green chemistry is that the manufacturer of a chemical needs to consider, what will happen to human life after that particular chemical is generated and introduced into the society? The fate of wastes generated during the production of that chemical or after its use is also a question of major concern. Twelve principles of green chemistry can be used to assess the greenness of a particular synthetic protocol (Anastas and Warner, 1998). These principles address several concerns, such as use of various solvents, amount of chemical wastes produced, use of catalyst and reagents from quantity and reusability point of view, atom economy (maximum incorporation of atoms of reactants), the number of chemical steps (energy efficiency), use of safer chemicals and reaction conditions, etc. Practically, it is very difficult for a new synthetic protocol to satisfy all the 12 principles of green chemistry. It is not expected also, but the more principles, a protocol satisfies, the developed process will be considered relatively more greener.

There are two alternative ways to categorize green chemical approaches-
- Synthesis via environment friendly synthetic pathway or process and
- To develop any new benign replacement, capable of achieving the desired performance on one hand without any negative human or ecological impact on the other.

This type of greener protocol can be achieved through the proper choice of starting materials (feed stocks), atom economic methodologies with a minimum number of chemical steps (atom economy), the use of appropriate greener solvents, reagents & reaction conditions, and efficient strategies for product isolation & purification. Thus, green chemistry has emerged as a promising discipline that permeates all aspects of synthetic chemistry. A major goal of this endeavor must be to maximize the efficient use of harmless (or less harmful) raw materials and to reduce the wastes produced in a particular process, simultaneously.

Therefore, the demands for green and sustainable synthetic methods in the fields of health care or fine chemicals combined with the pressure to produce these substances expeditiously and that too in an environmentally benign way, pose significant challenges to the synthetic chemical community. These objectives can be conveniently achieved through the development of aqueous synthetic protocols using microwave heating. It should be noted that rapid development of green organic chemistry is due to the recognition that environment friendly process and products will prove to be economical and efficient in the long-term as they circumvent the need for treating pollutants and by-products generated by conventional synthesis.

Microwave-assisted heating under controlled conditions is an invaluable technology because it not only often dramatically reduces reaction time, typically from days or hours to minutes or even seconds, but enhances the rate of reaction and also fulfills the major aims of green chemistry by reducing side reaction resulting into increased yield and improved reproducibility (Hayes, 2002).

Microwave chemistry has become a central tool in this fast paced, time sensitive field and it has blossomed into a useful technique for a variety of applications in organic synthesis where high yielding protocols and facility of purifications are highly desirable (Bradley, 2001). Faced with the increasing demands of novel drug targets, there is considerable current interest to accelerate this technology associated with combinatorial and computational chemistry.

The major problems for the use of microwave ovens are the lack of temperature control and also the flammability of organic solvent used in the reaction. Now a days, some modified/improved microwave instruments are available with both the controls; temperature and pressure so as to have monitoring of chemical reaction easier.

Pharma industries synthesizing Active Pharmaceutical Ingredients (API) are looking towards microwave technology as a hope to assist them for large-scale production and that too in limited time.

Microwave radiations have proved to be highly effective heating source for chemical reactions. It provides rapid and homogeneous heating, which has certain advantages such as reaction rate acceleration, milder reaction condition and higher chemical yield. In short, microwave enhanced chemical reactions are safer, faster, cleaner and more economical than conventional reactions. It helps in developing cleaner and greener synthetic routes.

KEYWORDS

- **Advantages**
- **Green Chemistry**
- **Limitations**
- **MAOS**
- **Microwave**
- **MORE**

REFERENCES

Agrawal, Y. K., & Joshipura, H. M. (2005). Indian Journal of Chemistry, 44B, 1649–1652.

Alajarin, R., Vaquero, J. J., Garcia Navio, J. L., & Alvarez-Builla, J. (1992). Synlett, 297–298.

Albrecht, C., Gauthier, S., Wolf, J., Scopelliti, R., & Severin, K. (2009). European Journal of Inorganic Chemistry, 967.

Ameta, C., Ameta, A., Ameta, R., Punjabi, P. B., & Ameta, S. C. (2011). Journal of Indian Chemical Society, 88, 1165–1185.

Ameta, C., Ameta, K. L., Sharma, B. K., & Ameta, R. (2014) Microwave-assisted Organic Synthesis: A Need of the Day. In Green Chemistry Fundamentals and Application, Ameta, S. C., & Ameta, R. (Eds.) New Jersey: Apple Academic Press, 283–215.

Amore, K. M., Leadbeater, N. E., Miller, T. A., & Schmink, J. R. (2006). Tetrahedron Letters, 47, 8583–8586.

Anastas, P. T., & Warner, J. C. (1998). Green chemistry, theory and practice. New York: Oxford University Press.

Ben-Alloum, A., Bakkas, S., & Soufiaoui, M. (1997). Tetrahedron Letters, 38, 6395–6396.

Bentiss, F., Lagrenée, M., & Barbry, D. (2000). Tetrahedron Letters, 41, 1539–1541.

Bogdal, D. (2005). Microwave-assisted organic synthesis. One hundred reaction procedures, Oxford, UK: Elsevier, 23–32.

Bosch, A. I., De La Crug, P., Diez-Barra, E., Loupy, A., & Langa. F. (1995). Synlett, 12, 1259–1260.

Bradley, D. (2001). Modern Drug Discovery, 4, 32–36.

Caddick, S. (1995). Tetrahedron, 51, 10403–10432.

Chakraborty, B., Sharma, P. K., Rai, N., & Sharma, C. D. (2012). Journal of Chemical Sciences, 124, 679–685.

Chakraborty, V., & Bordoloi, M. J. (1999). Journal of Chemical Research (S), 2, 118–119.

Clark, J. H. (1995). Chemistry of waste minimization, London: Chapman and Hall.

Corsaro, A., Chiacchio, U., Pistarà, V., & Romeo, G. (2004). Current Organic Chemistry, 8, 511–538.

Dallinger, D., & Kappe, C. O. (2007). Chemical Reviews, 107, 2563–2591.

Dandepally, S. R., & Williams, A. L. (2009). Tetrahedron Letters, 50, 1071–1074.

Davis, O.A., Hughes, M., & Bull, J.A. (2013). Journal of Organic Chemistry, 78, 3470–3475.

Deodhar, D. K., Tipnis, A. S., & Samant, S. D. (2010). Indian Journal of Chemistry, 49B, 1552–1555.

Desai, K. R. (2005). Green Chemistry: Microwave Synthesis, Mumbai: Himalaya.

Didem, O. (2013). Journal of Biobased Materials and Bioenergy, 7, 449–456.

Ericsson, C., & Engman, L. (2004). Journal of Organic Chemistry, 69, 5143–5146.

Feng, J. C., Liu, B., Dai, L., Yang, X. T., & Tu, S. J. (2001). Synthetic Communications, 31, 1875–1877.

Feng, J., Yang, L., Meng, Q., & Liu, B. (1998). Synthetic Communications, 28, 193–196.

Filip, S. V., Surducan, E., Vlassa, M., Silberg, L. A., & Jucan, G. (1996). Heterocyclic Communications, 2, 431–434.

Gadhwal, S., Baruah, M., & Sandhu, J. S. (1999). Synlett, 10, 1573–1574.

Gadhwal, S., Dutta, M. P., Boruah, A., Prajapati, D., & Sandhu, J. S. (1998). Indian Journal of Chemistry, 37B, 725–727.

Galema, S. A., Halstead, B. S. J., & Mingos, D. M. P. (1998). Chemical Society Reviews, 27, 213–232.

Galica, M., Kasprzyk, W., Bednarz, S., & Bogdal D. (2013). Chemical Papers, 67, 1240–1244.

Ganto, M. M., Lee, Y. C., & Kaye, P. T. (2011). Synthetic Communications, 41, 1688–1702.

Garrigues, B., Laurent, R., Laporte, C., Laporterie, A., & Dubac, J. (1996). Liebigs Annalen der Chemie, 5, 743–744.

Gedye, R., Smith, F., Westaway, K., Humera, A., Baldisera, L., Laberge, L., & Rousell, J. (1986). Tetrahedron Letters, 27, 279–282.

Giguere, R. J., Bray, T. L., Duncan, S. M., & Majetich, G. (1986). Tetrahedron Letters, 27, 4945–4948.

Goncalo, P., Roussel, C., Mélot, J. M., & Vébrel, J. (1999). Journal of the Chemical Society, Perkin Transactions 2, 10, 2111–2115.

Gonzalez-Arellano, C., Luque, R., & Macquarrie, D. J. (2009). Chemical Communications, 45, 1410–1412.

Gupta, R., Gupta, A. K., Paul, S., & Kachroo, P. L. (1995). Indian Journal of Chemistry, 34B, 61–62.

Hajipour, A. R., Mallakpour, S. E., & Imanzadeh, G. (1999). Journal of Chemical Research, 3, 228–229.

Hashemi, M. M., Rahimi, A., Karimi-Jaberi, Z., & Ahmadibeni, Y. (2005). Acta Chimica Slovenica, 52, 86–87.

Hayes, B. L. (2002). Microwave synthesis: Chemistry at the speed of light, NC: CEM Publishing.

Hayes, L. B. (2004). Aldrichimica Acta, 37, 66–76.

Hosseinzadeh, R., Sharifi, A., Tabar-Heydar, K., & Mohsenzadeh, F. (2002). Monatshefte für Chemie, 133, 1413–1415.

Hu, Y. L., Yu, J. H., Yang, S. Y., Wang, J. X., & Yin, Y. Q. (1999). Synthetic Communications, 29, 1157–1164.

Jacob, J. (2012). International Journal of Chemistry, 4, 29–43.

Jnaneshwara, G. K., Bedekar, A. V., & Deshpande, V. H. (1999). Synthetic Communications, 29, 3627–3633.

Ju, Y., Kumar, D., & Varma, R. S. (2006). Journal of Organic Chemistry, 71, 6697–6700.

Kad, G. L., Kaur, J., Bansal, P., & Singh, J. (1996). Journal of Chemical Research (S), 4, 188–189.

Kad, G. L., Singh, V., Kaur, K. P., & Singh, J. (1998). Indian Journal of Chemistry Section B, 37, 172–173.

Kaddar, H., Hamelin, J., & Benhaoua, H. J. (1999). Journal of Chemical Research, 12, 718–719.

Kanth, S. R., Reddy, G. V., Rao, V. V. V. N. S. R., Maitraie, D., Narsaiah, B., & Rao, P. S. (2002). Synthetic Communications, 32, 2849–2853.

Kappe, C. O., & Dallinger, D. (2006). The Nature Reviews Drug Discovery, 5, 51–63.

Kappe, C. O., & Stadler, A. (2005). Microwaves in organic and medicinal chemistry, Weinheim: Wiley-VCH.

Kaur, N., Sharma, P., Sirohi, R., & Kishore, D. (2012). Archives of Applied Science Research, 4, 2256–2260.

Kim, Y. H., Kim, J. K., Kwon, T. W., & Chung, S. K., & Kwon, P. S. (1997). Bulletin of the Korean Chemical Society, 18, 1118–1119.

Kimura, H., Mori, D., Harada, N., Ono, M., Ohmomo, Y., Kajimoto, T., Kawashima, H., & Saji, H. (2012). Chemical and Pharmaceutical Bulletin, 60, 79–85.

Kotha, S., Mandal, K., Deb, A. C., & Banerjee, S. (2004). Tetrahedron Letters, 45, 9603–9605.

Kumar, H. M. S., Subbareddy, B. V., Anjaneyulu, S., & Yadav, J. S. (1998). Synthetic Communications, 28, 3811–3815.

Kumar, P., & Gupta, K. C. (1996). Chemistry Letters, 8, 635–636.

Lai, Q. Y., Liao, R. S., Wu, S. Y., Zhang, J. X., & Duan, X. H. (2013). New Journal of Chemistry, 37, 4069–4076.

Lami, L., Casal, B., Cuadra, L., Merino, J., Ruiz-Hitzky, E., & Alvarez, A. (1999). Green Chemistry, 1, 199–204.

Li, C. J. (1993). Chemical Reviews, 93, 2023–2035.

Lidstrom, P., Tierney, J., Wathey, B., & Westman, J. (2001). Tetrahedron, 57, 9225–9283.

Liu, J. S., Cao, J. M., Li, Z. Q., Ji, G. B., & Zheng, M. B. (2007). Materials Letters, 61, 4409–4411.

Loupy, A. (Ed.) (2002). Microwaves in organic synthesis, Weinheim: Wiley-VCH.

Mahdi, M. A., Hassan, J. J., Hassan, Z., & Ng, S. S. (2012). Journal of Alloys and Compounds, 541, 227–233.

Meshram, H. M., Reddy, G. S., Sumitra, G., & Yadav, J. S. (1999). Synthetic Communications, 29, 1113–1119.

Navratilova, H., Kriz, Z., & Potacek, M. (2004). Synthetic Communications, 34, 2101–2115.

Pasha, M. A., & Nizam, A. (2012). Journal of Saudi Chemical Society, 16, 237–240.

Patil, B. S., Vasanthakumar, G. R., & Babu, V. V. S. (2002). Letters in Peptide Science, 9, 231–233.

Patrick, T. B., Gorrell, K., & Rogers, J. (2007). Journal of Fluorine Chemistry, 128, 710–713.

Paul, S., Nanda, P., Gupta, R., & Loupy, A. (2003). Synthesis, 2877–2881.

Polshettiwar, V., & Varma, R. S. (2008). Accounts of Chemical Research, 41, 629–639.

Puciova, M., Ertl, P., & Toma, S. (1994). Collection of Czechoslovak Chemical Communications, 59, 175–185.

Qiu, G., Dharmarathna, S., Zhang, Y., Opembe, N., Huang, H., & Suib, S. (2012). Journal of Physical Chemistry C, 116, 468–477.

Rajabi, F., & Saidi, M. R. (2004). Synthetic Communications, 34, 4179–4188.

Rama Rao, A. V., Gurjar, M. K., & Kaiwar, V. (1992). Tetrahedron: Asymmetry, 3, 859–862.

Rao, N. N., & Meshram, H. M. (2013). Tetrahedron Letters, 54, 1315–1317.

Reddy, D. S., Reddy, P. P., & Reddy, P. S. N. (1999). Synthetic Communications, 29, 2949–2951.

Runhua, D., Yuliang, W., & Yaozhong, J. (1994). Synthetic Communications, 24, 111–115.

Sagar, A. D., Patil, D. S., & Bandgar, B. P. (2000). Synthetic Communications, 30, 1719–1723.

Sarma, R., Sarmah, M. M., & Prajapati, D. (2012). Journal of Organic Chemistry, 77, 2018–2023.

Shaabani, A. (1998). Journal of Chemical Research (S), 10, 672–673.

Singh, P., Natani, K., Jain, S., Arya, K., & Dandia, A. (2006). Natural Product Research, 20, 207–12.

Soukri, M., Guillaumet, G., Besson, T., Aziane, D., Aadil, M., Essassi, E. M., & Akssira, M. (2000). Tetrahedron Letters, 41, 5857–5860.

Štefane, B., Požgan, F., Sosič, I., & Gobec, S. (2012). Tetrahedron Letters, 53, 1964–1967.

Van Atta, S. L., Duclos, B. A., & Green, B. D. (2000). Organometallics, 19, 2397–2399.

Varma, R. S. (1999). Green Chemistry, 1, 43–55.

Varma, R. S., & Saini, R. K. (1997). Tetrahedron Letters, 38, 4337–4338.

Verma, A., Prajapati, N., Salecha, S., Giridhar, R., & Yadav, M. R. (2013). Tetrahedron Letters, 54, 2029–2032.

Villemin, D., Labiad, B., & Loupy, A. (1993). Synthetic Communications, 23, 419–424.

Walia, A., Kang, S., & Silverman, R. B. (2013). Journal of Organic Chemistry, 78, 10931–10937.

Wu, J., Sun, L., & Dai, W. M. (2006). Tetrahedron, 62, 8360–8372.

Yang, Q., Wang, X. J., Li, Z. Y., Sun, L., & You, Q. D. (2008). Synthetic Communications, 38, 4107–4115.

Yu, H.-M., Chen, S.-T., Tseng, M.-J., Chen, S.-T., & Wang, K.-T. (1999). Journal of Chemical Research (S), 1, 62–63.

CHAPTER 2

INSTRUMENTATION

CHETNA AMETA, PURNIMA DASHORA, and RITU VYAS

CONTENTS

In early days, it was recognized that microwaves could heat water in a dramatic fashion. Cooking food with microwaves was discovered accidentally in the 1940 s. Domestic and commercial appliances for heating and cooking of foods began to appear in the 1950 s. The appliance called "Radarange" appeared in the market in 1947, and it was proposed for food processing. The first domestic microwave oven was introduced by Tappan Stove Company on 25 October 1955 but the widespread use of domestic microwave ovens occurred during the 1970s and 1980s. The first application of microwaves irradiation in chemical synthesis appeared in 1986.

The term 'microwave' is inseparably linked in our modern society to the rapid heating or warming of foodstuffs. What is fascinating from a chemical synthesis standpoint is the dynamic range of temperatures afforded by modern laboratory microwave instrumentation. To date, most of the efforts have been focused on elevated temperature transformations or reactions requiring heating. Synthetic transformations unachievable through conductive heating have recently been realized using microwaves as the energy source. Low temperature reactions via microwave energy have been recently introduced with the key point being that gentler reaction conditions. Gentler reaction conditions are especially important with respect to biochemical applications, where the preparation of peptides (Murray et al., 2005; Murray and Gellman, 2006), peptoids and oligosaccharides (Brun et al., 2006), etc. are of interest.

Heating water in a closed vessel well above its boiling point produces supercritical water. In this situation, the form of water is less polar and thus, it is more effective in dissolving organic substrates. In addition, the increased use of water in industrial settings is a popular notion in terms of green chemistry, as water is not only low cost but environmentally more benign than any other traditional organic solvent and when used in this context, it provides facile separation of the solvent, organic reactants and products.

2.1   MICROWAVES AS ENERGY SOURCE

Microwave radiations are basically electromagnetic radiation, which are widely used as a source of heating in organic synthesis. Microwaves have enough momentum to activate reactants to cross the energy barrier and lead a reaction to completion. Microwaves occupy a place in the electromagnetic spectrum between infrared waves and radio waves. They have wavelengths between 0.01 and 1 m, and operate in a frequency range between 0.3 and 30 GHz. The typical bands for industrial applications are 915 ± 15 and 2450 ± 50 MHz. The wavelength between 1 cm and 25 cm are extensively used for RADAR transmissions and the remaining wavelength range is used for telecommunications.

The entire microwave region is therefore not available for heating applications and the equipment operating at 2.45 GHz, corresponding to a wavelength of 12.2

cm, is quite commonly used. The energy carried by microwave at 2.45 GHz is 1 Joule per mole of quanta, which is relatively very small energy.

2.2 MICROWAVE APPARATUS

Most of the pioneering experiments in organic synthesis using microwaves were carried out in domestic microwave ovens. However, later on developments in microwave equipment technology have enabled researchers to use dedicated apparatus for organic reactions. Advance microwave system consists of a microwave source (Magnetron), a microwave cavity or an applicator (Multi-mode cavity or Single mode cavity), mode stirrer, sensors probe (thermocouples or IR sensor) and software with digital display.

2.2.1 MAGNETRON

Magnetron is a vacuum tube in which electrons are affected by magnetic and electric fields in such a way that they produce microwaves radiations of particular wavelength. A magnetron is a thermo-ionic diode that works on the principle of dielectric heating by converting part of the electric power into electromagnetic energy and the rest into heat energy.

A microwave cavity is a special type of resonator, which consists of a closed metal structure that confines electromagnetic fields in the microwave region of the spectrum. The structure is either hollow or filled with dielectric material. It acts similarly to a resonant circuit with extremely low loss at its frequency of operation, which results in quality factors up to the order of 10^6, compared to 10^2 for circuits made with separate inductors and capacitors at the same frequency. They are used in oscillators and transmitters to create microwave signals, and as filters to separate a signal at a given frequency from other signals, in microwave ovens.

The earliest description of a magnetron (the high-power generator of microwave power), as a diode with a cylindrical anode was published by Hull long back (1921a, 1921b) but it was developed practically by Randall and Boot at the University of Birmingham in England. On February 21, 1940, they verified their first microwave transmissions of 500 W at 3 GHz. Microwave techniques were developed just before and during Second World War, when most of the efforts were concentrated on the design and manufacture of microwave navigation and communications equipments for military use. Probably, the first announcement of a microwave oven was for a newly developed Radar range for airline use (Anonymous, 1946a; 1946b).

Microwave cavity or applicator is also known as reactor. The microwave chemistry apparatus is classified in following two categories:

- Single-mode or mono-mode apparatus
- Multi-mode apparatus

2.2.2 SINGLE-MODE APPARATUS

The single-mode cavity allows only a single-mode to enter the cavity by waveguide. A properly designed single-mode cavity or reactor can prevent the formation of "hot and cold" spots. This advantage is very important in organic synthesis since the actual heating pattern can be controlled. Therefore, higher reproducibility and predictability are achieved.

The basic difference of a single-mode apparatus is its ability to create a new standing wave pattern. This pattern is generated by the interference of fields having the same amplitude but different oscillating directions. This interface generates an array of nodes (zero microwave intensity) and antinodes (highest magnitude of microwave energy) (Fig. 2.1).

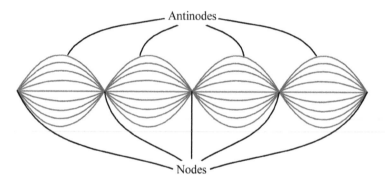

FIGURE 2.1   Nodes and antinodes standing wave pattern.

The design of a single-mode apparatus is governed by the distance of the sample from the magnetron. It should be appropriate to ensure that the sample is placed at the antinodes of the standing electromagnetic wave pattern (Fig. 2.2).

Several consequences of single-mode cavity design must be appreciated. First, microwave-absorbing materials placed inside such a cavity will absorb microwaves and as a consequence, the material is heated. The second consequence is more important. There are some specific positions inside the single-mode cavity, where items are placed so that the sample is heated. On the other hand, there are certain positions in the single-mode cavity, where no heating will occur.

One of the limitations of this apparatus is that anything placed inside this single-mode cavity can disrupt the standing wave pattern. Therefore, it is quite uncommon for single-mode cavities to be designed to accept more than one target object (such as a reaction vessel) to be heated at a time.

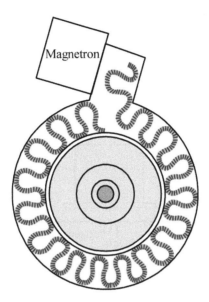

FIGURE 2.2   Single-mode heating apparatus.

Single-mode heating equipments are at present used for drug discovery on a small scale, and some combinatorial chemical applications. An advantage of a single-mode apparatus is its high rate of heating. This is all because the sample is always placed at the antinodes of the field, where the intensity of microwave radiation is the highest. On the contrary, the heating effect is averaged out in a multi-mode apparatus.

2.2.3   MULTI-MODE APPARATUS

An essential feature of a multi-mode apparatus is to deliberately avoid the generation of a standing wave pattern inside it (Fig. 2.3).

Basically, there are two approaches to achieve this goal:
(i)   The dimensions of the cavity must be carefully controlled, so as to avoid whole number multiples of the microwave full or half wavelength and
(ii)   Some means must be employed to physically disrupt any standing waves formed as a consequence of items placed in the cavity.

The shape of the wave and its movement is such that the microwave field is continually stirred, and therefore, the field intensity becomes homogeneous in all directions and all locations throughout the complete cavity.

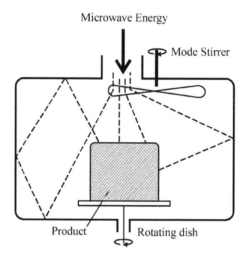

FIGURE 2.3   Multi-mode heating apparatus.

Since the field in a multi-mode cavity is continuously stirred and remains homogenized, anything placed inside the cavity will not affect permanently the field intensity distribution. It means that there are no limitations on the size and shape of the vessels placed inside the cavity. If that vessel or material physically fit inside the cavity and absorb microwaves, then only they are heated. This also means that multiple objects (multiple reaction vessels) can be heated simultaneously, as effectively as single objects (single reaction vessel). This is an added advantage of multi-mode cavity design. This characteristic makes a multi-mode heating apparatus useful for bulk heating and carrying out different chemical analysis processes such as ashing, extraction, etc. Multi-mode reactors mainly provide a field pattern with areas of high and low field strength of microwaves. These are commonly referred to as hot and cold spots, respectively. This nonuniformity of the field varying drastically with different positions of the sample leading to the drastic differences in heating efficiency. This drawback of a multi-mode apparatus is overcome by using a mode stirrer. Since the field is heterogeneous in nature, so the use of the apparatus for synthetic purposes requires mapping of the field, which involves determination of hot spots of high energy. This can be done using a filter paper sheet impregnated with a solution of cobalt chloride.

The use of multi-mode reactors has certain limitations. These are:

- The distribution of electric field inside the cavity results from multiple reflections off the walls and reaction vessel and as a consequence, it is heterogeneous in nature;
- The temperature cannot be simply and accurately measured; and
- The power is not tunable.

2.2.4 MODE STIRRER

The mode stirrer is a periodically moving metal vane (flat blades) that continuously changes the instantaneous field pattern inside the cavity. The shape of the vane as well as its movement is such that the microwave field is continuously stirred. Hence, the field intensity becomes homogeneous in all directions and all locations throughout the cavity. Now, the microwave absorbing materials can be placed anywhere inside the cavity, because the field is homogeneous throughout the cavity.

Some modifications of domestic microwave ovens have been suggested by various workers (Caddick, 1995) such as introduction of condensers by boring through the top of the oven (Villemin and Thibault-Starzyk, 1991) or reaction flasks being fitted with condensers (Plazl, 1994) and charged with precooled, microwave-inactive coolants, like xylene, carbon tetrachloride, etc.

In modern microwave reactors, digital thermometers (sensors and probes) are used for temperature control. Moreover, some sophisticated ovens even interface with computers for reaction monitoring (Barlow and Marder, 2003; Raner et al., 1995).

There are a few examples in the literature, which indicate that microwave heating was carried out in stirred tank reactors, for example the hydrolysis of sucrose by conventional and microwave heating (Plazl et al., 1995), the esterification of benzoic acid with 2-ethylhexanol (Pipus et al., 2002).

2.2.5 WORKING OF THE MICROWAVE OVEN

Microwaves are generated by a magnetron in a microwave oven. This magnetron is a thermo-ionic diode having an anode and a directly heated cathode. As the cathode is heated, electrons are released and these are attracted towards the anode. The anode is made up of an even number of small cavities, each of which acts as a tuned circuit. Thus, anode is a series of circuits, which are tuned to oscillate at a specific frequency or at its overtones.

A very strong magnetic field is induced axially through this anode assembly and as a result, the path of electrons bended as they travel from the cathode to the anode. As the deflected electrons pass through the cavity gaps, a small charge is induced into the tuned circuit, which results in the oscillation of the cavity. Alternate cavities are linked by two small wire straps, which ensure the correct phase relationship. This process of oscillation continues until the oscillation has achieved sufficiently high amplitude. It is then taken off by the anode via an antenna. The variable power available in domestic ovens is produced by switching the magnetron on and off according to the duty cycle.

Microwave dielectric heating is effective, when the matrix has a sufficiently large dielectric loss tangent (i.e., contains molecules possessing a dipole moment). It was reported that the use of a solvent is not always mandatory for the transport of heat (Andrews and Atkinson, 1984) and therefore, reactions may be performed under solvent-free conditions. It presents an alternative in the microwave chemistry (an environmentally benign technique), which avoids the generation of toxic residues, like organic solvents and mineral acids, and thus allows the attainment of high yields of products at reduced environmental costs and risk.

2.2.6 CONTINUOUS FLOW MICROWAVE REACTOR

Microwave-assisted chemistry is energy efficient, provides faster heating rates and enables rapid optimization of procedures as there is no direct contact between the chemical reactants and the energy source. From the early experiments in domestic ovens (Gedye et al., 1986; Giguere et al., 1986), the use of mono-mode or multi-mode (Loupy et al., 1998) instruments designed for organic synthesis has been implemented worldwide and continues to be developed (Strauss and Trainor, 1995). Although modern mono-mode or multi-mode instruments devoted for microwave-assisted organic synthesis are quite successful in small scale operations, efforts to process this technology in continuous flow (CF) reactors were not successful by the physical limitations of microwave heating, less penetration depth (only a few centimeters) and the limited dimensions of the standing wave cavity. Current technology has attempted to overcome these problems with the conventional instruments by using CF reactor, which pumps the reagents through a small heated coil that winds in and out of the cavity and external temperature monitoring using a fiber optic sensor. Some alternative methods, such as using a multi-mode batch (Khadilkar and Madyar, 2001; Stadler et al., 2003) or CF reactor (Cablewski et al., 1994; Chemat et al., 1998) have also been described (Fig. 2.4).

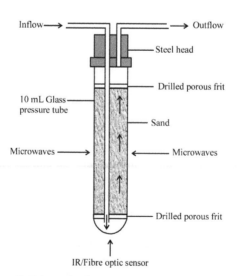

FIGURE 2.4   Flow cell (Schematic diagram).

The flow cell was designed to make optimum use of the cavity and to monitor the temperature of the flow cell directly using an inbuilt IR sensor. To this end, a standard pressure rated glass tube (10 mL) is fitted with a custom built steel head and it was filled with sand (10 g) between two drilled porous frits. This is to minimize dispersion and effectively create a lattice of microchannels. It is charged with solvent (5 mL volume), sealed using polytetrafluoroethylene (PTFE) washers and connected to an HPLC flow system with a backpressure regulator.

The flow cell was inserted into the cavity of a self-tunable single-mode microwave synthesizer. There it was irradiated and stabilized at the required reaction temperature through moderation of microwave power before the reagents is introduced into the reactor. This system has a number of advantages over other commercially available coils, like simple measurement of temperature of the flow cell, no additional requirement of expensive equipment, short of an HPLC pump and with a potential to carry out heterogeneous as well as homogeneous reactions simply by immobilizing a catalyst on the support in the glass tube (Bagley et al., 2005) (Fig. 2.5).

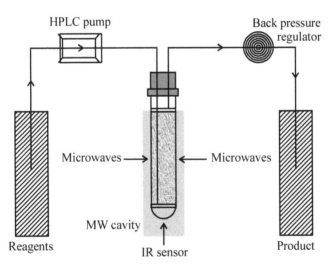

FIGURE 2.5 The CF reactor (Schematic diagram).

Moseley et al. (2008) reported a survey of some microwave reactors designed for scale-up by different manufacturers. The variety of instruments indicate that there is no satisfactory solution to the problem of microwave scale-up. Microwave chemistry is linearly scalable, from the level of a test tube to more than a liter. There is presently no commercial microwave scale-up solution. At present, commercially scale-up microwave reactor are not available, which is capable of meeting the needs of the pharmaceutical industry for the wide range of reactions, for example, for the proper process development and pilot scale.

2.3   PRINCIPLE

The basic mechanism behind the heating in a microwave oven is the interaction of charged particle of the reaction material with electromagnetic wavelength of a particular frequency. The heat is produced by electromagnetic irradiation either by collision or by conduction and some time by both.

All the wave energy changes its polarity from positive to negative with every cycle of the wave. This causes a rapid orientation and reorientation of molecule, resulting into heating by collision. If the charge particles of any material are free to travel through that material, a current will be induced, traveling in phase with the field. If these charge particles are bound within certain regions of the materials, the electric field component will cause them to move until opposing force balances this electric force (Adam, 2003; Blackwell, 2003; Bradley, 2001; Dzieraba and Combs, 2002; Johansson, 2001; Larhed and Hall berg, 2001; Sharma et al., 2002; Wathey et al., 2002).

2.3.1   HEATING MECHANISM

Materials may be heated using high frequency electromagnetic waves in microwave oven. This heating arises from the interaction of electric field component of the wave with charge particle in the material. Two basic mechanisms are involved in the heating of materials. These are:

- Dipolar polarization; and
- Conduction mechanism

2.3.2   DIPOLAR POLARIZATION

Dipolar polarization is responsible for the majority of microwave heating. It depends upon nature (Polarity) of solvents and reactants. Different electronegativities of individual atoms result in a permanent electric dipole in polar molecules. This dipole is sensitive to external electric fields and will attempt to align with them by rotation. This realignment is quite rapid for free molecule, but this is not the case as the instantaneous alignment is prohibited by the presence of other molecules in liquid. Therefore, a limit is placed on the ability of that dipole to respond to a electric field, which affects the behavior of the molecule with different frequencies. The dipole may react by aligning itself in phase with the electric field under low frequency irradiation and molecule will polarize uniformly and no random motion results. Some energy is gained in the molecule by this behavior, and some energy is lost in collisions and therefore, the overall heating effect is small.

On the other hand, the polar molecules will attempt to follow the electric field under high frequency irradiation, but intermolecular inertia stops any significant motion before the field has reversed. The dipoles do not have sufficient time to respond to this reversal of the field and so dipoles do not rotate. As no motion has been induced in the molecules, no energy transfers will take place. Therefore, as a consequence, there is no heating of polar molecules. In intermediate frequency, the electric field will be such that molecule is almost able to keep in phase but not quite with the field polarity. The microwave frequency is comparatively low enough that the dipoles have enough time to respond to the changing field and therefore, they do rotate, but high enough so that the rotation does not precisely follow the electric field. As the dipole is reoriented to align itself with the field, the field is already changing. As a result, phase difference causes energy to be lost from the dipole in random collisions to give rise to dielectric heating (Fig. 2.6).

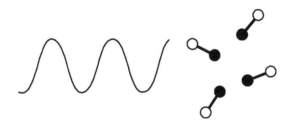

FIGURE 2.6   Dipolar mechanism.

2.3.3   CONDUCTION MECHANISM

The conduction mechanism generates heat through resistance to an electric current. The oscillating electromagnetic field generates an oscillation of electrons or ions in a conductor resulting in an electric current. This current faces internal resistance and consequently the conductor is heated (Fig. 2.7).

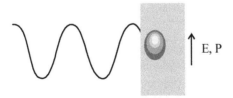

FIGURE 2.7   Conduction mechanism.

When the irradiated sample is an electrical conductor, the charge carriers (electrons, ions, etc.) move through the material under the influence of the external electric field and thus, cause a polarization. These induced currents will cause heating in the sample due to electrical resistance.

2.4   REACTION VESSELS

The reaction vessels used in microwave-assisted organic synthesis should preferably be made up of teflon, polystyrene or glass (Bose et al., 1994; Strauss and Trainor, 1995). These vessels should be tall and loosely covered with a capacity much greater than total volume of the reactants. Metallic containers are not advised for use as these are heated soon due to preferential absorption and reflection of rays.

2.5   REACTION MEDIUM

The reaction medium plays an important role in microwave chemistry, as the reactions are carried out in a solvent. The chosen solvent must have a dipole moment and it should have a boiling point higher than the desired reaction temperature as well as a high dielectric constant.

An excellent energy transfer medium for many types of reactions in a domestic oven is N,N-dimethyl formamide (DMF) as it has high boiling point (154°C) and high dielectric constant ($\varepsilon = 36.70$). Other solvents of choice are: formamide (b.p. 210°C, $\varepsilon = 24.6$), chlorobenzene (b.p. 131°C, $\varepsilon = 5.62$), 1,2-dichlorodenzene (b.p. 180°C, $\varepsilon = 9.93$), 1,2,4-trichorobenzene (b.p. 214°C, $\varepsilon = 2.20$), 1,2-dichloroethane (b.p. 84°C, $\varepsilon = 10.66$), ethylene glycol (b.p. 197.3°C, $\varepsilon = 37.7$), diglyme (b.p. 162°C, $\varepsilon = 7.30$) and triglyme (b.p. 216°C, $\varepsilon = 7.50$).

The presence of salts in polar solvents can frequently enhance microwave coupling.

Due to less dipole moment, hydrocarbon solvents such as benzene (b.p. 80°C, $\varepsilon = 2.27$), toluene (b.p. 111°C, $\varepsilon = 2.38$), o-xylene (b.p. 144°C, $\varepsilon = 2.57$), and cyclohexane (b.p. 80.7°C, $\varepsilon = 2.02$) are unsuitable as they poorly absorb microwave radiations. But the addition of small amount of polar solvents such as water or alcohol to these solvents, can lead to dramatic coupling effects. Hence, ethanol: toluene (1:4) mixture can be heated to boiling in few minutes in a microwave oven (Table 2.1).

TABLE 2.1   Solvents Commonly Used

Absorbance level	Solvents
High	Methanol; Ethanol; Propanols; Nitrobenzene; Formic Acid; Ethylene glycol; Dimethyl sulfoxide
Medium	Water; N, N-Dimethylformamide (DMF); NMP; Butanol; Acetonitrile; Hexamethylphospromide (HMPA); Acetone, Methyl ethyl ketone and other ketones; Nitromethane; 1,2-Dichloroethane; 1,2-Dicholorobenzene; 2-Methoxyethanol; Acetic acid; Trifluoroacetic acid
Low	Pentane, Hexane, and other hydrocarbons; Chloroform; Dichloromethane; Carbon tetrachloride; 1,4-Dioxane, Tetrahydrofuran (THF), Glyme, and other ethers; Ethyl acetate; Pyridine; Triethylamine; Benzene; Toluene; Chlorobenzene; Xylenes.

For solid state reactions, mineral oxides such as zeolite, alumina, silica, montmorillonite K10 clay, etc. are used as absorbents.

2.6 MICROWAVE EFFECT

There are two general classes of microwave effects:
- Specific microwave effects; and
- Non- thermal microwave effects.

Stuerga and Gaillard (1996) have discussed these effects and summarized the examples of these microwave effects in organic chemistry.

2.6.1 SPECIFIC MICROWAVE EFFECTS

These are the effects that cannot be easily emulated through conventional heating methods, for example, (i) Selective heating of specific reaction components; (ii) Rapid heating rates and temperature gradients; (iii) Elimination of wall effects; and (iv) Superheating of solvents. These specific effects are not controversial and support conventional explanations (i.e., kinetic effects) for the observed effects.

2.6.2 NON-THERMAL MICROWAVE EFFECTS

These effects have been proposed in order to explain unusual observations in microwave chemistry. As the name suggests, these effects are supposed not to require the transfer of microwave energy into thermal energy. Instead, the microwave energy itself directly couples to energy modes within the molecules or lattice. Non-thermal effects in liquids are almost certainly nonexistent (De la Hoz et al., 2005; Kappe, 2004) as the time for energy redistribution between molecules in a liquid is comparatively much less than the period of a microwave oscillation. Non-thermal effects in solids are still a topic of debate. It is likely that, through focusing of electric fields at particle interfaces, microwaves cause plasma formation and enhance diffusion in solids via second-order effects (Booske et al., 1992, 1997; Freeman et al., 1998). As a result, they may enhance solid-state sintering and diffusion processes (Whittaker, 2005). Although some attempts have been made to explain the effects, but debates are still on about nonthermal effects of microwaves that have been reported in solid-state phase transitions (Robb et al., 2002).

2.7 SAFETY

Safety should be of paramount concern with any chemical reaction especially where the irradiation of samples with electromagnetic radiation is carried out. The reaction enhancement and rapid heating benefits also requires additional safety measures on the other hand. Much of the chemistry is conducted within sealed vessels. Although the microwave instruments manufactures have addressed these issues through development of explosion proof reactors, shutdown mechanism for situations of overheating or sudden increase in pressure and venting mechanisms for closed vessel reactions. But still some more improvements are expected in large-scale reactors, as the microwave chemistry is expected to play a long inning in organic synthesis at industrial level.

KEYWORDS

- **Conduction**
- **Dipolar**
- **Microwave Effect**
- **Multi-Mode**
- **Polarization**
- **Single-Mode**

REFERENCES

Adam, D. (2003). Nature, 421, 571–572.

Andrews, J., & Atkinson, G. F. (1984). Journal of Chemical Education, 61, 177–178.

Anonymous (1946a). Electronics, 19, 178.

Anonymous (1946b). Elec. Eng., 65, 591.

Bagley, M. C., Jenkins, R. L., Lubinu, M. C., Mason, C., & Wood, R. (2005). Journal of Organic Chemistry, 70, 7003–7006.

Barlow, S., & Marder, S. R. (2003). Advanced Functional Materials, 13, 517–518.

Blackwell, H. E. (2003). Organic & Bimolecular Chemistry, 1, 1251–1255.

Booske, J. H., Cooper, R. F., & Dobson, L. (1992). Journal of Materials Research, 7, 495–501.

Booske, J. H., Cooper, R. F., & Freeman, S. A. (1997). Materials Research Innovations, 1, 77–84.

Bose, A. K., Mannas, M. S., Banik, B. K., & Robb, E. W. (1994). Research on Chemical Intermediates, 20, 1–11.

Bradley, D. (2001). Modern Drug Discovery, 4, 32–36.

Brun, M. A., Disney, M. D., & Seeberger, P. H. (2006). ChemBioChem, 7, 421–424.

Cablewski, T., Faux, A. F., & Strauss, C. R. (1994). Journal of Organic Chemistry, 59, 3408–3412.

Caddick, S. (1995). Tetrahedron, 51, 10403–10432.

Chemat, F., Esveld, E., Poux, M., & Dimartino, J. L. (1998). Journal of Microwave Power Electromagnetic Energy, 33, 88–94.

De la Hoz, A., Diaz-Ortiz, A., & Moreno, A. (2005). Chemical Society Review, 34, 164–178.

Dzieraba, C. D., & Combs, A. P. (2002). Microwave-assisted chemistry as a tool for drug discovery. Annual reports in medicinal chemistry, Academic Press, 37, 247–256.

Freeman, S. A., Booske, J. H., & Cooper, R. F. (1998). Journal of Applied Physics, 83, 5761–5772.

Gedye, R. N., Smith, F. E., Westaway, K., Ali, H., Baldisera, L., Laberge, L., & Rousell, J. (1986). Tetrahedron Letters, 27, 279–282.

Giguere, R. J., Bray, T. L., Duncan, S. M., & Majetich, G. (1986). Tetrahedron Letters, 27, 4945–4948.

Hull, A. W. (1921a). Journal of the American Institute of Electrical Engineers, 40, 715–723.

Hull, A. W. (1921b). Physical Review, 18, 31–57.

Johansson, H. (2001). American Laboratory, 33, 28–32.

Kappe, C. O. (2004). Controlled Microwave Heating in Modern Organic Synthesis. Angewandte Chemie International Edition, 43, 6250–6284.

Khadilkar, B. M., & Madyar, V. R. (2001). Organic Process Research and Development, 5, 452–455.

Larhed, M., & Hall berg, A. (2001). Drug Discovery Today, 6, 406–416.

Loupy, A., Petit, A., Hamelin, J., Texier-Boullet, F., Jacquault, P., & Mathe, D. (1998). Synthesis, 9, 1213–1234.

Moseley, J. D., Lenden, P., Lockwood, M., Ruda, K., Sherlock, J. P., Thomson, A. D., & Gilday, J. P. (2008). Organic Process Research and Development, 12, 30–40.

Murray, J. K., & Gellman, S. H. (2006). Journal of Combinatorial Chemistry, 8, 58–65.

Murray, J. K., Farooqi, B., Sadowsky, J. D., Scalf, M., Freund, W. A., Smith, L. M., Chen, J., & Gellman, S. H. (2005). Journal of the American Chemical Society, 127, 13271–13280.

Pipus, G., Plazl, I., & Lescovsak, S. T. (2002). Industrial and Engineering Chemistry Research, 41, 1129–1134.

Plazl, I. (1994). Acta Chimica Slovenica, 41, 437–445.

Plazl, I., Leskovek, S., & Koloini, T. (1995). The Chemical Engineering Journal, 59, 253–257.

Raner, K. D., Strauss, C. R., Traineer R. W., & Thorn, J. S. (1995). Journal of Organic Chemistry, 60, 2456–2460.

Robb, G., Harrison, A., & Whittaker, A. G. (2002). Phys Chem Comm, 5, 135–137.

Sharma, S. V., Ramasarma G. V. S., & Suresh, B. M. (2002). Indian Journal of Pharmaceutical Sciences, 64, 337–344.

Stadler, A., Yousefi, B. H., Dallinger, D., Walla, P., Van derEycken, E., Kaval, N., & Kappe, C. O. (2003). Organic Process Research and Development, 7, 707–716.

Strauss, C. R., & Trainor, R. W. (1995). Australian Journal of Chemistry, 48, 1665–1692.

Stuerga, D. A. C., & Gaillard, P. (1996). Journal of Microwave Power and Electromagnetic Energy, 31, 101–113.

Stuerga, D. A. C., & Gaillard, P. (1996). Journal of Microwave Power and Electromagnetic Energy, 31, 87–100.

Villemin, D., & Thibault-Starzyk, F. (1991). Journal of Chemical Education, 68, 346.

Wathey, B., Tierney, J., Lidrom, P., & Westman, J. (2002). Drug Discovery Today, 7, 373–380.

Whittaker, A. K. (2005). Chemistry of Materials, 17, 3426–3432.

CHAPTER 3

OXIDATION

CHETNA AMETA, POONAM KUMAWAT, and
ABHILASHA TRIPATHI

CONTENTS

3.1   INTRODUCTION

Oxidation is a chemical reaction that involves the transfer of electrons. Specifically, it means the side that gives away electrons. Oxidation involves the loss of electrons or hydrogen or gain of oxygen or increase in oxidation state. Any chemical reaction, in which the oxidation numbers (oxidation states) of the atoms are increased is termed as an oxidation reaction. Oxidation-reduction reactions (or redox) reactions are chemical reactions that involve a transfer of electrons between two species. Oxidation-reduction reactions are quite vital for many biochemical reactions and industrial processes. The electron transfer system in cells and oxidation of glucose in the human body are some well known examples of redox reactions. Redox reactions are used to reduce ores to obtain metals, to produce electrochemical cells, to convert ammonia into nitric acid for fertilizers, to coat compact discs, etc.

Microwave-assisted organic synthesis (MAOS) is a new and rapidly developing area in synthetic organic chemistry. This synthetic technique has been based on the empirical observations that some organic reactions proceed much faster and with higher yields under microwave irradiation compared to conventional heating. In many cases, reactions that normally require many hours at reflux temperature under classical conditions can be completed within few minutes or even seconds in a microwave oven, even at comparable reaction temperatures. While different hypotheses have been proposed to account for the effect of microwaves on organic reactions, the reason for such dramatic acceleration effects still remains largely unknown. Regardless of the exact origin of the microwave effect, it is extremely efficient and applicable to a very broad range of practical syntheses.

Over the past few years, a considerable number of reactions have been developed in which inorganic solid supports such as alumina, silica gel, montmorillonite, etc. appeared to be useful in terms of mildness of conditions, yield and convenience. Microwave-assisted solvent-free synthesis in organic reactions has been of growing interest as an efficient, economic and clean procedure.

3.2   HYDROCARBONS

Luu et al. (2009) have converted efficiently some allylbenzenes from essential oils (Allylbenzenes, such as safrole and eugenol, are the main ingredients of several essential oils in plants) into the corresponding benzaldehydes in good yields by a two-step green reaction. This pathway is based on a solventless alkene group isomerization by KF/Al_2O_3 to form the corresponding 1-arylpropene and a subsequent solventless oxidation of the latter to the corresponding benzaldehyde by $KMnO_4$/ $CuSO_4.5H_2O$. Microwave irradiation provides products in very short reaction times (<15 min).

where X = H, -OCH$_3$, -O-CH$_2$-O-; Y = H, -OCH$_3$, -O-CH$_2$-O-.

An environmentally benign and simple versatile method for hydroboration-oxidation of alkenes, dienes and alkynes under microwave irradiation has been explored by Jayakumar et al. (2008). They carried out a simple and facile hydroboration-oxidation of alkenes by N,N-dimethylaniline-borane using microwave irradiation. Hydroboration of various alkenes with N,N-dimethylaniline-borane was carried out in dry THF under specially designed microwave apparatus. It was designed to accommodate magnetic stirrer and refluxing condenser with mercury bubbler. Hydroboration was followed by oxidation with H$_2$O$_2$/NaOH, which resulted in the formation of corresponding alcohols in 90–95% yield.

The ease of recovery of amine after hydroboration reaction and the possibility of recycling makes this method environmentally friendly.

Manktala et al. (2006) reported that the allylic oxidation of alkene can be carried out successfully keeping other functional groups intact in the presence of catalytic quantities of SeO$_2$, using urea-hydrogen peroxide (UHP). A comparative study has also been carried out keeping all other conditions identical. A decrease in reaction time to 40 sec was observed with enhanced yields (75–85%). It takes 1 h or more in conventional heating with relatively low yields (58–75%).

R	Time (s)	Yield (%)
CH$_2$CH$_2$OH	40	80
CH$_2$CH$_2$OAc	40	85
CH$_2$CH$_2$Br	40	75
= O	40	80

Chen et al. (2008) observed that oxidation of functionalized internal alkynes with DMSO in the presence of I_2, gives 1,2-diaryldiketones under microwave irradiation with good yields. They developed this microwave enhanced, simple and efficient process for synthesis of α-diketones using DMSO/I_2. The most attractive feature of this method was the short reaction times, good yields, low-cost and easy preparation.

It was observed that the yields of product increase with increasing mol % of iodine.

I (mol %)	Yield (%)
10	19
20	21
30	64
40	74
50	95

Selenium dioxide oxidizes 1, 2-diarylethanones to corresponding diones in about 8 h. The same oxidation was carried out under microwave radiations by Shirude et al. (2006) by using dimethylsulfoxide as solvent, the reaction time was reduced considerably (30 to 90 sec).

3.3 ALCOHOLS

In recent years, oxidation processes have received much attention, especially in the search for selective and environmentally friendly oxidants. The oxidation of alcohols to carbonyl compounds is an elementary transformation in organic chemistry since carbonyl compounds are widely used as intermediates; both in manufacturing and research. Although the oxidation of organic compounds under nonaqueous conditions has become an effective technique for modern organic synthesis; however, the methods still suffer some disadvantages including the cost of preparation, long reaction time and tedious work up procedures.

A rapid microwave oxidation protocol for the oxidation of alcohol to carbonyl compound has been reported by Varma (2001). Different types of oxidizing reagents have been used with microwave irradiations. Alcohols were converted into corresponding carbonyl compounds in the presence of montmorillonite K10 clay-supported iron (III) nitrate (Clayfen), Oxone-alumina, IBD-alumina, 35% MnO_2 doped silica, CrO_3-alumina or $CuSO_4$-alumina under microwave irradiation and solvent-free conditions in few minutes (0.25–3.5 min).

Ghorbani-Vaghei et al. (2007) reported an efficient rapid and mild methodology for the oxidation of primary and secondary alcohols to the corresponding carbonyl compounds in presence of N,N,N',N'-tetrabromobenzene-1,3-disulfonamide (TBBDA) or poly(N-bromobenzene-1,3-disulfonamide) (PBBS) using microwave irradiation under solvent-free conditions.

where R_1 = Aryl, Alkyl; R_2 = H, Alkyl.

The formation of carbonyl compounds from alcohols most likely involves a Br^+ transfer from reagent to substrate. This oxidation method has an additional advantage that it does not require highly specialized equipment and expensive reagents. The reaction is fast and easy to perform under mild reaction conditions, there is no side product formation and it is completed in very less reaction time. A wide range of aliphatic and benzylic alcohols can be converted to their corresponding aromatic aldehydes and ketones.

Benzyl alcohol, p-nitrobenzyl alcohol and 2,4-dicholorobenzyl alcohol react very fast, whereas other benzylic alcohols required a longer reaction time. The nature of substituents on the aromatic ring affects the rate of the reaction. It was observed that the introduction of an electron withdrawing substituent into the aromatic ring increases the yield compared to electron donating substituent. Aliphatic alcohols were less reactive than aromatic alcohols and therefore, the reaction was observed to be slower. The oxidation of secondary alcohols was slower and the yield was also lower than with primary alcohols. It was interesting to observe that primary and secondary alcohols were oxidized to aldehydes and ketones without overoxidation to carboxylic acids.

Aliphatic, benzylic and allylic alcohols were also rapidly oxidized without any overoxidation to carboxylic acids. Secondary carbinols were slowly oxidized. These reactions are highly chemoselective. The method was found to be highly selective for primary aromatic alcohols. In a mixture of benzyl alcohol and benzhydrol, 85% of benzaldehyde and 10% of benzophenone was obtained.

Primary and secondary alcohols were oxidized into their corresponding ketones and carboxylic acids within 10–20 min using 30% aqueous H_2O_2 in the presence of sodium tungstate and tetrabutylammonium hydrogen sulfate (TBAHS) as a phase-transfer catalyst under microwave irradiation (Bogdal and Lukasiewicz, 2000). It is an environmentally benign and safe protocol. Its advantages are a simple reaction set-up, use of commercially available 30% aqueous hydrogen peroxide and catalysts, high product yields, shorter reaction times as well as the elimination of solvents and inorganic solid supports.

Primary alcohols $C_6H_{13}OH$ and $C_7H_{16}OH$ were converted into $C_5H_{11}COOH$ (75% yield) and $C_7H_{16}COOH$ (80% yield), respectively only in 20 min while secondary alcohol $CH_3CHOHC_6H_{13}$ was oxidized to $CH_3COC_6H_{13}$ (94% yield) in 10 min.

Galica et al. (2013) also reported microwave-assisted oxidation of alcohols by hydrogen peroxide catalyzed by tetrabutylammonium decatungstate (TBADT). Catalytic activity of tetrabutylammonium decatungstate (VI) was observed in the oxidation of some selected alcohols with hydrogen peroxide using 1,2-dichloroethane/ water or acetonitrile/ water as a solvent system. The most efficient reaction system was found to be acetonitrile/water. The advantage of this environmentally benign and safe procedure was a simple reaction set-up. Moreover, the use of a microwave-pressurized reactor reduces the reaction time by half compared with the time necessary for a conventional reactor.

The oxidation of native starch by hydrogen peroxide has been carried out by Lukasiewicz et al. (2005) using microwaves as an energy source. The reaction results in oxidation of some primary hydroxyl group into corresponding aldehyde or carboxylic group. The changes in the oxidation level observed at the same time for microwave as compared to conventional method could be the good alternative for other polysaccharide oxidation processes conducted at both; the laboratory and industrial scales.

An efficient methodology for oxidation of alcohols and polyarenes using cetyltrimethylammonium bromochromate (CTMABC) or tetramethylammonium fluorochromate (TMAFC) in CH_2Cl_2 as solvent under microwave irradiation has been reported (Mohammadi et al., 2008, 2009). This method offers some advantages like simplicity, mild operation condition, no side product and a very short reaction time. Apart from these, a wide range of substrates can be used and the reduced reagent $CH_3(CH_2)_{15}N^+(CH_3)_3CrO_2Br^-$ and $(CH_3)_4N^+CrO_2F^-$ could also be recycled after oxidation.

CTMABC had been a very well suited oxidizing reagent in microwave synthesis, as it is an ionic and magnetically retrievable material.

Oxidation of organic compounds to their corresponding carbonyl compounds has also been studied by Mohammadi (2013) under microwave irradiation. It is an efficient, clean and mild methodology for oxidation using cetyltrimethylammonium chlorochromate (CTMACC). These reactions were completed in short times with excellent yields.

A facile, efficient and selective solvent-free synthesis of ketones from secondary alcohols with tert-butylhydroperoxide (TBHP) as the oxidant under microwave irradiation has been achieved, where the copper (II) 2,4-alkoxy-1,3,5-triazapentadienato complexes (CATAPD) worked as efficient catalysts providing high yields (even upto 100%) (Figiel et al., 2010).

They observed that these complexes were remarkably active catalysts and the oxidation was fast, selective, require a small amount of catalyst and proceeds in the absence of any additional solvent or additives; thus, making it a green catalytic process.

A simple and efficient protocol for microwave-assisted solvent-free oxidation of hydrobenzoins to benzoins or benzils, benzoins to benzils, and alcohols to the corresponding aldehydes or ketones, using N-bromosuccinimide over neutral alumina has been also reported (Khurana and Arora, 2008).

It was observed that hydrobenzoin was selectively oxidized to benzoin (99% yield), when exposed to microwaves for 20 sec or to benzyl in quantitative yields, when exposed for 2 min using NBS and neutral Al$_2$O$_3$. It is a quite simple and efficient solvent-free protocol for the microwave accelerated oxidation of hydrobenzoins, benzoins and alcohols with N-bromosuccinimide–neutral alumina. The reaction conditions were optimized by carrying out reactions of hydrobenzoin as a model substrate and changing the molar ratios of NBS and Al$_2$O$_3$ besides variable microwave irradiation.

Silica gel supported bis(trimethylsilyl) chromate in dry media also provided a fast, efficient and simple method for oxidation of alcohols to corresponding carbonyl compounds under microwave irradiation (Heravi et al., 1999).

Bis (trimethylsilyl) chromate supported on silica gel can serve as an excellent oxidant for the oxidation of various types of alcohols under microwave irradiation without any solvent. This method decreases the time of reactions dramatically compared to ordinary condition. The products can be isolated by addition of dichoromethane to the crude and filtration of reaction mixture. The high reactivity and selectivity of the supported reagent under microwave irradiation avoid the use of an excess of the oxidant, which may cause an overoxidation and other possible side reactions.

The oxidation of benzylic alcohols to carbonyl compounds is one of the fundamental reactions in organic synthesis. However, the majority of methods using various oxidizing reagents commonly suffer from disadvantages such as difficulty in manipulation, longer reaction time and utilization of toxic reagents. The [hydroxy(tosyloxy)iodo]benzene is one of the most versatile reagent and when it is coupled with microwave, it provides a rapid and convenient way to prepare corresponding carbonyl compounds in solvent-free conditions with ecofriendly conditions and fast reaction rates, making this protocol a valuable alternative to the other existing methods. Therefore, an efficient method for the oxidation of benzylic alcohols by [hydroxy(tosyloxy)iodo]benzene (HTIB, Koser's reagent) under solvent-free microwave irradiation conditions has been described (Lee et al., 2004).

Palombi et al. (1997) reported the oxidation of secondary (linear and cyclic) and benzylic alcohols to the corresponding carbonyl compounds using t-butyl hydroperoxide, 3Å molecular sieves (MS) and microwave irradiation under solvent-free conditions. Under the same conditions, α,β-unsaturated alcohols are converted into α,β-epoxyalcohols in regio- and diastereoselective way. The reactions provided the corresponding carbonyl compounds in good yields.

Under microwave irradiations, benzimidazolium dichromate (BIDC) in CCl_4 was used as a selective reagent for the oxidation of benzylic and allylic alcohols to the corresponding carbonyl compounds with 73–97% yields (Meng et al., 1998).

Pyridinium chlorochromate has also been used (Chakraborty and Bordoloi, 1999) under microwave irradiation for the oxidation of alcohols to the corresponding carbonyl functions with an efficient and mild methodology. Microwave irradiation of alcohols with silica supported active manganese dioxide in solvent-free condition provides rapid and selective oxidation of alcohols to the corresponding carbonyl compounds (Varma et al., 1997).

$$\begin{array}{c} R_2 \\ \diagdown \\ CH{-}OH \\ \diagup \\ R_1 \end{array} \xrightarrow[\text{MW, 20-60 s}]{MnO_2.\text{Silica}} \begin{array}{c} R_2 \\ \diagdown \\ C{=}O \\ \diagup \\ R_1 \end{array}$$

(67-96%)

where R_1 = C_6H_5; R_2 = H; the yield was 88% in 20 sec.

There is an increasing need as well as interest in developing some processes that minimize production of toxic waste due to environmental concerns. In this context, the combination of supported reagents and microwave irradiation without solvent has proved to be of prime importance in synthetic organic chemistry with a wide range of reactions in short times, high conversion and selectivity. However, some of these reagents and catalysts have limitations like availability of the reagent, difficult work-up, longer reaction time, toxicity and high cost of the reagents. Thus, Hashemi et al. (2005) used iodic acid as an oxidant, which is milder and more selective with moderate oxidizing power in aqueous acidic medium and has lower toxicity to humans.

$$\underset{R}{\overset{R'}{>}}CHOH \xrightarrow[MW]{HIO_3/K-10\ clay\ or\ silica\ gel} \underset{R}{\overset{R'}{>}}C=O$$

where R and R' = H, Aryl, Heteroaryl.

They carried out oxidation of benzyl alcohol successfully by using montmorillonite K10 and silica gel supported iodic acid under microwave irradiation without solvent and that too with simplicity of performance and with high yield (92%).

Rajabi et al. (2013) also used supported iron oxide nanoparticles on aluminosilicate catalyst as efficient and easily recoverable material in the aqueous selective oxidation of alcohols to their corresponding carbonyl compounds using hydrogen peroxide under both; conventional and microwave heating. This procedure also has an easy work-up, simplicity and the utilization of mild reaction conditions as well as high selectivity toward aldehydes. In addition, the supported iron oxide nanoparticles could be easily recovered from the reaction mixture and reused several times without any loss in activity.

A remarkable decrease in reaction time from 4 h to 15 min with 98% yield in presence of microwave irradiation has been reported without any significant loss of activity and metal leaching.

The solvent-free potassium permanganate promoted oxidation of alcohols into the corresponding carbonyl derivatives has been examined by Luu et al. (2008). Secondary alcohols were oxidized very efficiently to the corresponding ketones at ambient temperature by $KMnO_4$ absorbed on a fourfold molar amount of copper (II) sulfate pentahydrate. The reaction rate was enhanced considerably by ultrasonic irradiation also, but drastically in the presence of microwave irradiation, may be due to synergistic effect. Hydroquinone can be converted 100% into p-benzoquinone within 4.5 min under the microwave irradiations. Benzylic alcohols were most efficiently oxidized to the corresponding benzaldehydes under heterogeneous reaction conditions.

It was observed that the most efficient oxidant was $KMnO_4$ absorbed on a four-fold molar amount of $CuSO_4.5H_2O$ (100% yield), but attempts were made to oxidize 2-heptanol, under solvent-free conditions, by $KMnO_4$ alone (i.e., in the absence of the support of an inorganic salt hydrate) were absolutely unsuccessful. Various inorganic salts were tried and yielded varied amounts of the product. The better supports include nickel sulfate (90%), zinc sulfate (74%), and cobalt sulfate (41%) while other supports were not that interesting like magnesium sulfate (12%), calcium sulfate (11%) and barium chloride (3%). Zeolite HZSM-5 was used as a catalyst for the oxidation of alcohols to the corresponding carbonyl compound with chromium trioxide under solvent-free conditions and microwave irradiation (Heravi et al., 1999).

Varma et al. (1998) used alumina-supported iodobenzene diacetate (IBD) as an oxidant for the rapid oxidation of alcohols to carbonyl compounds under microwave irradiation. This reaction takes 1–3 min and provides excellent yields.

Under microwave irradiation, the oxidation of variety of benzylic alcohols to carbonyl compounds using clay supported ammonium nitrate (clayan) was carried out by Meshram et al. (2006). The selectivity and solvent-free conditions were the important features of this procedure. It is known that nitrates in the presence of clay produced nitrosonium ion (NO$^+$) and therefore, this reaction may proceed via nitrosonium intermediate.

They demonstrated that it is a rapid, highly selective and environmentally benign procedure for oxidation. Easy preparation of reagent, self-destroying nature and dry conditions made this method more attractive.

Lee et al. (2005) developed a facile and environment friendly synthetic procedure to oxidize benzyl alcohols into the corresponding benzaldehydes under microwave exposure. They were initially interested in utilization of neat nitric acid in the oxidation of benzyl alcohols under solvent-free condition and microwave irradiation. However, oxidation of benzyl alcohols with neat nitric acid under microwave irradiation proved to be unpractical because the reaction mixture scattered. After examining various metal nitrate reagent systems, the sodium nitrate/p-toluenesulfonic acid system was found to be the most suitable in terms of cleanness and effectiveness for the oxidation of benzyl alcohols under these conditions.

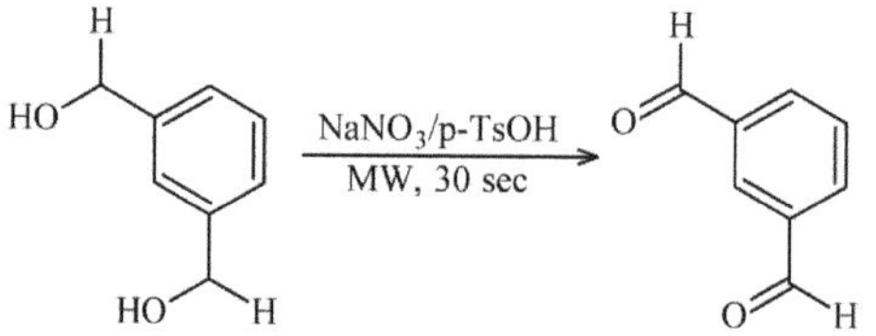

(N-Heterocyclic carbene)-Pd (NHC-Pd) has been used as a precatalyst for the anaerobic oxidation of secondary alcohols. It was reported that the use of this complex allows for a drastic reduction in the reaction times and catalyst loading, when compared to the unsaturated counterpart. This catalytic system was compatible with the use of microwave dielectric heating. Domino Pd-catalyzed oxidation-arylation reactions of secondary alcohols have also been carried out (Landers et al., 2011).

Pd-catalyzed anaerobic oxidation:

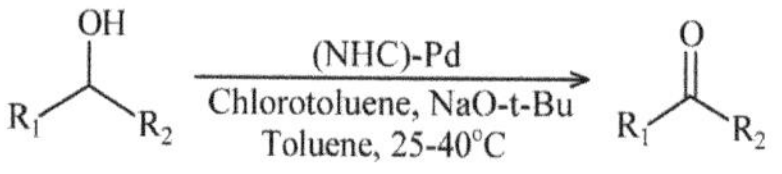

Pd-catalyzed domino oxidation-arylation:

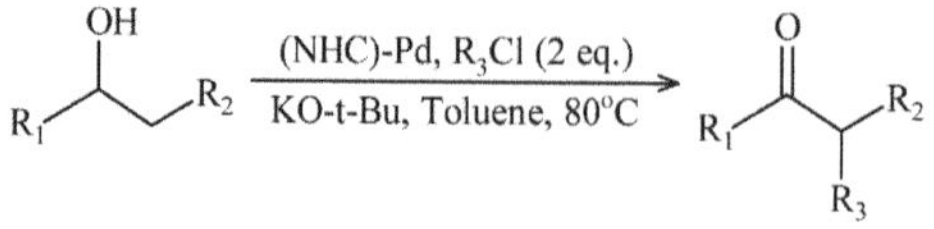

The conversion of a α-hydroxyketones into 1,2-diketones using three different oxidants under microwave irradiation with good yield (78–94%) was reported by Mitra et al. (1999).

Use of sodium tungstate as catalyst in 30% aqueous hydrogen peroxide oxidizes primary alcohols to the corresponding carboxylic acids under microwave irradiation.

Primary alcohol   Carboxylic acid

3.4 ALDEHYDES

The oxidation of aldehydes to carboxylic acids is an important transformation in organic synthesis and several methods have been developed to accomplish this conversion (Srivastava and Venkataramani, 1998). Oxidation of benzaldehydes with oxone (2 $KHSO_5$, $KHSO_4$, K_2SO_4)/wet-alumina under microwave irradiation in solventless system is a rapid, manipulatively simple, inexpensive and selective protocol. (Nikje and Bigdeli, 2004).

$$Ar-CHO \xrightarrow{\text{Oxone/Wet-alumina}} ArCOOH$$

Shie and Fang (2003) carried out oxidation of some aromatic aldehydes with I_2 in ammonia water and in situ cycloadditions with NaN_3/$ZnBr_2$ under microwave irradiation at 80 °C for 10 min to afford 5-aryl-1,2,3,4-tetrazoles in 70–83% yields. This microwave-accelerated reaction in aqueous media is quite safer and more efficient than conventional heating using prolonged reflux (17–48 h) at a high temperature (>100 °C) where R = C_6H_5, 4-MeOC$_6$H$_4$, 4-O$_2$NC$_6$H$_4$, 2-Furyl, 2-Thienyl.

They worked out such reactions with alcohols also along with aldehydes (2007).

Use of hydrogen peroxide and KMnO$_4$ as oxidants for H$_5$PV$_2$Mo$_{10}$O$_{40}$ catalyzed oxidation of aromatic aldehydes to the corresponding carboxylic acids under mild conditions has been reported by Shojaei et al. (2011). This system provides an efficient, convenient and practical method for the oxidation of aromatic aldehydes.

Aromatic aldehydes possessing a wide variety of substituents were smoothly converted to carboxylic acids in excellent yields by stoichiometric bismuth (III) nitrate pentahydrate in presence of aerial oxygen in a microwave-oven under solvent-free conditions. This method is also environmentally benign being solvent-free, operationally simple, highly chemoselective and employing no bulky work-up procedures, along with wide general applicability, short reaction times, mild reaction conditions, large scale-ups and very good yields (Mukhopadhyay and Datta, 2008).

Preyssler's anion, [NaP$_5$W$_{30}$O$_{110}$]$^{14-}$, was used as catalyst for the oxidation of aromatic aldehydes to related carboxylic acids using hydrogen peroxide as an oxidizing agent, under microwave irradiation or at 70°C. Oxidation of aldehydes is a surface type reaction and with Preyssler's anion, it produces the highest yields (Bamoharram et al., 2006).

where R = 4-Cl, 4-Br, 4-Nitro, 2-Methyl, 3-Methyl, 4-Methyl, 2-Hydroxy, 3-Hydroxy, 4-Hydroxy.

In most of the reactions, benzaldehydes with electron withdrawing substituents are thought to be hard-oxidizing aldehydes, but it was observed that product yields are highest in the presence of Preyssler's catalyst even with electron withdrawing substituents. The effects of various substituents on the yields of carboxylic acids have been reported.

The order of the efficiency in the term of yield for this oxidation is:

$4\text{-Cl} > 4\text{-Br} > 4\text{-NO}_2 > 4\text{-CH}_3 > 3\text{-CH}_3 > 2\text{-CH}_3$

Only traces of the products were obtained, when hydroxyl group was present on any position. It was observed that not only the nature of the substituent is important, but its position also. Electron withdrawing substituents result in higher yields than electron donating substituents in oxidation of these aldehydes.

3.5 ARENES

The oxidation of toluene to the benzoic acid with $KMnO_4$ under normal conditions takes 10–12 h reflux, which takes only 5 min under microwave irradiation and the yield is 40% (Surati et al., 2012).

Oussaid and Loupy (1997) used $KMnO_4$ impregnated on alumina under microwave activation in dry media to oxidize arenes into ketones within 10–30 min instead of several days under conventional conditions.

Magtrieve™ is a DuPont's trademark for the oxidant based on tetravalent chromium dioxide (CrO_2). Lukasiewz and Bogdal (2002) used Magtrieve™ as an oxidant, because it has been proved to be a useful oxidant in some reactions including the oxidation of alcohols. Magtrieve™ is very well suited reagent as an oxidant for microwave synthesis, because it is an ionic and magnetically retrievable material.

It carries a benefit of converting of electromagnetic energy efficiently into heat according to dielectric heating mechanism.

The satisfactory yields in this reaction supported Magtrieve™ as a useful microwave-working oxidant. It is because of the fact that its ionic structure and magnetic properties of Magtrieve™ is strongly coupled with microwave irradiation. However, a toxic transition metal oxidant chromium has been used in this process, even then it is environment friendly because of short reaction time, easy set-up and easy separation of oxidant. Additionally, the possible recycling of this oxidant put it in the list of methods of oxidation as a powerful and green tool in modern organic synthesis.

The oxidation of some arenes with the alkyl side groups has been carried out by hydrogen peroxide under microwave irradiation. Tungstoboric acid was used as the activator of hydrogen peroxide. The catalyst was used under both homogeneous and heterogeneous conditions (Lukasiewicz et al., 2006). Mohammadi et al. (2008) carried out oxidation of alcohols and polyarenes with tetramethylammonium fluro-chromate under microwave irradiation.

Zhang et al. (2012) studied catalytic oxidation of benzene in two different heating modes, microwave heating and conventional electric furnace heating. The catalyst had better catalytic activity for the oxidation of benzene under microwave heating than electric furnace heating. It was also observed that benzene acquires high oxidation efficiency due to the local hot spots and dipole polarization effect of microwave and stable bed reaction temperature. The effects of copper (Cu)-manganese (Mn) mass ratio, doping dose of cerium (Ce) and calcination temperature on the catalytic activity of Cu-Mn-Ce/molecular sieve catalyst were also investigated in catalytic oxidation of benzene with microwave heating.

3.6   PHENOLS

Phenols were readily oxidized to corresponding quinones by hydrogen peroxide, in the presence of catalyst Fe (III). Both were supported on alumina. The reaction was carried out under microwave irradiation and it reduces the time of reaction and the yield of product was also improved. (Sahu et al., 2009).

Oxidation of phenol and 1,4-dihydroxybenzene to 1,4-benzoquinone, using hydrogen peroxide supported on alumina in the presence of Fe (III) under microwave irradiation (2.8 and 1.5 min) gave 48% and 70% yields, respectively.

Gomez-Lara et al. (2000) used MnO_2 or HNO_3 on bentonite clay for the oxidation of some hydroquinones to the corresponding quinones under microwave irradiations. These reactions were completed in short times with good yields (100%).

3.7 LACTAMS

Taherpour and Mansuri (2005) reported a rapid conversion of lactams to cyclic imides using peracetic acid (CH_3-CO_3H) and manganese chloride ($MnCl_2$) in ethyl acetate as solvent and under microwave irradiation (5 min) and obtained good yields. Cyclic imides are very useful materials in the chemical industry.

2-Pyrolidinone was converted into succinimide with 70% yields. Succinimide is used as growth stimulants for plants and/or as starting materials for the synthesis of heterocycles.

3.8 SULPHIDES

A variety of alkyl, aryl and cyclic sulfides are rapidly oxidized to the corresponding sulfoxides in the presence of PhI(OAc)$_2$-alumina with high yield upon microwave exposure (Varma, 1999).

Varma (2006) has reported the use of urea and hydrogen peroxide as an oxidant for converting alkyl cyanide, sulfides, o-hydroxybenzaldehyde and pyridine into an amide, sulphoxide or sulphone, catechon and pyridine-oxide, respectively. The reactions were carried out on recyclable mineral support like alumina, silica, clay, etc.

Using solid reagent systems, copper (II) sulfate–alumina, or oxone–wet alumina symmetrical and unsymmetrical benzoins can be rapidly oxidized to benzils in high yields under the influence of microwaves (Varma et al., 1999).

when R, = R$_1$ = C$_6$H$_5$, 88% yield was obtained in 2 min.

3.9 ENAMINES

Benhaliliba et al. (1998) successfully oxidized β,β-disubstituted enamines into ketones and formamides with KMnO$_4$/Al$_2$O$_3$ in domestic (255 W, 82 °C) as well as in focused (330 W, 140 °C) microwave ovens under solvent-free conditions. Oxidation yields were higher under focused microwave oven as compared to domestic oven as well as classical heating.

3.10 MISCELLANEOUS

Heravi et al. (1998) have reported direct oxidative deprotection of different trimethylsilyl ethers to their corresponding carbonyl compounds with 75–92% yields using montmorillonite K10 supported bis(trimethylsilyl)chromate in dichloromethane.

Villemind and Hammadi (1995) used DMSO in presence of KSF for the oxidation of epoxide into α-hydroxyketones under microwave irradiation. The α-hydroxyketones were produced in good yields (55–90%).

SeO$_2$/t-BuOOH adsorbed on SiO$_2$ has been employed as a selective reagent for the oxidation of allylic methyl groups to trans-α,β-unsaturated aldehydes under microwave irradiation. The product was obtained in 68–85% yield (Singh et al., 1997).

Varma and Kumar (1999) reported the use of phenyliodine (III) bis(trifluoroacetate) (PIFA) or elemental sulfur for oxidation of 1,4-dihydropyridines to pyridines under microwave irradiations. Dealkylation at the 4-position in the cases of ethyl, isopropyl and benzyl substituted dihydropyridine derivatives with PIFA is circumvented by an alternative general procedure using elemental sulfur, which provides pyridines in good yield (68–90%).

Khurana et al. (2008) reported solvent-free oxidation of organic halides using aqueous hydrogen peroxide under controlled microwave irradiation for the synthesis of carbonyl compounds while Varma and Dahiya (1998) reported that the claycop (Copper (II) nitrate on clay)–hydrogen peroxide is an efficient and selective oxidizing reagent for a variety of compounds under solvent-free conditions using microwave irradiation. The reaction gives 82–83% yields in 60–90 sec.

$$\begin{array}{c} R_1 \\ \diagdown \\ \diagup \\ R_2 \end{array} CH{-}R_3 \xrightarrow[\text{MW}]{\text{Claycop-H}_2\text{O}_2} \begin{array}{c} R_1 \\ \diagdown \\ \diagup \\ R_2 \end{array}C{=}O$$

where R_3 = COOH, NH_2, CN, Br.

Hashemi et al. (2004) synthesized aromatic aldehydes and ketones from alkyl halides without solvent using wet montmorillonite K10 supported iodic acid as oxidant.

Oxidation is an important step in converting a functional group to another and it is most desired in organic synthesis. Microwave irradiation helps in carrying out such reactions in a rapid manner with a variety of oxidants and at some places with specificity along with no possibility of over oxidation.

KEYWORDS

- **Alcohol**
- **Aldehyde**
- **Arenes**
- **Hydrocarbon**
- **Oxidation**
- **Phenol**

REFERENCES

Bamoharram, F. F., Roshani, M., Alizadeh, M. H., Razavi, H., & Moghayadi, M. (2006). Journal of Brazilian Chemical Society, 17, 505–509.

Benhaliliba, H., Derdour, A., Bazureau, J., Texier-Boullet, J., & Hamelin, J. (1998). Tetrahedron Letters, 39, 541–542.

Bogdal, D., & Lukasiewicz, M. (2000). Synlett, 1, 143–145.

Chakraborty, V., & Bordoloi, M. (1999). Journal of Chemical Research (S), 118–119.

Chen, M., Zhao, Q., She, D.-B., Yang, M.-Y., Hui, H.-H., & Huang, G.-S. (2008). Journal of Chemical Sciences, 119, 347–351.

Figiel, P. J., Kopylovich, M. N., Lasri, J., da Silva, M. F. C. G., da Silva, J. J. R. F., & Pombeiro, A. J. L. (2010). Chemical Communications, 46, 2766–2768.

Galica, M., Kasprzyk, W., Bednarz, S., & Bogdal, D. (2013). Chemical Papers, 67, 1240–1244.

Ghorbani-Vaghei, R., Veisi, H., & Amiri, M. (2007). Journal of Chinese Chemical Society, 54, 1257–1260.

Gomez-Lara, J., Gutierrez-Perez, R., Penieres-Carrillo, G., Lopez-Cortes, J., G., Escudero-Salas, A., & Alvarez-Toledano, C. (2000). Synthetic Communications, 30, 2713–2720.

Hashemi M. M., Rahimi, A., & Ahmadibeni, Y. (2004) Acta Chimica Slovenica, 51, 333–336.

Hashemi, M. M., Rahimi, A., Karimi-Jaberi, Z., & Ahmadibeni, Y. (2005). Acta Chimica Slovenica, 52, 86–87.

Heravi, M. M., Ajami, D., & Tabar-Hydar, K. (1999). Synthetic Communications, 29, 163–166.

Heravi, M., M., Ajami, D., Tabar-Hydar, K., & Ghassemzadeh, M. (1999). Journal of Chemical Research (S), 334–335.

Heravi, M., M., Ajami, D., Tabar-Hydar, K., & Mojtahedi, M. M. (1998). Journal of Chemical Research (S), 620–621.

Jayakumar, S. V., Srinivas, K. A., Hiriyanna, S. G., & Pati, H. N. (2008). Rasayan Journal of Chemistry, 1, 326–330.

Khurana, J. M., & Sanjay, K. (2008). Academic Journal, 137–139.

Khurana, J. M., & Arora, R. (2008). Arkivoc, (XIV), 211–215.

Landers, B., Berini, C., Wang, C., & Navarro, O. (2011). Journal of Organic Chemistry, 76, 1390–1397.

Lee, J. C., Lee, J. Y., & Lee, S. J. (2004). Tetrahedron Letters, 45, 4939–4941.

Lee, J. C., Lee, J. Y., & Lee, J. M. (2005). Bulletin of the Korean Chemical Society, 26, 1300–1302.

Lukasiewicz, M., Bogdal, D., & Pielichowski, J. (2006). Molecular Diversity, 491–493.

Lukasiewicz, M., & Bogdal, D. (2002). Sixth International Electronic Conference on Synthetic Organic Chemistry (ECSOC–6).

Luu, T. X. T., Christensen, P., Duus, F., & Le, T. N. (2008). Synthetic Communications, 38, 2011–2024.

Luu, T. X., Lam, T. T., Le, T. N., & Duus, F. (2009). Molecules, 14, 3411–3424.

Manktala, R., Dhillon, R. S., & Chhabra. B. R. (2006). Indian Journal of Chemistry, 45B, 1591–1594.

Meng, Q., Feng, J., Bian, N., Liu, B., & Li, C. (1998). Synthetic Communications, 28, 1097–1102.

Meshram, H. M., Muralidhar, B., Eeshwaraiah, B., Ramesh Babu, K., Aravind, D., & Yadav, J. S. (2006). Indian Journal of Chemistry, 45B, 500–502.

Mitra, A. K., De, A., & Karchaudhuri, N. (1999), Journal of Chemical Research (S), 246–247.

Mohammadi, M. K., Ghammamy, S., & Imanieh, H. (2008). Bulletin of Chemical Society of Ethiopia, 22, 449–452.

Mohammadi, M. K. (2013). Open Journal of Synthesis Theory and Applications, 2, 87–90.

Mohammadi, M. K., Ghammamy, S., & Imanieh, H. (2009). Chinese Journal of Chemistry, 27, 1501–1504.

Mohammadi, M. K., Ghammamy, S., and Imanieh, H. (2008). Bulletin of Chemical Society of Ethiopia, 22, 1011–3924.

Mukhopadhyay, C., & Datta, A. (2008). Catalysis Communications, 9, 2588–2592.

Nikje, M. M. A., & Bigdeli, M. A. (2004). Phosphorus, Sulfur, and Silicon and the Related Elements, 179, 1445–1448.

Oussaid, A., & Loupy, A. (1997). Journal of Chemical Research (S), 342–343.

Palombi, L., Bonadies, F., & Scettri, A. (1997). Tetrahedron, 53, 15867–15876.

Rajabi, F., Pineda, A., Naserian, S., Balu, A. M., Luqueb, R., & Romerob A. A. (2013). Green Chemistry, 15, 1232–1237.

Sahu, A., Rawal, M. K., Sharma, V. K., & Parashar, B. (2009). Rasayan Journal of Chemistry, 2, 536–538.

Shie, J. J., & Fang, J. M. (2003). Journal of Organic Chemistry, 68, 1158–1160.

Shie, J. J., & Fang, J. M. (2007). Journal of Organic Chemistry, 72, 3141–3144.

Shirude, S. T., Patel, P., Giridhar, R., & Yadav, M. R. (2006). Indian Journal of Chemistry, 45B, 1080–1085

Shojaei, A. F., Rezvani M. M., & Heravi, M. (2011). Journal of the Serbian Chemical Society, 76, 1513–1522.

Singh, J., Sharma, M., Kad, G. L., & Chhabra, B. R. (1997). Journal of Chemical Research (S), 264–265.

Srivastava, R. G., & Venkataramani, P. S. (1998). Synthetic Communications, 18, 2193–2200.

Surati, M. A., Jauhari, S., & Desai, K. R. (2012). Archives of Applied Science Research, 4, 645–661.

Taherpour, A. A., & Mansuri, H. R. (2005). Turkish Journal of Chemistry, 29, 317–320.

Varma, R. S., & Dahiya, R. (1998). Tetrahedron Letters, 39, 1307–1308.

Varma, R. S. (1999). Green Chemistry, 1, 43–55.

Varma, R. S. (2001). Pure and Applied Chemistry, 73, 193–198.

Varma, R. S. (2006). Indian Journal of Chemistry, 45B, 2305–2312.

Varma, R. S., Naicker, K. P., Kumar, D., Dahiya, R., & Liesen, P. J. (1999). Journal of Microwave Power and Electromagnetic Energy, 34, 113–123.

Varma, R. S., Saini, R. K., & Dahiya, R. (1997). Tetrahedron Letters, 38, 7823–7824.

Varma, R. S., & Kumar, D. (1999), Journal of the Chemical Society, Perkin Transactions 1, 1755–1758.

Varma, R. S., Saini, R. K., & Dohiya, R. (1998). Journal of Chemical Research (S), 120–121.

Villemin, D., & Hammadi, M. (1995), Synthetic Communications, 25, 3141–3144.

Zhang, Y. C., Bo, L. L, Wang, X. H., Liu, H. N., & Zhang, H. (2012). Huan Jing Ke Xue, 2759–2765.

CHAPTER 4

REDUCTION

SANGEETA KALAL, KIRAN MEGHWAL, MEENAKSHI JOSHI, and
PINKI B. PUNJABI

CONTENTS

4.1 INTRODUCTION

Reduction is a process, which either involves the gain of electrons/hydrogen or loss of oxygen/decrease in oxidation state. An oxidation-reduction reaction occurs in our everyday life and is very vital for some of the basic functions of life. Some examples include photosynthesis, respiration, combustion, corrosion and rusting.

Microwave radiations can be used to carry out a wide variety of reactions in short time, high yield and sometimes, even without solvent. This approach has proved to be very beneficial since the conventional process always has some or the other disadvantages. In conventional heating, reactants are slowly activated by a conventional external heating source. Thus, it is a slow and inefficient method for transferring energy into the reacting system. Microwaves couple directly with the molecules of the entire reaction mixture, leading to a rapid rise in the temperature. Here, only the contents of reaction vessel are heated and not the vessel itself; thus, it provides better homogeneity and selective heating of polar molecules. The microwave irradiation provides enhanced reaction rate and improved product yield in chemical synthesis and it is proving quite successful in the formation of a variety of bonds.

Reduction of organic functional groups plays an important role in organic synthesis. However, its application is often limited and is poorly compatible with the synthesis of sensitive products because of the harsh conditions used in reduction reactions. Therefore, it is an important issue to improve the currently existing reduction reaction schemes towards faster, more selective, economically and ecologically acceptable procedures. In this context, the application of microwave heating has been shown to have a strong beneficial impact on a large variety of synthetic procedures involving reduction. Reaction times can be drastically reduced with an overall milder reaction conditions, higher yields and pure products. In numerous examples, the application of microwave irradiation has also allowed the use of milder and more environmentally acceptable reagents (Bose et al., 1994; Lidstrom et al., 2001; Wathey et al., 2002). The use of microwave irradiation is thus expected to improve reductions of number of functional groups.

4.2 MULTIPLE BOND REDUCTION

Berthold et al. (2002) reported use of formate salts such as ammonium formate and triethylammonium formate as hydrogen source with N-butyl-N'-methylimidazolium hexaflourophosphate, ([bmim][PF$_6$] for catalytic transfer hydrogenation of different homo- or heteronuclear organic compounds at 150°C with 92–98% yield.

$$\text{Ph} \diagdown \diagup \text{Y} \quad \xrightarrow[\substack{\text{5 eq. Et}_3\text{N, 5 eq. HCO}_2\text{H} \\ \text{[bmim] [PF}_6\text{], MW, 150°C}}]{\text{10\% Pd/C (50 mg/mmol substrate)}} \quad \text{Ph} \diagdown \diagup \text{Y}$$

where Y = Ph, COR$_1$, COOR$_1$.

Wolfson et al. (2009) carried out the reduction of carbon-carbon double bonds in the presence of glycerol with palladium supported on carbon (0.2–0.3 mol% Pd) while bulky olefines (cis-stilbene and 1,1-diphenylethylene) and linear aliphatic alkenes (1-hexene and 1-octene) were partially reduced through this methodology under microwave irradiation (Tavor et al., 2010).

where R$_1$R$_2$ = –(CH$_2$)$_4$–; R$_1$ = Alkyl, Phenyl; R$_2$ = H, Alkyl, Phenyl.

One pot reaction for the synthesis of aryl substituted pyrrole derivatives from but-2-ene-1,4-diones and but-2-yne-1,4-diones via Pd/C-catalyzed hydrogenation of the carbon-carbon double bond/triple bond followed by amination-cyclization has been studied by Rao et al. (2004).

formate as a hydrogen source and ethylene glycol as the microwave energy transfer agent for catalytic transfer hydrogenation. Reduction of double bonds and hydrogenolysis of several functional groups were carried out safely and rapidly (3–5 min) at about 110–130°C with 10% Pd/C as an efficient catalyst. When 10% Pd/C was used as the catalyst, the β-lactam ring cleavage proceeds by hydrogenolysis of the N–C bond of 4-aryl-2-azetidinones. But under these conditions, raney nickel as a catalyst did not cause cleavage of this β-lactam ring but only alkene group was reduced.

1,4-Dihydropyridines have been used in the reduction of carbon-carbon double bond without solvent under microwave irradiation (Torchy et al., 2002).

where R_1 = Ph; A = NO_2, CHO, $COCH_3$, COPh.

They observed that efficiency of the reaction is dramatically dependent on the steric effects in the dyhydropyridines and on the electronic effects in the olefins.

Desai and Danks (2001) investigated that a combination of formate bound to an ion- exchange resin and Wilkinson's catalyst can be used in the transfer hydrogenation of electron deficient alkene. Reactions were completed upon microwave irradiation in 30 sec and the products were obtained in quantitative yields (80–95%).

Akisanya et al. (2000) observed that (1-azabuta-1,3-diene)tricarbonyliron (0) complexes with sodium borohydride do not reduce C-C multiple bond but only C=N bond under thermal condition. On the other hand, both the reductions take place in the presence of microwave irradiation. When the reaction was performed with sodium borodeuteride, 1,2,3-trideutero secondary amines were obtained. The complexes are totally inert to reduction by sodium borohydride under thermal conditions.

(90%)

where R = Aryl; R_1 = H, CH_3.

4.3 CARBONYL GROUP REDUCTION

Varma and Saini (1997) reported that alumina supported sodium borohydride can be used for the reduction of carbonyl compounds to corresponding alcohols under microwave irradiation, using solventless conditions giving 62–93% yield.

where R = CH_3; R_1 = C_2H_5.

White and Kittredge (2005) reported a microwave-assisted reduction of cyclo-hexanone by sodium borohydride that is supported on SiO_2. The reaction was completed in less than 3 min while Feng et al. (2001) reported the reduction of ester to the corresponding alcohols, using potassium borohydride/lithium chloride and microwave irradiation under solvent-free condition. The reactions were generally completed in 2–8 min, with good to excellent yields (55% to 95%).

A simple and efficient protocol for microwave-assisted reduction of carbonyl compounds to alcohols and carboxylic acids is by Cannizzaro reaction. The reaction was achieved using a mixture of an aldehyde, sodium hydroxide, and basic alumina. Microwave irradiation of only 15 sec was reported to give almost quantitative yield of both; alcohol and the carboxylic acid (Sharifi et al., 1999).

Gracia et al. (2009) investigated the transfer hydrogenation of carbonyl compounds to their corresponding alcohols using supported Pt and Pd nanoparticles on Al-SBA-15 materials under microwave irradiation with short times of reaction (15–30 min).

An efficient way for microwave-assisted reduction of liquid and solid ketones using sodium borohydride without solvent has also been reported. An excess amount of $NaBH_4$ without alumina or any support was used for the reduction of acetophenone and the yield of the product was 81% under microwave irradiation. It was concluded that the powder of excess $NaBH_4$ disperses the aggregation of ketone and made a useful contact successfully to provide an efficient reaction, but in case of benzophenone, the reduction needs more time and $NaBH_4$ to complete the reaction. This lower conversion may be due to some steric reasons (Chen et al., 1999).

Microwave heating involves direct absorption of energy by those functional groups that bear ionic conductivity or a dipole rotation effect. Then this energy is released to the surrounding solution. This absorption of energy causes the functional groups to have high reactivity with surrounding reactant. The rapid transfer of heat via microwave irradiation reduces the reaction time, which is more evident in case of solid-state reactions as the heat conducted in solid state was slower than in solution phase. It was reported that the solid supported reduction of ketone under microwave irradiation proceed very well not only with inorganic alumina support but also with nonmetal support.

Solid-state reduction of solid ketones with $NaBH_4$ under microwave is-

where R = Phenyl, X = Br, OH, Phenyl.

When cellulose was used as a support, then the yield was about 90–98% in 5 min
Solid-state reduction of liquid ketones with $NaBH_4$ under microwave is-

where R_1 = Phenyl, R_2 = CH$_3$, C$_2$H$_5$, when celite, cellulose MgSO$_4$ or SiO$_2$ was used as a support, the yield was almost 98% in 5 min.

SiO$_2$ has been used as a supporting agent to initiate the microwave-assisted reduction of chalcones. This procedure improves the chemoselectivity of reduction of chalcone.

The chemoselective reduction of chalcone using NaBH$_4$/Support.

When silica was used as a support, then the ratio of the yields (A:B) was 98:2 in 5 min while it was only (A:B) was 1.4:1 on using cellulose support.

Meerwin-Ponndorf-Verley (MPV) reduction of carbonyl compounds to alcohols is an important reaction in organic chemistry. Posner et al. (1977) carried out the efficient reduction of carbonyl compounds using isopropyl alcohol/diisopropyl carbinol and alumina under conventional heating, that has now been translated to a solvent less and expeditious reduction that uses aluminum alkoxides under microwave irradiation conditions and the reaction was completed in 2 min (Barbry and Torchy, 1997).

The reduction of carbonyl compounds like ketones and α,β-unsaturated cyclic ketones takes place very quickly and efficiently with excellent yield of products by using NaBH$_4$/MgSO$_4$.7H$_2$O under microwave irradiation. The reduction of carbonyl compounds using this catalyst is a convenient method because of its ready availability, mild, simple and cost effective properties.

The procedure of reduction involves a mixture of NaBH$_4$ and MgSO$_4$.7H$_2$O. Carbonyl compound was added to this mixture is well grounded with a pestle and then irradiated in a microwave oven for an appropriate time. After completion of the reaction, the mixture was diluted with water and the product was collected by filtration and recrystallized from toluene (Hossein et al., 2008).

Only product

Wolff–Kishner reduction of carbonyl groups is generally conducted with hydrazine hydrate in the presence of a strong base and at elevated temperatures (about

200 °C) for hours. Soriano (1993) reported a simple Wolff–Kishner reduction of isatin under mild condition. However, this method requires a 3–4 h time and the base, sodium ethoxide, should be freshly prepared, which presents a potential danger. It was observed that the reaction time was greatly decreased, when the reaction was carried out in household microwave oven.

The Wolff-Kishner reaction is commonly used for the reduction of carbonyl group to the corresponding hydrocarbons. Parquet and Lin (1997) investigated the microwave-assisted Wolff-Kishner reduction reaction. The reaction was performed in two steps. The first step involved the conversion of the carbonyl group into the hydrazone with 55% hydrazine in ethylene glycol through microwave irradiation for 30 sec. Second step is the reduction step, where KOH in ethylene glycol was used and reaction mixture was irradiated for 10 sec. In this reaction, 32% yield of product was obtained.

The two-step synthesis has many advantages:
- A very short reaction time with no need for special microscale glassware.
- Mild experimental conditions. (Hot oil baths and heating mantles are not required.)
- Reagents are easy to handle.

Several β-trimethylsilyl carbonyl compounds were reduced by sodium borohydride in short reaction time and with good yields. It involves simple mixing of β-trimethylsilyl aldehyde or ketone with 10% $NaBH_4$/neutral alumina and irradiating the mixture in a teflon container in a conventional microwave oven for 2–3 min. The product was extracted with CH_2Cl_2 after cooling. They obtained excellent yields (60–100%) (Zadmard et al., 1998).

Varma et al. (1998) performed a crossed Cannizzaro reaction using solventless method. Here, aldehydes react with 2 eq. paraformaldehyde and 2 eq. barium hydroxide, $Ba(OH)_2.8H_2O$ under microwave irradiation for 0.25–2 min at 900 W. Yield of the alcohols was 80–99%, whereas 1–20% yield of carboxylic acid as a by product was also obtained.

$$RCHO + (CH_2O)_n \xrightarrow[\text{MW or in oil bath}]{Ba(OH)_2.\ 8H_2O} RCH_2OH + RCOOH$$

4.4 REDUCTIVE AMINATION

The reductive amination of carbonyl compounds can be carried out using sodium cyanoborohydride, sodium triacetoxyborohydride or $NaBH_4$ coupled with sulfuric acid. These reagents involve the use of corrosive acids and/or produce waste stream. The environmentally benign methods using wet montmorillonite K10 clay supported sodium borohydride facilitated by microwave irradiation has been developed by Varma and Dahiya (1998). Clay montmorillonite K10 not only behaves as a Lewis acid but it also provides water from its interlayers that enhance the reducing ability of $NaBH_4$.

$$R_1\text{-}\underset{R}{C}=O + H_2N\text{-}R_2 \xrightarrow[\text{2 min}]{\text{Clay, MW}} R_1\text{-}\underset{R}{C}=N\text{-}R_2$$

$$\downarrow \begin{array}{l} NaBH_4\text{-Clay} \\ H_2O, MW \end{array}$$

$$R_1\text{-}\underset{R}{CH}\text{-}\underset{H}{N}\text{-}R_2$$

78-97% Yield

where (i) R = i-Pr, Ph, o-HOC_6H_4, p-$MeOC_6H_4$, p-$NO_2C_6H_4$; R_1 = H; R_2 = Ph
  (ii) R and R_1 = -$(CH_2)_5$-; R_2 = Ph
  (iii) R and R_1 = -$(CH_2)_6$-; R_2 = n-Pr
  (iv) R = p-ClC_6H_4; R_1 = H; R_2 = o-HOC_6H_4
  (v) R = R_1 = Et; R_2 = Ph
  (vi) R = n-C_5H_{11}; R_1 = Me; R_2 = Morpholine, Piperidine
  (vii) R = i-Pr; R_1 = H; R_2 = n-$C_{10}H_{21}$
  Loupy et al. (1996) reported the Leuckart reductive amination of carbonyl compounds with ammonium formate or formamide. The reaction was carried out under solvent-free condition with microwave irradiation (60 W). Excellent yields of N-alkylated formamides (75–97%) were produced in reaction time of about 30 min.

$$R^1\text{-}\underset{R^2}{\overset{O}{C}} \xrightarrow[\text{MW, 60 W, 30 min}]{HCO_2NH_4 \text{ or } HCO_2H, HCONH_2} R^1\text{-}\underset{R^2}{\overset{NHCHO}{C}}$$

(75-97%)

4.5 NITRO GROUP REDUCTION

Quinn et al. (2010) carried out reduction of heteroaromatic and aromatic nitro groups to amines using Pd/C or Pt/C catalyst, which is extremely effective with 1,4-cyclohexadiene as the hydrogen transfer source. In general, the reactions were completed within 5 min at 120 °C. Pt/C is effective and it gives little or no dehalogenation in case substrate contains potentially labile aromatic halogens.

The selective and rapid reduction of nitro group into corresponding amine using sodium hypophosphite under microwave irradiation was reported by Meshram et al. (2000). This reaction showed best result in terms of yields and purity, when the substrates were irradiated by microwaves under solvent-free conditions. The reaction was chemoselective and was not affected by functional groups like CN, OH, COOH, $CONH_2$ and halogens.

A variety of nitroarenes such as electron rich and electron poor are reduced to corresponding aromatic amines using Al_2O_3 supported sodium hydrogen sulfide under microwave irradiation with high yields. Kanth et al. (2002) observed that the substituents at different positions of nitrobenzene have no influence on the rate of reaction and yields of products.

Farhadi et al. (2011) reported efficient and selective reduction of aromatic nitro compounds into the corresponding amine by using nanosized $SmFeO_3$ as a heterogeneous catalyst, propan-2-ol as a hydrogen donor (reducing agent) and KOH as a promoter under microwave irradiation. The method was found to be highly regio- and chemoselective method for aromatic nitro compound. The advantage of this catalytic method is that it is fast, clean, inexpensive and high yielding. In addition, the nanosized $SmFeO_3$ catalyst could be reused without loss of any activity.

Vanier (2007) reported that the hydrogenation of various substrates (alkenes, alkynes, ethers, nitro compounds, etc.) can be carried out using gaseous hydrogen into a sealed reaction system under microwave irradiation in short time with moderate temperature (80–100°C) with 50 psi of hydrogen.

The reduction of aromatic nitro compound to corresponding amine by using microwave irradiation is a fast, easy and environment friendly method with good yields of product in short reaction time, in the presence of zinc dust-ammonium chloride and solvent-free condition. The time of the reaction was generally 8–15 min and the product was obtained in good to excellent yields (82–95%).

where R = p-OCH$_3$, -NH$_2$, o-CH$_3$, p-OCH$_3$, o-NH$_2$, p-NH$_2$, p-CH$_3$, p-Br, p-Cl.

The obvious advantages of this microwave method over the conventional reduction methods are simple experimental conditions, short time with high yield of the substituted aromatic amines, avoiding the use of organic solvents and strong acid medium and less expensive materials (Abdulla and Suliman, 2011).

4.6 ARYL HALIDE REDUCTION

The reductive dehalogenation of aryl halide by triethylsilane with catalytic amount of palladium chloride under microwave irradiation is a simple, efficient and selective method. The silicon hydrides are nontoxic and stable and hence, these are effective reducing agents.

No reaction was observed under same conditions without palladium. 4-Bromoanisole in tetrahydrofuran gave only 21% yield. The low yield in this solvent was due to the low dielectric constant value of THF. Most of the aryl chlorides react moderately but no reaction was observed with aryl fluoride.

The reduction of 4-iodoanisole by triethylsilane catalyzed by palladium chloride has been observed by Villemin and Nechab (2000).

The reduction reaction of an aromatic halide is also useful for destruction of highly toxic polyhalogenated aromatic hydrocarbons (PCBs) (Safe and Hutzinger, 1987).

4.7   NITRILE GROUP REDUCTION

Kumar and Das (2013) observed that solid supported ruthenium can be used as a heterogeneous catalyst for the reduction of aromatic, α,β-unsaturated and aliphatic nitriles to primary amides at 130–140°C for 0.5–2 h in microwave. The catalyst was found to be very stable under moisture and microwave irradiation, easily separable from the reaction mixture and recyclable up to ten times without any loss of catalytic activity.

60–99% Yield

4.8   REDUCTION OF OXIMES

Bandgar et al. (1999) investigated the conversion of aldoximes into nitrile using natural kaolinitic clay EPIC^R, EPZG^R or EPZ10^R as a heterogeneous catalyst and microwave irradiation under solvent-free condition. This reaction provides 52–95% yield.

4.9   AZIDE GROUP REDUCTION

Azido compounds were reduced into their corresponding amines within 10–20 min using nickel boride (Ni$_2$B) in the presence of acidic media, MeOH-HCl (1 m) under microwave irradiation (70 W). The corresponding amine was obtained in good yield (93–84%) in 1 min (Shankaraiah et al., 2009).

where R = H, OH.

Bartoli et al. (2008) reported that NaI is a useful regent for reduction of azide into primary amine in the presence of $CeCl_3.7H_2O$ (Lewis acid) system. The rate and yield of the reaction was improved by employing microwave-assisted procedure.

Stankovic et al. (2010) carried out a new synthetic protocol for the $LiAlH_4$-promoted reduction of nonactivated aziridines under microwave conditions. It involves ring opening of 2-(phenoxymethyl) aziridines and conversion into corresponding β-amino alcohols. In this reaction, 2-(phenoxymethyl)aziridines were treated with $LiAlH_4$ under microwave irradiation giving 1-methoxypropan-2-amines.

4.10 BIOREDUCTION

Bioreduction of aromatic aldehydes into corresponding aromatic alcohols using an extract of *Aloe vera* in aqueous suspension under microwave irradiations is a green chemical method. This reaction was performed in a very short time (Leyva et al., 2012).

where R_1 = H, OH; R_2 = OCH_3.

4.11 MISCELLANEOUS

A rapid conversion of benzyl alcohols into N-benzyl-substituted amides can be achieved under microwave irradiation using benzonitriles in the presence of catalytic amount of Envirocatalyst EPZG in solvent-free conditions. EPZG is a novel supported reagent catalyst based on montmorillonite K10 clay. It is selective, eco-friendly, reusable and hence provides cost effective and safe approach for production in industries.

The conversion of alcohol to amide via nitrile is known as Ritter reaction (Ritter and Minieri, 1948). This method is very useful in peptide synthesis or in preparation of amines through the hydrolysis or reduction of initially formed amides. The reaction time was greatly decreased, when the reaction was carried out under microwave irradiation. The classical Ritter reaction, which functions only on tertiary alcohol does not provide good yield of amide (9%). This was explained because of difficulty in contact between active site of catalyst and bulky reactant (Clark and Macquarie, 1997). However, the results with different types of benzyl alcohols are quite good with both; electron releasing as well as electron withdrawing substituents in the starting material. The advantages of the method are high yield, use of solvent-free technique, short reaction time and easy isolation of product. The only disadvantage of this method is that it is suitable only for conversion of benzyl alcohols.

The reaction involving a mixture of benzyl alcohol, nitrile and Envirocat EPZG catalyst was stirred under microwave irradiation in a microwave reactor for a given time. After cooling to room temperature, this is further treated with CH_2Cl_2. The catalyst can be removed by filtration and the residue can be recrystallized from ethanol-water (Veverkova and Toma, 2005).

where R_1 = H, Cl, $(CH_3)_3C$; R_2 = H, Cl; R_3 = H, F, Cl, CH_3O, $(CH_3)_2N$.

Mojtahedi et al. (1999) obtained high yields of β-amino alcohols from the reaction of primary and secondary amines with epoxides in the presence of montmorillonite K10 clay, under solvent-free conditions and microwave irradiation.

Seltzman and Berrang (1993) reported the reduction of iodine-substituted aryl nitro compounds to the corresponding anilines with nickel boride (Ni_2B) under microwave irradiation (70 W).

This reaction proceeds in the presence of iodo and ortho carbo-alkoxy group, which is otherwise a problematic reaction by conventional methods.

Reduction is an up hill process and the reduced compound is higher in energy than its counter part, that is, oxidized substrate. Reduced functional group plays an important role in the activity of that compound. Microwave-assisted reduction provides a convenient route to reduce many functional groups with reasonably good yields and that too in relativity shorter time.

KEYWORDS

- **Aryl Halides**
- **Azide**
- **Carbonyl Compounds**
- **Multiple Bond Reduction**
- **Nitrile Group**
- **Nitro Compounds**
- **Reductive Amination**
- **Wolff-Kishner Reduction**

REFERENCES

Abdulla, J. M., & Suliman, M. M. (2011). Iraqi National Journal of Chemistry, 43, 418–423.

Akisanya, J., Danks, T. N., & Garman, R. N. (2000). Journal of Organometallic Chemistry, 603, 240–243.

Bandgar, B. P., Sadavarte, V. S., & Sabu, K. R. (1999). Synthetic Communications, 29, 3409–3413.

Banik, B. K., Barakat, J. K., Wagle, D. R., Manhas, M. S., & Bose, A. K. (1999). Journal of Organic Chemistry, 64, 5746–5753.

Barbry, D., & Torchy, S. (1997). Tetrahedron Letters, 38, 2959–2960.

Bartoli, G., Di, A. G., Giovannini, R., Giuli, S., Lanari, S., Paoletti, M., & Marcantoni, E. (2008). Journal of Organic Chemistry, 73, 1919–1924.

Berthold, H., Schotten, T., & Honig, H. (2002). Synthesis, 11, 1607–1610.

Bose, A. K., Manhas, M. S., Banik, B. K., & Robb, E. W. (1994). Research on Chemical Intermediates, 20, 1–11.

Chen, S., Yu, H., Chen, S., & Wang, K. (1999). Journal of the Chinese Chemical Society, 46, 509–511.

Clark, J. H., & Macquarie, D. J. (1997). Organic Process & Research Development, 1, 149–162.

Desai, B., & Danks, T. N. (2001). Tetrahedron Letters, 42, 5963–5965.

Farhadi, S., Siadatnasab, F., & Kazem, M. (2011). Journal of Chemical Research (S), 35, 104–108.

Feng, J., Liu, B., Dai, L., Yang, X., & Tu, S. (2001). Synthetic Communications, 31, 1875–1877.

Gracia, M. J., Campelo, J. M., Losada, E., Luque, R., Marinas, J. M., & Romero, A. A. (2009). Organic & Biomolecular Chemistry, 7, 4821–4824.

Hossein, A. A., Katayoun, M., & Hossein, M. G. (2008). Synthetic Communications, 38, 3414–3421.

Kanth, S. R., Reddy, G. V., Rao, V. V. V. N. S. R., Maitraie, D., Narsaiah B., & Rao, P. S. (2002). Synthetic Communications, 32, 2849–2853.

Kumar, S., & Das, P. (2013). New Journal of Chemistry, 37, 2987–2990.

Leyva, E., Moctezuma, E., Santos-Diaz, M. D. S., & Loredo-Carrillo, S. E. (2012). Revista Latinoamericana Química, 40, 140–147.

Lidstrom, P., Tierney, J. & Westman, J. (2001). Tetrahedron, 57, 9225–9283.

Loupy, A., Monteux, D., Petit, A., Aizpurua, J. M., Dominguez, E., & Palomo, C. (1996). Tetrahedron Letters, 37, 8177–8180.

Meshram, H. M., Ganesh, Y. S. S., Sekhar, K. C., & Yadav, J. S. (2000). Synlett, 7, 993–994.

Mojtahedi, M. M., Saidi, M. R., & Bolourtchian, M. (1999). Journal of Chemical Research (S), 2, 128–129.

Parquet, E., & Lin, Q. (1997). Journal of Chemical Education, 74, 1225.

Posner, G. H., Runquist, A. W., & Chapdelaine, M. J. (1977). Journal of Organic Chemistry, 42, 1202–1208.

Quinn, J. F., Bryant, C. E., Golden, K. C., & Gregg, B. T. (2010). Tetrahedron Letters, 51, 786–789.

Rao, H. S. P., Jothilingam, S., & Scheeren, H. W. (2004). Tetrahedron, 60, 1625–1630.

Ritter, J. J., & Minieri, P. P. (1948). Journal of the American Chemical Society, 70, 4045–4048.

Safe, S., & Hutzinger, O. (Ed.) (1987). Polychlorinated Biphenyls (PCBs): Mammalian and Environmental Toxicology, Berlin: Springer, 15–48.

Seltzman, H. H., & Berrang, B. D. (1993). Tetrahedron Letters, 34, 3083–3086.

Shankaraiah, N., Markandeya, N., Espianoza-Moraga, M., Arancibia, C., Kamal, A., & Santos, L. S. (2009). Synthesis, 13, 2163–2170.

Sharifi, A., Mojtahedi, M. M., & Saidi, M. R. (1999). Tetrahedron Letters, 40, 1179–1180.

Soriano, D. S. (1993). Journal of Chemical Education, 70, 332.

Stankovic, S., Dhooghe, M., & Kimpe, N. D. (2010). Organic & Biomolecular Chemistry, 8, 4266–4273.

Tavor, D., Popov, S., Dlugy, C., & Wolfson, A. (2010). Organic Communications, 3, 70–75.

Torchy, S., Cordonnier, G., Barbry, D., & Eynde, J. J. V. (2002). Molecules, 7, 528–533.

Vanier, G. S. (2007). Synlett, 1, 131–135.

Varma, R. S., & Dahiya, R. (1998). Tetrahedron Letters, 54, 6293–6298.

Varma, R. S., & Saini, R. K. (1997). Tetrahedron Letters, 38, 4337–4338.

Varma, R. S., Naicker, K. P., & Liesen, P. J. (1998). Tetrahedron Letters, 39, 8437–8440.

Veverkova, E., & Toma, S. (2005). Chemical Papers, 59, 8–10.

Villemin, D., & Nechab, B. (2000). Journal of Chemical Research (S), 9, 432–434.

Wathey, B., Tierney, J., Lidstrom, P., & Westman, J. (2002). Drug Discovery Today, 7, 373–380.

White, L. L., & Kittredge, K. W. (2005). Journal of Chemical Education, 82, 1055–1056.

Wolfson, A., Dlugy, C., Shotland, Y., & Tavor, D. (2009). Tetrahedron Letters, 50, 5951–5953.

Zadmard, R., Saidi, M. R., Bolourtchian, M., & Nakhshab, L. (1998). Phosphorus, Sulfur, and Silicon and the Related Elements, 143, 63–66.

CHAPTER 5

SUBSTITUTION

MEENAKSHI SINGH SOLANKI, SHIKHA PANCHAL, and RITU VYAS

CONTENTS

5.1   INTRODUCTION

In a substitution reaction, a functional group in a particular chemical compound is replaced by another group. Substitution reaction is also known as displacement reaction or replacement reaction. Substituted compounds are those chemical compounds, where one or more hydrogen atoms of a core structure have been replaced with a functional group like alkyl, hydroxy, halogen, etc. Organic substitution reactions are classified in several main organic reaction types depending on:

(i)   whether the reagent that brings about the substitution is an electrophile or a nucleophile.

(ii) whether a reactive intermediate involved in the reaction is a carbocation, a carbanion or a free radical.

(iii) whether the substrate is aliphatic or aromatic. Detailed understanding of a reaction type helps to predict the product outcome in a reaction. It is also helpful for optimizing a reaction with regard to variables such as temperature and choice of solvent.

Ju et al. (2006) carried out microwave synthesis of various azides, thiocyanates, and sulfones in an aqueous medium by reacting alkyl halides or tosylates with alkali azides, thiocyanates, or sulfinates at 110°C for 20 min in the absence of any phase transfer catalyst, and a variety of reactive functional groups are tolerated. This is a practical, rapid, and efficient microwave synthesis method. 1.3 eq. of alkali azides, thiocyanates, or sulfinates are required to react with alkyl monohalide or alkyl monotosylates to give azides, thiocyanates, and sulfones, respectively.

$$R-X + M^+-Nu \xrightarrow[\text{MW, (70-100 W)}]{H_2O} R-Nu$$

where R = Alkyl, Benzyl ; X = Br, Cl, I, OTs; M = K, Na; Nu = SCN, N_3

2.6 equivalents of alkali azides, thiocyanates or sulfinates are required to react with alkyl dihalide or alkyl ditosylates.

$$X-(CH_2)_n X + M^+-Nu \xrightarrow[\text{MW}]{} Nu-(CH_2)_n Nu$$

where n = 4; X = Br, OTs; M = Na, K; Nu = N_3, SCN

Different substitutions have been carried out under microwave irradiation reactions and excellent yields (85–95%) of various products have been reported.

5.2   ACYLOXYLATION

Onogi et al. (2012) carried out reaction of methyl (1R,2S,4R,5R)-2-amino-4,5-di-bromocyclohexanecarboxylate in presence of K_2CO_3 and dimethylacetamide solvent (DMA) at 60°C for 0.5 h to give unstable intermediate methyl (1S,2R,4S,5R)-7-aza-5-bromo-bicyclo[2.2.1]heptane-2-carboxylate via intermolecular cycloamination, which further reacts at 140°C for 1.5 h to form methyl (1S,2R,4S,5S)-7-aza-5-hy-droxy-bicyclo[2.2.1]heptane-2-carboxylate as final product under microwave irradiation. In this acyloxylation reaction, unusual endo-selectivity occurs due to 7-az-abicyclo[2.2.1]heptane skeleton.

5.3   AMINATION

Amination of aryl halides has been an important and frequently required reaction for the synthesis of some interesting compounds containing N-aryl moiety. Arylamines are very common and important entities in many pharmaceuticals, agrochemicals, photography, xerography, pigments, and natural products. Aromatic nucleophilic substitution of activated aryl halides with different amines is the conventional method for the preparation of substituted arylamines. But traditional method involves long reaction time, less friendly solvent like DMF, DMSO and vigorous reaction condition.

Meng et al. (2013) carried out amination reaction of fluorobenzenes in N-meth-ylpyrrolidinone (NMP) with weak base (K_2CO_3) in absence of any catalyst under microwave irradiation. It was observed that leaving ability of fluorine increases in the presence of additional halogen atom(s). m-substituted fluorine gave higher acti-vation as compared to o-substituted. 1,2,3-Trifluorobenzene, 1,2,4-trifluorobenzene and 1,2,4,5-tetrafluorobenzene produced the corresponding regioselective mono-substituted products in good to excellent yields (55–92%).

Pichowicz et al. (2010) reported nucleophilic aromatic substitution of 3,4,5-tri-halopyridines with 1–1.1 eq. of primary and secondary amines for efficient syn-thesis of 4-amino-3,5-dihalopyridines using N-methyl-2-pyrrolidone solvent under microwave irradiation at 220 °C in about 60 min. The reaction is also applicable to electron-rich arylamines.

Yaunner et al. (2012) reported synthesis of piperidinyl pyridines by piperidination of halopyridines under microwave irradiation by Buchwald-Hartwig reaction, Ullmann reaction and nucleophilic aromatic substitution ($S_N Ar$). A comparative study of these methodologies show that the Ullmann reaction was most effective for less reactive halopyridines, while uncatalyzed $S_N Ar$ was sufficient for more reactive ones. Reaction was carried out at 150°C for about half an hour in ethylene glycol.

where X = Halogen group.

Angrish et al. (2005) reported a rapid and efficient method for aromatic nucleophilic substitution reaction of chloronitrobenzene with number of substituted amines in presence of 1-butyl-3-methylimidazolium tetrafluoroborate ionic liquid (green reaction media) to give N-substituted nitrobenzene under microwave irradiation.

where X = 2-NO_2, 3-NO_2, 4-NO_2; Amines = HN⟨ ⟩N–CH_3; NH⟨ ⟩N–Ph;

NH⟨ ⟩N–CH_2–Ph; [bmim] [BF_4] = [–N⊕N⧺] [BF_4]

Traditional aromatic nucleophilic substitution ($S_N Ar$) of N-nucleophiles is limited to only activated aromatic substrates. Petit and Vo-Thanh (2011) developed a rapid and highly efficient procedure for the synthesis of N-arylamines by reacting aryl halides with various amines such as secondary aliphatic amines, cycloaliphatic and aromatic amines by aromatic nucleophilic substitution using tris(dioxa-3,6-heptyl)amine (TDA) as phase transfer catalyst, solvent-free microwaves heating. This method is successfully applied to the amination reaction of activated as well as unactivated aryl halides. Good to excellent yields (42–90%) were obtained in very short reaction times. This reaction works out best through an elimination-addition mechanism.

where X = F, Cl, Br.

where X = F, Cl.

The N-arylindole subunit is an important species in many biologically active and pharmaceutically important compounds, which display antiestrogen, analgesic, antiallergic, antimicrobial and neuroleptic activity. Xu and Fan (2008) carried out microwave-assisted nucleophilic aromatic substitution reaction of indoles with fluoro- and chloro-substituted aryl halides to synthesize N-arylindoles via $C(sp^2)$–$N(sp^2)$ bond formation in the presence of K_2CO_3 or Cs_2CO_3 under microwave irradiation (420 W). Reaction was completed within 25–40 min with good yields (46–94%).

where R_1 = NO_2, CN; R_2 = CH_3; R_3 = CH_3, NO_2; X = F, Cl, Br.

Microwave-assisted aromatic nucleophilic substitution or Suzuki coupling were carried out of mono- or di-substituted aminopyrimidine derivatives by Luo et al. (2002).

5.4   AZA-PRINS REACTION

Parchinsky et al. (2011) reported synthesis of symmetrically and unsymmetrically substituted chiral, racemic 1-azaadamantanes via $BF_3.OEt_2$-promoted aza-Prins reaction under microwave irradiation. Reaction of (±)-[(4-methylcyclohex-3-en-1-yl) methyl]amine with an equimolar amount of an aldehyde in the presence of 1.0 equiv. of $BF_3.OEt_2$ at 180°C for 1 h gives predominantly bicyclic piperidines whereas in presence of excess amount of the aldehydes, symmetrically substituted 1-azaadamantanes are the major products. The yield of the products ranges from 52–83%.

where R = COOEt, Phenyl 4–$CH_3C_6H_4$–, 4–ClC_6H_4–.

It also reacts with aromatic aldehydes in presence of benzene as solvent for 30 min to give N-benzylidene-1-(4-methylcyclohex-3-en-1-yl)methanamine, which further reacts to give unsymmetrical substituted 1-azaadamantanes under same conditions.

where Ar = Phenyl, 4–ClC_6H_4–, 4–CH_3–C_6H_4–, 4–$CH_3OC_6H_4$–, 4–$C_2H_5OC_6H_4$–.

5.5 CYANATION

Cyanation of aryl halides (aryl iodides and aryl bromides) was completed within 20 min using $K_4[Fe(CN)_6]$ as a cyanide source, water as the solvent and tetrabutylammonium bromide (TBAB) in presence of palladium catalyst using microwave heating (Velmathi and Leadbeater, 2008).

where X = Halogen group (I, Br).

5.6 HALOGENATION

Recently, Wannberg et al. (2013) reported an easy and two step method for conversion of aryl alcohols to aryl fluoride using nonafluorobutylsulfonates (ArONf) as a reagent by one-pot procedure to give moderate to good yield (41–85%) at 180°C in 30 min under microwave-assisted heating. This fluorination reaction is palladium-catalyzed.

Heropoulos et al. (2007) reported bromination reaction (ultrasound and microwave-assisted) of various alkylaryls with NBS (N-bromosuccinimide), either neat or in water with diverse chemoselectivity. Thus, only ring substitution occurs in water with ultrasound, whereas with microwaves both; side-chain α-bromination as well as ring substitution occurs. In presence of water, microwaves promote side-chain α-bromination and ring substitution, both. With neat reactant, side-chain α-brominated form was obtained as the major product.

Bissember and Banwell (2009) have observed that the microwave irradiation of certain chloro-, bromo-, trifluoromethanesulfonyloxy- and nonafluorobutanesulfonyloxy substituted quinolines in the presence of acetic anhydride and sodium iodide leads to iodinated products via a transhalogenation process. As a result, corresponding iodides were obtained in high yield. Under similar conditions, related products of pyridines and isoquinolines can also be achieved.

where X = Cl, Br, OTf or ONf.

It is known that chemically modified graphite is an economical material with important applications in the synthesis of graphene, but yet only two methods, that is, oxidation and fluorination have been found due to its extreme chemical inertness, which can give modified graphite with high yield and large throughput. Zheng et al. (2012) have used a microwave sparks assisted halogenation reaction to synthesize large quantities of highly modified graphite. It was observed that the resulting graphite halide could easily be exfoliated into monolayer graphene in organic solvents and after thermal annealing of this graphene halide, the structure and electronic properties of the original graphene could also be recovered. They also reported that the graphene halide can be further modified by a variety of organic functional groups.

Upadhyay and Jursic (2011) have proposed an efficient, safe, high yielding and rapid microwave-assisted method for the preparation of protected p-bromomethyl and p-dibromomethylanilines, which act as new protection groups for alcohol, thiol, and amine. This method involves microwave-assisted N-bromosuccinimide (NBS) radical bromination of readily available N-protected p-toluidine. It was observed that the microwave-assisted radical bromination was superior to the conventional NBS radical bromination.

Pingali et al. (2011) have developed an ecofriendly, green chemical benzyl mono- and di-bromination synthetic protocol, which was found superior to the classic carbon tetrachloride bromination procedure due to less reaction time and high yield. The microwave reaction was performed in diethyl carbonate as media, which is not being used in the case of conventional heating. It was found that both; the solvent and the brominating reagent, N-bromosuccinimide, were recyclable. The target compounds were prepared in less than two hours.

Bernard et al. (2009) have studied microwave-assisted solvent-free brominations of organic substrates with tetrabutylammonioum tribromide (TBATB). They observed that the reactions were facile and the products were obtained in high yields within very short reaction times.

Microwave induced brominations with TBATB were successfully carried out under solvent-free conditions using different types of organic substrates. This process was found better than the existing methods of bromination due to facile nature of reactions, along with very short reaction times and simple purification process. The products were also obtained in moderate to high yields.

Lee et al. (2004) have reported a novel and direct method for the synthesis of α-halocarbonyl compounds using sequential treatment of carbonyl compounds with [hydroxy(tosyloxy)iodo]benzene. This method was further followed by magnesium halides under solvent-free microwave irradiation conditions. This microwave-assisted procedure was found better than the existing methods due to the ease of manipulation, fast reaction rates, and formation of cleaner products under neutral reaction conditions.

where X = Br, Cl, I.

5.7   KNOEVENAGEL REACTION

Xu et al. (2008) studied three-component Knoevenagel nucleophilic aromatic substitution reaction of 4-halobenzaldehydes with cyanoacetic acid esters/cyanoacetamides, and cyclic secondary amines in alcoholic medium under microwave irradiation giving products in 5 min. This reaction also gave high yield in a short time. The yield of product ranges from 72–90%.

where n = 0–1; X = F, Cl, Br; Y = CH$_2$ or O, NR; Z = COOEt, CO$_2$Me, CONH$_2$.

5.8 N-ALKYLATION

4-Bromo-N-alkylnaphthalimides undergo aromatic nucleophilic substitution reaction with amines, alkoxides and thiols under microwave irradiation in the presence of KF/Al$_2$O$_3$ under solvent-free conditions to afford a number of fluorescent 4-substituted-1,8-naphthalimide dyes. This is an efficient method for C-N, C-O and C-S bond formation by applying suitable nucleophiles. Adducts were produced in good to excellent yields (70–95%) and relatively in short times (Bardajee, 2013).

Karuehanon et al. (2012) reported the microwave-assisted S$_N$Ar reaction of 2,4,6-trichloro-1,3,5-triazine with various unprotected amino acids for the synthesis of C_3-symmetrical polycarboxylate ligand. These products can be used as structural directing units in metal–organic frameworks. A domestic microwave oven was used as the heating device and the reactions were performed in water. Comparison between the reactions performed under conventional heating and microwave irradiation was made. It was observed that the microwave method takes 20 minutes less time to afford the desired ligands than the conventional method.

where R$_1$ R$_2$ NH = Valine, Tyrosine, Aspartic acid, Glutamic acid.

5.9 NITRATION

4-Hydroxy-3-nitrobenzaldehyde reacts with nitromethane to give 2-nitro-4-(2'-nitroethenyl)phenol.

This product has been isolated from marine sources. This metabolite was synthesized first in 1964 in Japan and discovered later on in 1991 in India as a natural product. Bose et al. (2004) further reported a new method for synthesis of 2-nitro-4-(2'-nitroethenyl)phenol from p-hydroxy cinnamic acid under microwave irradiation.

In this reaction, ipso-attack (substitution) of a carboxyl group by a nitro group takes place.

Sana et al. (2012) have carried out solvent-free and microwave-assisted nitration reactions, which underwent smoothly in the presence of group V and VI metal salts with high regio-selectivity for anilides, moderately and nonactivated aromatic compounds. Good to excellent yields were obtained under solvent-free conditions and the nitration was completed in few minutes under microwave irradiation. It was observed that when the ortho position was engaged, p-nitro derivatives were obtained, while o-nitro derivatives were obtained when para position was engaged. The nitration of aromatic carbonyl and related compounds was observed with good yield and the reaction times were about 3–5 min.

where X = OH, NH$_2$, NHCOPh, NHCOCH$_3$, CHO, COCH$_3$, COPh, COOH; Y = EWG or EDG; Catalyst = $(NH_4)_6 Mo_7O_{24}.4H_2O$, K_2CrO_4, $Na_2WO_4.2H_2O$, $BiNaO_3$, BiN_3O_9.

Bismuth nitrate pentahydrate induced nitration of eugenol has been investigated by Canales et al. (2011) in detail. They have studied 20 five different conditions and found that microwave induced solvent-free reaction was the best. 5-Nitroeugenol was the sole product in all the cases and no oxidized or isomerized products were detected.

Yadav et al. (2012) studied an easy-to-complete, microwave-assisted electrophilic nitration of phenol using $Cu(NO_3)_2$ in acetic acid, which was green chemical in nature.

Bose et al. (2006) have carried out nitration of phenolic compounds, which was found feasible with a mixture of calcium nitrate and acetic acid (as an efficient nitration agent) under microwave irradiation. This method was found green chemical as the calcium salts, the inorganic by products, were useful agrochemicals rather than waste.

where R = CHO, COCH$_3$.

Paik and Jung (2012) have examined the effect of ligands and solvents in Cu catalyzed Ullmann-type reaction using 10 mol% of CuI, 20 mol% of ligand, and 1 eq. of nitrite salt with 4-iodoanisole as a model substrate under microwave irradiation for 10 min. The desired compound was obtained in 84% and 83% yields in presence of N,N'-dimethylethylenediamine and trans-N,N'-dimethyl-1,2-cyclohexanediamine, respectively. However, desired product could not be obtained in good amount in presence of other ligands.

5.10   PAAL–KNORR REACTION

Minetto et al. (2004) carried out synthesis of tetrasubstituted pyrroles (and trisubstituted furans) in a three-step reaction of β-ketoester with an aldehyde followed by oxidation to give number of substituted 1,4-dicarbonyl compounds, which is further cyclized by microwave-assisted with the Paal–Knorr reaction.

where PCC = Pyridinium chlorochromate.

5.11   SONOGASHIRA CROSS-COUPLING

Sajith and Muralidharan (2012) reported $PdCl_2(PCy_3)_2$ as an efficient catalyst for the copper and amine-free Sonogashira cross-coupling reactions of 2-halo-3-alkyl imidazo[4,5-b]pyridines (where halogen group is I/Br/Cl) using tetrabutyl ammonium acetate as an activator and NMP as a solvent under microwave conditions. This reaction gives good to excellent yield of product (70–95%).

where R_1 = Aryl, Alkyl.

5.12   SULFONATION

Maza et al. (2011) have reported the efficient and fast O-sulfonation of heparin oligosaccharide intermediates by using microwaves. Here, high yielding sulfonation of a series of oligosaccharides using $SO_3 \cdot Me_3N$ complex has been observed. Application of microwave irradiation dramatically reduced reaction times, allowing for the rapid and high yielding sulfonation of products ranging from di- to hexasaccharides.

About 200 times excess sulfuric acid was required for monosulfonation in conventional method, while in presence of microwaves, not only the less amount of sulfuric acid was required but the reaction was also completed in less time (Umrigar et al., 2007). The activation energy required was also less and selectivity and conversion were enhanced. Molar ratio and 2-N conversion were found as S:2-N = 1:1, S:2-N = 2:1, O:2-N = 2:1 and 87, 81 and 54%, respectively.

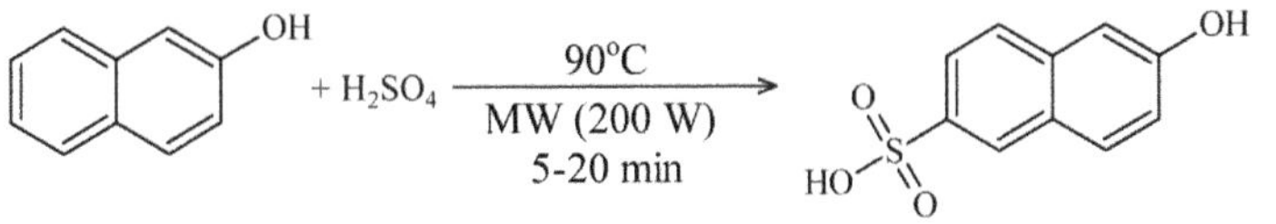

2-Hydroxynaphthalene-
6-sulfonic acid
(Schaeffer's acid)

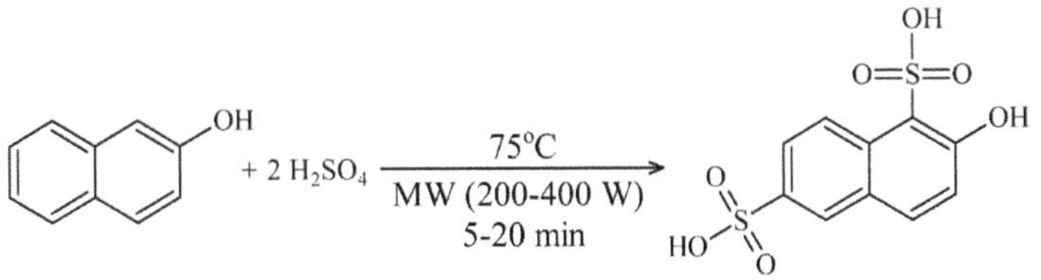

2-Hydroxynaphthalene-
1,6-disulfonic acid

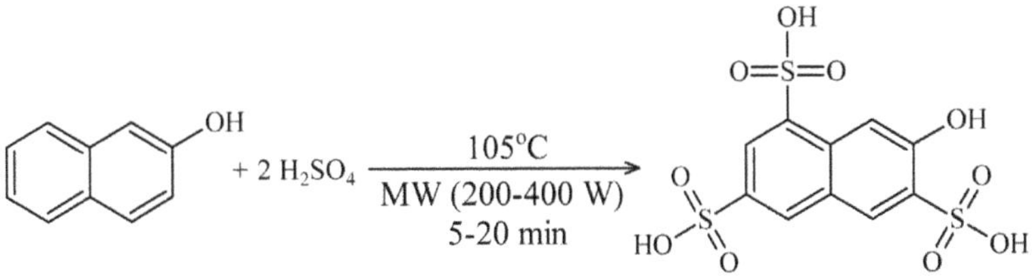

2-Hydroxynaphthalene-
3,8-disulfonic acid

One possible way of polymer waste management is the chemical modification of polymeric materials. Sulkowski et al. (2013) used expanded polystyrene waste (EPS), as the reference material, which was converted into polymeric flocculants by the sulfonation reaction. Under conventional heating and microwave conditions, poly(styrenesulphonate) acids (EPSS) were obtained from EPS during the sulfonation process with sulfuric acid as the sulfonation agent and Ag_2SO_4 as the catalyst. It was observed that the sulfonation process performed under microwave irradiation gave same amount of product as in case of conventional conditions but the reaction time was substantially reduced from 1–1.30 h to 15 min.

Stuerga et al. (1993) have examined the sulfonation of naphthalene and found that the substitution preferentially occurred in 2-position under microwaves at temperatures higher than 130 °C; where as the final product contained almost equimolar mixture of 1- and 2- naphthalene sulfonic acids under conventional heating conditions at temperature lower than 130 °C.

5.13 SUZUKI-MIYAURA CROSS-COUPLING

A facile synthesis of 5-dialkylamino-6-aryl-(2H)-pyridazin-3-one from 5,6-dichloropyridazinone was carried out by Cao et al. (2008) under microwave irradiation by using a palladium catalyzed Suzuki-Miyaura cross-coupling of 6-chloro-5-dialkylaminopyridazinone with various arylboronic acids (3 eq.) as the key transformation.

An efficient microwave-assisted palladium/copper comediated C-8 direct alkenylation of purines with styryl bromides has been developed by Vabre et al. (2011). Combined with subsequent nucleophilic substitution, it provides an easy access to new 6,8,9-trisubstituted purines (48–79% yield). The procedure is regioselective, rapid, functional group tolerant and compatible with other related azoles.

5.14 THIOCYANATION

Murthy et al. (2011) reported an efficient, solvent-free and microwave promoted procedure for regioselective thiocyanation of substituted anilines and indoles both; on and off the solid surface of acidic alumina. Anilines and indoles gave unique p-thiocyano substituted anilines and 3-thio-cyano substituted indoles, respectively.

where R = H, –CH$_3$, –CH(CH$_3$)$_2$, Cl, –OCH$_3$, –COOH, Br, -CN; R$_1$ & R$_2$ = H, –CH$_3$.

In absence of alumina, reaction time ranges from 4.5–9 min and yield of product ranges from 45–65% whereas in presence of alumina, reaction time was reduced to approximately half (2.5–4 min) and yield was also far better (64–90%).

5.15   TSUJI-TROST REACTION

Liao et al. (2005) carried out microwave-assisted allylic substitution with various carbon and heteronucleophiles in an 1-ethyl-3-methylimidazolium tetraborate ionic liquid/water ([emim] BF$_4$/H$_2$O) containing catalyst system of Pd(OAc)$_2$/TPPTS (3,3',3''-phosphanetriyltris(benzene sulfonic acid trisodium salt) in 1:4 ratio. The ionic liquid/water containing catalyst system can be recycled 8 times.

where R = H, Ph,

Appukkuttan et al. (2004) investigated three-component reaction of alkyl halides, sodium azide, and alkynes in presence of copper (I) as a catalyst for 10–15 min using water as a medium under microwave irradiation to give number of 1,4-disubstituted-1,2,3-triazoles with complete regioselectivity (100%) and yield of 90% or more.

5.16   MISCELLANEOUS

Joshi et al. (2006) reported synthesis of some symmetrically and asymmetrically substituted acylic enediynes. In conventional heating method, preparation of such asymmetrical and symmetrical substituted acyclic enediyne takes 14–18 h, while same reaction was completed in 3–5 min using microwave irradiation.

Alcohol/phenol/thiophenol reacts with propargyl bromide in presence of anhydrous K_2CO_3 in dry DMF or NaH in dry THF to give substituted alkyne. The yield of substituted alkyne range from 65–85%. The reaction takes 8–10 h to complete. These alkynes are used in the synthesis of enediynes.

where R = Phenyl, Naphthyl, o-CHO-C_6H_4-.

Reaction between substituted alkyne (1 eq. each) and cis-dichloroethylene (1 and 0.5 eq.), under catalytic amount of Pd(PPh$_3$)$_4$, CuI and n-BuNH$_2$ in benzene give chloro-en-yne precursors (60–62% and 6–8%, respectively) and symmetrical enediyne (9–13% and 61–65%, respectively). This reaction was completed in 3–5 min under microwave exposure.

Chloro-en-yne precursor further reacts with substituted alkyne under identical reaction conditions to give desired enediyne as the final product in 52–66% yield.

Yield = 52-66%

Nitsche and Klein (2012) recently investigated that alkyl- and benzylamines react with carbon disulfide and chloroacetic acid at 100–120°C to give corresponding rhodanines in three steps, one-pot procedure under microwave irradiation in 40 min.

Another microwave-assisted one-pot, one step reaction was developed using bis(carboxymethyl)trithiocarbonate in water for the synthesis of N-arylrhodanines from anilines at 160°C within 15 min.

When R is the alkyl group in this reaction, the yield (58–77%) was more than in the other reaction, where yield ranges from 5–16% whereas with aryl group in a reaction, yield was 5–19% and 5–54%, respectively.

Hou et al. (2012) carried out methylation of cassava starch using thiocarbamide as an active catalyst, dimethyl carbonate (DMC) as a solvent and methylating reagent by applying microwave irradiation as an energy source. Degree of substitution was high (DS = 0.6) within 4 min.

The S_N^H [AE, (addition-elimination)] and S_N^H [AO, (addition-oxidation)] reactions of 5-bromopyrimidine with pyrroles and indoles for the synthesis of 4-(1R-pyrrol-2-yl)-and 4-(1R-indol-3-yl)-5-(hetero)aryl-substituted pyrimidines by combination of the Suzuki-Miyaura cross-coupling and nucleophilic aromatic substitution of hydrogen reactions was carried out by Verbitskiy et al. (2012a).

Combination of the Suzuki–Miyaura cross-coupling and nucleophilic aromatic substitution of hydrogen is a versatile method and it was carried out by Verbitskiy et al. (2012b). This method was used for the synthesis of 4-(thiophen-2-yl)-, 5-(thiophen-2-yl)-, and 4,5-di(thiophen-2-yl) substituted pyrimidines from the commercially available 5-bromopyrimidine.

Bapna et al. (2012) reported microwave-assisted synthesis of hydantoins and their derivatives, 3-(substituted)-5,5- diphenylimidazolidine-2,4-dione and 1-(substituted)-3,5,5- triphenylimidazolidine-2,4-dione by Mannich reaction. The synthesized compounds are good antiarrhythmic, anticonvulsant and antitumor agents and also have anti-HIV, antibacterial and antifungal activity.

Song et al. (2012) studied synthesis of a number of fluorinated thieno[2,3-d]pyrimidine derivatives containing 1,3,4-thiadiazole by the cyclization of 2-aminothiophene-3-carbonitrile with trifluoroacetic acid (TFA), chlorination and nucleophilic substitution reaction using microwave heating.

Radi et al. (2013) carried out the synthesis of 1,4-disubstituted pyrazolo[3,4-d] pyrimidines with allopurinol, tetra-n-butylammonium fluoride (TBAF) mediated by N1-functionalization, which was followed by C4 nucleophilic substitution under microwave-assisted heating. These derivatives have biological significance.

Alen et al. (2008) reported a reaction of 3,5-dichloropyrazinones with 2-bromo-4,6-disubstituted aniline to give 3-anilino-pyrazinones. A microwave-assisted Buchwald–Hartwig type cyclization converted various 3-anilino-pyrazinones easily to the substituted pyrazino[1,2-a]benzimidazol-1(2H)ones (tricyclic structures) using palladium as catalyst with moderate to good yield (27–78%).

Jia et al. (2006) compared microwave irradiated and conventionally heated reaction between 2-aminoarylketone or 2-aminoarylaldehyde and carbonyl compounds in the presence of p-toluene sulfonic acid to give various poly-substituted quinolines by Friedlander condensation under solvent-free condition. Yield of various products under microwave heating ranges from 85–96% in 0.25–1.0 min whereas in conventional heating, the yield is almost same in 3–10 min.

Salmoria et al. (1998) carried out microwave-assisted aromatic nucleophilic substitutions reaction between disubstituted-benzenes and nucleophile piperidine or potassium t-butoxide in homogeneous medium such as dimethyl sulfoxide or dimethylformamide at atmospheric pressure to give substituted product. The rate was 2.7–12 time faster under MW condition, than under conventional condition.

where Z = Nucleophile; X = Leaving group; G = Any group (electro withdrawing or electron donating)

Some 5-methyl-1,2-disubstituted benzimidazoles derivatives have been synthesized in an efficient and rapid solid-phase method with help of the phosphonium linker. The phosphonium linker was prepared by reaction between polymer-supported triphenylphosphine and 4-fluoro-3-nitrobenzyl iodide, which underwent aromatic substitution with primary amines, followed by one pot reaction with aldehydes in the presence of $SnCl_2 \cdot 2H_2O$ under microwave irradiation. The products were isolated from resin using NaOH to give high purity and good overall yield (Rios et al., 2013).

A rapid S_N2' allylic substitution of Baylis–Hillman acetates with ethyl(triphenyl phosphoranylidene) acetate was reported by Yadav et al. (2007) to give ethyl-5-aryl or alkyl-(E)-pent-4-enoates with high (E)-stereoselectivity under microwave irradiation.

Tu et al. (2007) carried out microwave-assisted multicomponent reactions of α, β-unsaturated ketone, methane-1,2-dicarbonitrile, and primary amine under microwaves to afford N-substituted 2-aminopyridines. Reactions were controlled by the basicity of amine and the nature of solvent.

Chan et al. (2009) observed nucleophilic substitution of halogen group from polycyclic aromatic halides with various N-, S-, and Se-nucleophiles to afford substituted phenanthrenes, anthracenes, acenaphthenes, and fluorenes under microwave heating to have good to excellent yields in a short reaction time.

where Nu = R_2NH, RONa, RSNa, RSeH.

5-Bromopyrimidine reacts with bithiophene/2-phenylthiophene/[2,2′:5′,2″] terthiophene in the presence of air-stable [PdCl(C_3H_5)dppb] complex as catalyst at 150°C for 16 h to form 5-(het)aryl substituted pyrimidines by C-C coupling and 5-bromo-4-(het)aryl-pyrimidines have been prepared from the same starting materials via S$_N$H (Nucleophilic aromatic substitution of hydrogen)- reaction catalyzed by a Lewis acid (Verbitskiy et al., 2013).

C-C Coupling

Major   Minor

S$_N$H Reaction

Two new unsymmetric derivatives of 1,2-bis-(5-phenyloxazol-2-yl)benzene (o-POPOP) under microwave exposure were synthesized by Lliashenko et al. (2011). In this reaction, fluorine was replaced by nucleophile such as hydroxyl ion or cyclic secondary amine. This reaction appears to be significantly more efficient as compared with conventional thermal activation.

A series of 1-[4-(3-substituted-acryloyl)-phenyl]-pyrrole-2,5-diones was synthesized and characterized by Aravind and Ganesh (2013). All the compounds were synthesized from 1-(4-acetyl-phenyl)-pyrrole-2,5-dione and substituted benzaldehydes by using Claisen-Schmidt condensation reaction under conventional heating and microwave irradiation. The starting compound 1-(4-acetyl-phenyl)-pyrrole-2,5-dione is the precursor for the preparation of 1-[4-(3-substituted-acryloyl)-phenyl]-pyrrole-2,5-diones and it was obtained by a reaction involving maleic anhydride and 4-aminoacetophenone in diethyl ether. The yield of product was 58–71% under conventional heating in 7–8 h, which increases to 84–90% in 3–5 min under microwave irradiation.

Upadhyay et al. (2010) prepared several new N-[(4-oxo-2-substituted aryl-1, 3-thiazolidine)-acetamidyl]-5-nitroindazoles from N-(arylidene amino acetamidyl)-5-nitroindazoles. This reaction is completed with 2–4 min under microwave irradiation while it takes about 4–7 h conventional heating.

Satyanarayana et al. (2013) synthesized 3-substituted-2H-pyrido[1,2-a]pyrimidin-2-ones via microwave induced cyclocondensation of methyl 2-(acetoxy(phenyl) methyl)acrylate with 2-aminopyridines under solvent-free conditions in the presence of a catalyst and with good yields and high selectivity.

Rat et al. (2011) reported simultaneous formation of glycosylation, esterification and a butyl ether resulting in substitution of a glucuronic acid trisaccharide. This substitution was easily performed in one step under microwave irradiation.

Microwave-assisted synthesis has also been used for synthesizing carbonyl complexes of some metals. Jung et al. (2009) reported microwave-assisted ligand substitution reactions of $Os_3(CO)_{12}$. The labile complex $Os_3(CO)_{11}(NCMe)$ was prepared in a remarkably short period of time with high yield without the need for a decarbonylation reagent such as trimethylamine oxide. Compounds of the type $Os_2(\mu-O_2CR)_2(CO)_6$ were also synthesized by microwave heating. The process includes the first instance with aromatic carboxylate ligands. The one step preparation of these complexes is quite straight and does not require an air-free reaction environment. The compounds are produced in higher yield than previously reported by the microwave irradiation of $Os_3(CO)_{12}$ for less than 15 min in acetic or propionic acid. This method may also be used to prepare the new compound $Os_2(\mu-O_2CH)_2(CO)_6$ by reaction with formic acid with a higher yield, when 1,2-dichlorobenzene was used as a cosolvent. Irradiation of a mixture of $Os_3(CO)_{12}$ in excess of benzoic acid in 1,2-dichlorobenzene gives another new complex $Os_2(\mu-O_2CC_6H_5)_2(CO)_6$. When the molar ratio of benzoic acid to $Os_3(CO)_{12}$ is 1:1, then the major product is the trinuclear cluster $Os_3(\mu-H)(\mu-O_2CC_6H_5)(CO)_{10}$ instead of a dinuclear compound. This represents the first instance, where a cluster of this type has been produced directly from $Os_3(CO)_{12}$ instead of a multistep procedure.

The substitution reaction has got an important place in synthetic organic chemistry as the substitution of a functional group by another group will change the properties (physical, chemical and biological) abruptly and therefore, one can synthesize different organic compounds with diverse activities through these substitution reactions. Microwaves assist in achieving this goal in a cleaner process and limited time.

KEYWORDS

- **Amination**
- **Cyanation**
- **Halogenation**
- **Knoevenagel Reaction**
- **Nitration**
- **Paal-Knorr Reaction**
- **Sulfonation**

REFERENCES

Alen, J., Robeyns, K., De Borggraeve, W. M., Meervelt, L. V., & Compernolle, F. (2008). Tetrahedron Letters, 64, 8128–8133.

Angrish, C., Kumar, A., & Chauhan, S. M. S. (2005). Indian Journal of Chemistry, 44B, 1515–1518.

Appukkuttan, P., Dehaen, W., Fokin, V. V., & Eycken, E. V. (2004). Organic Letters, 6, 4223–4225.

Aravind, K., & Ganesh, A. (2013). Der Pharma Chemica, 5, 261–264.

Bapna, M., Parashar, B., Sharma, V. K., & Chouhan, L. S. (2012). Medicinal Chemistry Research, 21, 1098–1106.

Bardajee, G. R. (2013). Dyes and Pigments, 99, 52–58.

Bernard, A., Kumar, A., Jamir, L., Sinha, D., & Sinha, U. B. (2009). Acta Chimica Slovenica, 56, 457–461.

Bissember, A. C., & Banwell, M. G. (2009). Journal of Organic Chemistry, 74, 4893–4895.

Bose, A. K., Ganguly, S. N., Manhas, M. S., Rao, S., Speck, J., Pekelny, U., & Pombo-Villars, E. (2006). Tetrahedron Letters, 47, 1885–1888.

Bose, A. K., Ganguly, S. N., Manhas, M. S., Srirajan, V., Bhattacharjee, A., Rumthas, S., & Sharma, A. H. (2004). Tetrahedron Letters, 45, 1179–1181.

Canales, L., Bandyopadhyay, D., & Banik, B. K. (2011). Organic and Medicinal Chemistry Letters, 1, 9. doi: 10.1186/2191–2858–1–9.

Cao, P., Qu, J., Burton, G., & Rivero, R. A. (2008). The Journal of Organic Chemistry, 73, 7204–7208.

Chan, S.-C., Jang, J. P., & Cherng, Y.-J. (2009). Tetrahedron, 65, 1977–1981.

Heropoulos, G. A., Carvotto, G., Screttas, C. G., & Steele, B. R. (2007). Tetrahedron Letters, 48, 3247–3250.

Hou, C., Chen, Y., & Li, W. (2012). Carbohydrate Research, 355, 87–91.

Jia, C.-S., Zhang, Z., Tu, S.-J., & Wang, G.-W. (2006). Organic & Biomolecular Chemistry, 4, 104–110.

Joshi, M. C., Joshi, P., & Rawat, D. S. (2006). Archive for Organic Chemistry, 16, 65–74.

Ju, Y., Kumar, D., & Varma, R. S. (2006). The Journal of Organic Chemistry, 71, 6697–6700.

Jung, J. Y., Newton, M. L., Powell, C. B., & Powell, G. L. (2009). Journal of Organometallic Chemistry, 694, 3526–3528.

Karuehanon, W., Fanfuenha, W., Rujiwatra, A., & Pattarawarapan, M. (2012). Tetrahedron Letters, 53, 3486–3489.

Lee, J. C., Park, J. Y., Yoon, S. Y., Bae, Y. H., & Lee, S. J. (2004). Tetrahedron Letters, 45, 191–193.

Liao, M. C., Duan, X. H., & Liang, Y. M. (2005). Tetrahedron Letters, 46, 3469–3472.

Lliashenko, R. Yu., Gorobets, N. Yu., & Doroshenko, A. O. (2011), Tetrahedron Letters, 52, 5086–5089.

Luo, G., Chen, L., & Poindexter, G. S. (2002). Tetrahedron Letters, 43, 5739–5742.

Maza, S., de Paz, J. L., & Nieto, P. M. (2011). Tetrahedron Letters, 52, 441–443.

Meng, X., Cia, Z., Xiao, S., & Zhou, W. (2013). Journal of Fluorine Chemistry, 146, 70–75.

Minetto, G., Raveglia, L. F., & Taddei, M. (2004). Organic Letters, 6, 389–392.

Murthy, Y. L. N., Govindh, B., Diwakar, B. S., Nagalakshmi, K., & Venu, R. (2011). Journal of the Iranian Chemical Society, 8, 292–297.

Nitsche, C., & Kelin, C. D. (2012). Tetrahedron Letters, 53, 5197–5201.

Onogi, S., Higashibayashi, S., & Sakurai, H. (2012). Tetrahedron Letters, 53, 3710–3712.

Paik, S., & Jung, M. G. (2012). Bulletin of Korean Chemical Society, 33, 689–691.

Parchinsky, V., Shumsky, A., & Krasavin, M. (2011). Tetrahedron Letters, 52, 7161–7163.

Petit, A., & Vo-Thanh, G. C. (2011). 15th International Electronic Conference on Synthetic Organic Chemistry (ECSOC-15).

Pichowicz, M., Crumpler, S., Mc Donald, E., & Blagg, J. (2010). Tetrahedron Letters, 66, 2398–2403.

Pingali, S. R. K., Upadhyay, S. K., & Jursic, B. S. (2011). Green Chemistry, 13, 928–933.

Radi, M., Bernardo, V., Vignaroli, G., Brai, A., Biava, M., Schenone, S., & Botta, M. (2013). Tetrahedron Letters, 54, 5204–5206.

Rat, S., Bosco, M., Tavernier, M. L., Michaud, P., Wadouachi, A., & Kovensky, J. (2011). Comptes Rendus Chimie, 14, 307–312.

Rios, N., Chavarria, C., Gil, C., & Porcal, W. (2013). Journal of Heterocyclic Chemistry, 50, 720–726.

Sajith, A. M., & Muralidharan, A. (2012). Tetrahedron Letters, 53, 5206–5210.

Salmoria, G. V., Dall'Oglio, E., & Zucco, C. (1998). Tetrahedron Letters, 39, 2471–2474.

Sana, S., Reddy, K. R., Rajanna, K. C., Venkateswarlu, M., & Ali, M. M. (2012). International Journal of Organic Chemistry, 2, 233–247.

Satyanarayana, S., Praveen, K. K., Laxmi Reddy, P., Narender, R., Narasimhulu, G., & Subba Reddy, B. V. (2013). Tetrahedron Letters, 54, 4892–4895,

Song, X. J., Duan, Z. C., Shao, Y., & Dong, X. G. (2012). Chinese Chemical Letters, 23, 549–552.

Stuerga, D., Gonon, K., & La Uemant, M. (1993). Tetrahedron, 49, 6229–6234.

Sulkowski, W. W., Wolinska, A., Sulkowska, A., Nowak, K., & Bogdal, D. (2013). e-Polymers, 8, 65–71.

Tu, S., Jiang, B., Zhang, Y., Jia, R., Zang, J., Yao, C. & Shi, F. (2007). Organic & Biomolecular Chemistry, 5, 355–359.

Umrigar, V. M., Chakraborty, M., & Parikh, P. A. (2007). Industrial & Engineering Chemistry Research, 46, 6217–6220.

Upadhyay, A., Srivastava, S. K., & Srivastava, S. D. (2010). European Journal of Medicinal Chemistry, 45, 3541–3548.

Upadhyay, S. K., & Jursic, B. S. (2011). Synthetic Communications, 41, 3177–3185.

Vabre, R., Chevot, F., Legraverend, M., & Piguel, S. (2011). The Journal of Organic Chemistry, 76, 9542–9547.

Velmathi, S., & Leadbeater, N. E. (2008). Tetrahedron Letters, 49, 4693–4694.

Verbitskiy, E. V., Cheprakova, E. M., Slepukhin, P. A., Kodess, M. I., Ezhikova, M. A., Pervova, M. G., Rusinov, G. L., Chupakhin, O. N., & Charushin, V. N. (2012b). Tetrahedron Letters, 68, 5445–5452.

Verbitskiy, E. V., Cheprakova, E. V., Slepukhin, P. A., Subbotina, J. O., Schepochkin, A. V., Rusinov, G. L., Chupakhin, O. N., & Charushin, V. N. (2013). Tetrahedron, 69, 5164–5172.

Verbitskiy, E. V., Rusinov, G. L., Charushin, V. N., Chupakhin, O. N., Cheprakova, E. M., Slepukhin, P. A., Pervova, M. G., Ezhikova, M. A. & Kodess, M. I. (2012a). European Journal of Organic Chemistry, 33, 6612–6621.

Wannberg, J., Wallinder, C., Unlusoy, M., Skold, C., & Larhed, M. (2013). Journal of Organic Chemistry, 78, 4184–4189.

Xu, F., & Fan, L.-L. (2008). Zeitschrift Fur Naturforschung, 63B, 298–302.

Xu, H., Yu, X., Sun, L., Lu, J., Fan, W., Shen, Y., & Wang, W. (2008). Tetrahedron Letters, 49, 4687–4689.

Yadav, J. S., Reddy, S. B. V., Basak, A. K., & Narsaiah, A. V. (2007). Journal of Molecular Catalysis A, 274, 105–108.

Yadav, U., Mande, H., & Ghalsasi, P. (2012). Journal of Chemical Education, 89, 268–270.

Yang, G., Shen, C., Zhang, L., & Zhang, W. (2011). Tetrahedron Letters, 52, 5032–5035.

Yaunner, R. S., Barrios, J. C., & Da Silva, J. F. M. (2012). Applied Organometallic Chemistry, 26, 273–276.

Zheng, J., Liu, H.-T., Wu, B., Di, C.-A., Guo, Y.-L., Wu, T., Yu, G., Liu, Y.-Q., & Zhu, D.-B. (2012). Scientific Reports, 2, doi: 10.1038/srep00662.

CHAPTER 6

ALKYLATION AND ARYLATION

SANYOGITA SHARMA, NEELAM KUNWAR, and SURESH C. AMETA

CONTENTS

6.1   INTRODUCTION

Alkylation/arylation is the transfer of an alkyl/aryl group from one molecule to another. This group may be transferred as an alkyl carbocation, a free radical, a carbanion or a carbene (or their equivalents). Alkylating/arylating agents are quite commonly used in chemistry because the alkyl group is the most universal group encountered in organic molecules. Many biological target molecules or their synthetic precursors are composed of an alkyl chain/aryl group with specific functional groups in a specific order. Selective alkylation/arylation, or adding parts to the chain with the desired functional groups, is important from synthesis point of view. In context of oil refining, alkylation refers to a particular alkylation of isobutane with olefins. For upgrading of petroleum, alkylation produces synthetic C_7–C_8 alkylate, which is a finest unification stock for gasoline (Stefanidakis and Gwyn, 1993).

6.2   ALKYLATION

Alkylation is generally carried out by Friedel-Crafts reaction.

This Lewis acid catalyzed electrophilic aromatic substitution allows the synthesis of alkylated products via the reaction of arenes with alkyl halides or alkenes. Since alkyl substituents activate the arene substrate, polyalkylation may occur. A valuable, two-step alternative is Friedel-Crafts acylation followed by a carbonyl reduction.

Alkylation is an important alteration that regularly employs toxic and hazardous reagents such as alkyl iodide (Johnstone and Rose, 1979) or dialkyl sulfate (Basak, et al., 1998). The use of unconventional reagents has been scarce due to the ruthless conditions required with dialkylcarbonate or methanol as alkylating agents.

Dimethylcarbonate (DMC) is a nontoxic and environmentally safe reagent that can be used in organic synthesis as a green substitute for toxic reagents such as phosgene in carbonylation reactions, and dimethylsulfate (DMS) and methyl chloride in methylation reaction (Tundo, 1991). N-Methylation of indoles with methyl iodide (Reineeke et al., 1972; Nunomoto et al., 1990) and dimethyl sulfate (Stouton and Topham, 1953) in presence of variety of bases, such as NaH (Buchi and Mak, 1977), KOH (Ottani et al., 1998) and NaOH are classical methods to form N-methylated indole derivatives. However, use of this method has several disadvantages. Methyl iodide has a very low boiling point (40 °C), causing air emission problem, and it is a suspected carcinogen also while dimethyl sulfate is highly toxic (LD_{50} orally in rats is 440 mg/kg). The by products generated by these methylating agents can cause waste disposal problems. In view of these disadvantages, an alternate methylating reagent and an efficient process for large scale manufacturing is highly desirable. Recently, dimethyl carbonate has been reported as a methylating reagent for phenols and NH-containing heteroaromatic compounds in conjunction with 1,8-diazabicy-clo[5,4,0]undec-7-ene (DBU) under conventional thermal heating with long reaction times. It was found that the change from conventional convection heating to microwave energy increases the speed of the reaction (Shieh et al., 2001).

In last few years, there has been a growing interest in the use of microwave heating in organic synthesis. Microwave heating is a valuable tool for synthetic chemists. It is capable of improving product yields and enhancing the rate of reactions as well as being a safe and convenient method for heating reaction mixtures to elevated temperatures. Domestic microwave ovens are being replaced by scientific microwave apparatus. Apart from being safer, these new instruments allow for accurate control of key parameters such as initial microwave power, reaction temperature and, in the case of sealed vessel reactions, the internal pressure.

Chemists would often use microwave irradiation as an alternate for heating reaction mixtures to high temperatures in sealed tubes in an attempt to complete their reactions. However, microwave heating has a much more valuable role to play in the preparative chemist's portfolio. It is possible to carry out reactions at modest reaction temperatures and still find reasonable improvements in rate of the reaction as well as yield of the product. Methylation of phenols and carboxylic acids using microwave energy in a commercial microwave oven without temperature control was reported by Rajabi and Saidi (2004). They used 1,8-diazabicyclo[5.4.0]undec-7-ene (DBU) as a base but in a catalytic amount.

The reactivity of dimethyl carbonate has been studied by several groups since the early 1980's. Their work was motivated by the report of the first reaction that fulfills green chemistry ecological standards for the synthesis of DMC based on the oxicarbonylation reaction of methanol (Delledonne et al., 1995). This fact was readily recognized and since the initial report, an intense research activity has been directed all over the globe towards innovative applications of DMC and its higher homologs. However, the use of DMC as a methylating reagent often requires high temperatures and long reaction times. As a result, autoclaves, sealed tubes (Barcelo et al., 1991; Shimizu and Lee, 1998) or the use of asymmetrical carbonates (Perosa et al., 2000) is required. Recently, Shieh et al. (2002) discovered that DBU can function as an effective nucleophilic catalyst for carboxylic acid esterification.

6.2.1   C-ALKYLATION

Runhua et al. (1994a; 1994b) have reported phase transfer promoted microwave-assisted C-alkylation of active methylenes.

$$R_1 \diagdown R_2 \xrightarrow[\substack{R_1, SPh, CH_3CO; R_2 = CO_2Et \\ R = Alkyl, allyl, benzyl}]{\substack{RX, KOH-K_2CO_3, PTC, UV, \\ 3-5 \ min, 58-83\%}} R_1 \diagup\diagdown R_2 \atop R$$

The effect of microwave irradiation on phase transfer assisted ether synthesis was reported by Yuncheng et al. (1993). It was observed that the effect was beneficial, leading to dramatically shorter reaction times with similar or better yields of isolated products. Thus, the reaction of benzyl chloride with ethanol was significantly accelerated using microwaves.

$$\xrightarrow[85\%, \ 5 \ min]{CH_3CH_2OH, \ PTC, \ MW}$$ OEt

Alkylation of active methylenes has also been successfully carried out using microwave irradiation. Treatment of ethyl phenylsulphonylacetate with alkyl halide in the presence of potassium carbonate and phase transfer catalyst leads to the alkylated product in 83% yield after 2–3 min of microwave irradiation as reported by Yuliang and Yaozhong (1992).

$$PhSO_2 \diagdown CO_2Et \xrightarrow[MW, \ 3 \ min, \ 83\%]{K_2CO_3, \ PTC, \ BuBr} \underset{PhSO_2 \diagdown CO_2Et}{\overset{Bu}{|}}$$

Giguere and Herberich (1991) reported a rapid and efficient preparation of allyldiphenylphosphine oxide via phosphinite allylation. Ordinarily the reaction proceeds by stirring the reagents at room temperature overnight but the reaction time can be reduced to 1 min using microwave radiation.

$$\diagup\diagdown_{Br} \xrightarrow[1-3 \ min]{Ph_2P-OEt, \ MW} \overset{O}{\underset{Ph_2P}{\overset{||}{\diagdown}}}\diagup\diagdown$$

Beam et al. (2000) observed that dilithiated 1-tetralone oxime was prepared in excess lithium diisopropylamide and condensed with a variety of esters followed by acid cyclization of C-acylated intermediates to substituted 4,5-dihydronaphth-[1,2-c]-isoxazoles.

Active methylenes also react with epoxides in the presence of aluminum oxide, lithium chloride and potassium fluoride under microwave for 5 min (Abenhdim et al., 1994). The main product is a lactone formed by epoxide ring opening followed by cyclization.

They reported a good selectivity with 58–83% yield for monoalkylated products and to ensure efficient mixing of the reagents, the reactions can be carried out in toluene.

6.2.2 N-ALKYLATION

N-Alkylation of aniline derivatives is an important reaction in organic synthesis, and it provides access to valuable building blocks that are used as intermediates or additives in the preparation of some dyes (Suwanprasop et al., 2004), fluorescence probes (Kim et al., 2002; Stauffer and Hartwig, 2003), agrochemicals (Montgomery, 1993) and pharmaceuticals (Negwer, 1994). Some methods for the direct N-monoalkylation of anilines are already available but, when alkylbromides or alkylchlorides are used as alkylating agents, the reactions proceed slowly and several polyalkylation/halogenated by products are observed (Onaka et al., 1985). The most commonly employed alkylating agents are alcohols in the presence of Raney Ni (Ainsworth, 1956), Al_2O_3 (Matsuhashi and Arata, 1991) and SiO_2 supported catalysts (Rusek, 1991) under different reaction conditions.

Aldehydes and ketones react with aniline derivatives in the presence of hydrogen and a catalyst to yield N-alkylanilines (Fache et al., 1996; Freifelder, 1971). N-Alkylation of aliphatic amines under microwave irradiation is a well-documented process (Caddick, 1995; Loupy et al., 1998; Varma, 1999; Lidstrom et al., 2001; Wathey et al., 2002). Jiang et al. (1996) reported the N-alkylation of anilines with alcohols over Raney nickel under microwave irradiation. More recently, Khadilkar and Jaizinghani (1999) carried out the direct monobenzylation of aniline with benzylchloride on alumina supported potassium carbonate. They reported a fast and efficient metal-free method for the N-alkylation of anilines under microwave irradiation. Romera et al. (2004) reported potassium iodide catalyzed monoalkylation of anilines, where they obtained 54–98% yield.

where R_1 = H, Me, Bn; R_2 = H, NO_2, CF_3, Halo, OMe; R_3 = Primary and secondary Alkyl; X = Br, I, OTs.

Barbry and Torchy (1996) reported that the primary and secondary amines are rapidly N-methylated by methanal and formic acid under microwave irradiation.

Grabowska et al. (2008) prepared a composite oxide $ZnAl_2O_4$ by microwave-assisted hydrothermal treatment of a precursor mixture of hydroxides obtained by precipitation of aluminum and zinc nitrates. Various studies show that $ZnAl_2O_4$ is nanosized and is a micro/mesoporous material with large a surface area (140 m²/g). The gas phase catalytic methylation of 4-hydroxypyridine in the presence of the $ZnAl_2O_4$ catalyst was performed in a continuous process at atmospheric pressure in the temperature range of 240–360 °C. A mixture of O- and N-alkylated products, namely 4-methoxypyridine and N-methyl-4-pyridone were obtained. The alkylation of 4-hydroxypyridine with methanol at 345 °C offered 87.6% selectivity towards N-methyl-4-pyridone with about 89% 4-methoxypyridine conversion.

The ionic liquid, tributylmethylammonium methyl carbonate, has been used as a catalytic base for benign N-methylation of indole with dimethyl carbonate. The reaction was optimized under microwave exposure to give 100% conversion and selectivity to N-methylindole. The incredibly short reaction time of 3 min, ease of workability, and high selectivity have major inference for the synthesis of a extensive series of pharmaceutical intermediates. It has also been shown that the ionic liquid can be generated in situ from tributylamine (Glasnov et al., 2012).

Shieh et al. (2001) observed that the 1,8-diazabicyclo[5.4.0]undec-7-ene (DBU) is a novel and active catalyst for the methylation reaction of phenols, indoles, and benzimidazoles with dimethyl carbonate under mild conditions. The rate enhancement is achieved by applying microwave irradiation, which can be further accelerated by incorporating tetrabutylammonium iodide. Combining these, very slow chemical transformations taking several days can be performed efficiently in high yield and that too within minutes.

where Y = C, N.

Lee et al. (2000) reported that a solution of acid chlorides in DMF solvent on reflux gives N,N-dimethylamides and in excellent yields.

$$R\overset{O}{\underset{Cl}{\|}} \quad \xrightarrow[\text{Reflux}]{\text{CHO–NMe}_2} \quad R\overset{O}{\underset{NMe_2}{\|}}$$

A new method for N-methylation of aromatic diamines using environmentally safe and less toxic methylation reagent dimethyl carbonate (DMC) was developed by Sharma et al. (2010). The effect of various functional groups on the aromatic ring has been investigated. This method provides the desired product in high yields with high purity under microwave irradiation.

$$\text{R} \underset{NH_2}{\overset{NH_2}{\bigcirc}} + (CH_3O)_2CO \xrightarrow[\text{100\% Power}]{K_2CO_3,\ DMF} \text{R} \underset{NH_2}{\overset{H_3C-N-CH_3}{\bigcirc}} + 2\ CO_2 + CH_3OH$$

Lourenco et al. (2013) reported N-alkylation reaction of amine functionalized phenylene moieties in crystal-like mesoporous silica. A potassium iodide catalyzed method commonly used for the selective N-monoalkylation of aniline, was adapted and optimized to the N-monoalkylation reactions of the amine functionalized periodic mesoporous phenylene-silica (NH_2-PMO) under microwave irradiation with conservation of the ordered mesostructure and the crystal-like molecular scale periodicity of the material. This functionalization opens a path for the preparation of new materials with different amino-alkyl groups specially designed for a desired application.

Jakopin and Dolen (2010) developed an easy and efficient method of N-alkylation. It was applied successfully to the alkylation of several substituted saccharins. Subsequent t-butoxide induced condensation of these products led to the formation of fused polycyclic compounds, which could be a privileged framework in the field of medicinal chemistry because of their favorable hydrophilic nature and straightforward functionalization. Ding et al. (1994) also reported the rapid N-alkylation of saccharin using a series of alkyl halides under microwave exposure with silica gel or alumina support.

A greener process for direct mono-N-alkylation of aromatic amines by alkyl halides was reported under microwave irradiation in water without any catalyst (Marzaro et al., 2009).

$$\text{R} \underset{}{\overset{NH_2}{\bigcirc}} + R_1X \xrightarrow[\text{MW, 150°C}]{H_2O} \text{R} \underset{}{\overset{NHR_1}{\bigcirc}}$$

N-Alkylations also proceed smoothly under microwave conditions; thus, irradiation of a mixture of saccharin and alkyl halide on silica gel or alumina leads to the isolation of the product in 91% yield (Jinchang et al., 1994).

(91%)

The N-alkylation of some secondary amides, 2-(1H-azol-1-yl)-N-(substituted phenyl) acetamides was carried out under different conditions; basic, microwave and Mannich reaction (Rajput and Gore, 2012).

where R = H, 2-CH$_3$, 4-CH$_3$, 2-Cl, 4-Cl.

N-Methylation is common among peptide natural products and it has a considerable impact on both the conformational states and physical properties of cyclic peptides. White et al. (2011) reported a method for the selective, on-resin N-methylation of cyclic peptides to manufacture compounds with drug-like film permeability and oral bioavailability. The selectivity and degree of N-methylation of the cyclic peptide was determined by backbone stereochemistry. The permeabilities of the N-methyl variants were confirmed by computational studies.

A solvent-free synthesis of N-methyl and N,N-dimethylsulfonamides has been achieved by Malik et al. (2008). The N-methylated products were obtained by treating the primary and secondary sulfonamides with Me$_3$S$^+$OI$^-$ and KOH under microwave irradiation on alumina support.

$$R_1\ SO_2-NH_2 \xrightarrow[\text{MW, 130°C, 4 min}]{Me_3S^+OI^-,\ KOH,\ Al_2O_3} R_1\ SO_2Me_2$$

The N-methylation of backbone amides in synthetic peptides is an important method for improving properties such as bioavailability, stability, as well as structural preferences, and therefore, it is an attractive design strategy (Roodbeen et al., 2012), However, the synthesis of N-methylated peptides can be challenging as the nucleophile in the acylation step is a sterically hindered secondary amine. They have systematically evaluated the use of microwave heating, different coupling conditions and the role of steric effects on coupling yields.

where R = Side chain of Ala, Cys, Phe, Ile, Arg, Thr, Val or Trp; R_1 = Ala, Cys, Asp, Asp, Phe, Ile, Pro, Arg, Thr or Val.

An alternative procedure has been developed by Yuncheng and Yulin (1992). It was observed that alkylation of carboxylic acids with alkyl halides in the presence of a phase transfer catalyst can also lead to high yields of esters with 10 min irradiation. The reaction of hexanoic acid with benzyl bromide and PTC in a sealed tube gave the ester in good yield (72%).

6.2.3 O-ALKYLATION

Microwave-assisted O-alkylation reactions of carboxylic acids, such as aryloxyacetic acids, 4-chlorobenzoic acid, (un)substituted furoic acids, and benzofuroic acid, with (un)substituted ω-haloacetophenones in dry media under phase transfer catalysis were carried out. 2-Oxo-2-arylethyl carboxylates were synthesized by this method in high yield using tetrabutylammonium bromide as catalyst. O-Alkylation of diverse carboxylic acids has played an important role in preparation of esters. O-Alkylation of various carboxylic acids with (un)substituted ω-haloacetophenones is a major source of 2-oxo-2-arylethyl carboxylates (Shi et al., 1995), which are important intermediates for organic synthesis and biologically active compounds because of their antibacterial, anesthetic, anticonvulsive and plant-growth regulating activities. The present methods for obtaining 2-oxo-2-arylethyl carboxylates are performed by the reaction in toxic solvents, such as acetonitrile and chlorobenzene, at reflux temperature for at least 12 h, giving only average yields. Furthermore, the carboxylic acids have to be converted into their potassium salts before reaction with ω-haloacetophenones (Fang et al., 1998; Shi et al., 1995). All this can be avoided by using microwave radiations.

Methylation of carboxylic acids and phenolic compounds with dimethyl carbonate (DMC) in the presence of a catalytic amount of $BF_3 \cdot OEt_2$, DBU, or KOH, in good yields was reported by Rajabi and Saidi (2004) under solvent-free conditions and microwave irradiation. Guerrero and Rivero (2008) have discovered that 1,2-dimethylimidazole (DMI) can also function (like DBU) as a nucleophilic catalyst reacting with DMC or diethyl carbonate (DEC) to form a more active alkylating reagent for the alkylation of phenols and esterification of carboxylic acids. The yields of the alkylated products were about 98–99%.

Belov et al. (2011) evaluated the potential of N,N-dimethylformamide dimethylacetal (DMF-DMA) as a methylating agent for a variety of p-substituted phenols under microwave irradiation. The reaction was completed within 30 min if substituent is an electron-withdrawing group while it is over in 60 min, if substituent is an electron-donating group. Enamino-ketone formation and esterification were obtained with carboxylic acid and ketone functional group, respectively.

Tetramethylammonium chloride (Me_4NCl) has been used as an alternative-methylating agent for phenols under microwave-assisted conditions by Maras et al. (2008). Its chemical behavior was tested in a reaction with 2-naphthol in the presence of various bases and solvents. This method was then applied under heterogeneous conditions using 1,2-dimethoxyethane (DME) or toluene for the O-methylation of a series of phenolic compounds with 10–90% yield. They found that a number of simple phenols can be methylated in the presence of K_2CO_3, while some other less-reactive phenols need the presence of the more reactive Cs_2CO_3.

$$Ar-OH + Me_4NCl \xrightarrow[\text{MW, 145°C}]{K_2CO_3 \text{ or } Cs_2CO_3/DME} Ar-OMe + Me_2N$$

where ArOH = Phenols, Naphthol.

The preparation of industrially important alkyl aryl ethers was reported by Saidi and Rajabi (2003). A number of phenolic compounds such as naphthols, phenols, and hydroxyl coumarins were O-methylated with trimethyl phosphite or trimethyl phosphate under microwave irradiation and solvent-free condition in almost quantitative yields. Reaction of 2-naphthol with trimethyl phosphate gave mixture of

2-methoxynaphthalene and 1-methyl-2-methoxynaphthalene while the reaction with trimethyl phosphite gave mostly 2-methoxynaphthalene. This method is extremely proficient for the methylation of phenolic compounds with an easy experimental procedure and environmental friendly conditions.

O-Methylation is of outstanding importance in structural polysaccharide chemistry. A novel method for the methylation of polysaccharides using microwave irradiation was described by Singh et al. (2003). Seed gum from *Cyamopsis tetragonolobus* (Guar) was fully methylated with dimethyl sulfate and sodium hydroxide using 100% microwave power in 4 min with 68% yield. The completely methylated seed gum was further hydrolyzed by 70% formic acid followed by 0.5N H_2SO_4 under full microwave power.

Singh et al. (2011) modified the sago starch and evaluated its efficacy as tablet disintegrant. Cross-linked carboxymethylated sago starch (CMSS) was synthesized with degree of substitution (DS = 0.31) using native sago starch and monochloroacetic acid with sodium hydroxide under microwave radiations. It was further evaluated as disintegrant in Ondansetron based tablets. The results revealed that CMSS could be used as disintegrant in tablet formulation in concentration dependent manner.

$$St-OH + CH_2Cl-COOH \xrightarrow[MW]{NaOH} St-O-CH_2-COOH + NaCl + H_2O$$

O-Methylation of phenolic compounds can be efficiently carried out by tetramethylammonium chloride in diglyme or polyethyleneglycol (PEG) at temperatures of 150–160 °C and in the presence of either NaOH or K_2CO_3. The benzylation and methylation of phenols occur, where the benzylation product was always predominating. With allyl-substituted phenols as substrates and using NaOH as a base, it was feasible to achieve both the alkylation and the double-bond isomerization of the allyl group to obtain (E/Z)-propenyl-substituted methyl and benzyl aryl ethers in a single preparative step (Maras et al., 2010).

Carboxylic esters have been readily prepared by a tribromolanthanoid mediated deetherification reaction (Yulin and Yunchang, 1994). Thus, microwave irradiation of benzyl ether with a carboxylic acid in the presence of $LnBr_3$ (Ln = La, Nd, Sm, Dy and Er) led to the isolation of ester in 61–84% yield. The reactions are dependent upon the presence of the $LnBr_3$, because in its absence, the yields are significantly reduced.

$$Ar{\diagup}OR \xrightarrow[\text{MW, 1.5-2 min}]{\text{LnBr}_3,\ \text{RCOOH}} Ar{\diagup}O{\diagup}\underset{O}{\overset{}{C}}R$$

When a similar transformation was attempted using a dialkylether in place of an aryl alkyl ether, yields obtained were much lower. An alternative microwave-assisted method for promoting esterification uses phase transfer catalysis. Irradiation of potassium acetate with alkyl halides for 1-2 min (Aliquat 336, 5–10%) gives the product in good to excellent yields (> 92%) (Bram et al., 1990).

$$CH_3COOK + RX \xrightarrow[\text{MW}]{\text{Alumina}} CH_3COOR$$

where R = n-C_8H_{17}, n-$C_{16}H_{33}$; X = Cl, Br.

Liu et al. (2008) reported that nonspecific proteolytic digestion of glycoproteins is an established technique in glycomics and glycoproteomics. Glycoproteins are digested to small glycopeptides having one to six amino acids residues in the presence of *pronase E.* Unfortunately; the long digestion times (1–3 days) limit its use. They used controlled microwave irradiation to accelerate the proteolytic cleavage of glycoproteins mediated by *pronase E.* When glycoproteins were digested in presence of *pronase E.*, it produced glycopeptides within 5 min under microwave irradiation, the glycopeptides having one or two amino acids were the major products. The sodiated forms of glycopeptides were methylated using methyl iodide. This controlled methylation procedure resulted in quaternization of the amino group of the N-terminal amino acid residue. The methylated products of glycopeptides containing two or more amino acid residues were more stable than those containing only a single Asparagine (Asn) residue.

6.2.4 S-ALKYLATION

Villemin and Alloum (1990) have reported a useful procedure for the preparation of sulphones by microwave-assisted alkylation of sodium phenylsulphinate. Thus irradiation of a mixture of the sulphinate and benzyl halide adsorbed onto alumina led to the isolation of the desired product in 40–99% yield.

$$\text{PhCH}_2\text{Cl} \xrightarrow[\text{MW, 5 min}]{\text{PhSO}_2^-\text{Na}^+,\ \text{Al}_2\text{O}_3} \text{PhCH}_2\text{SO}_2\text{Ph}$$

(43-99%)

The solvent less S-methylation of thiols and O-methylation of phenols and naphthols with dimethyl sulfate (DMS) supported on basic alumina was reported by Heravi et al. (2005).

$$R-SH \xrightarrow[\text{MW, 2-5 min}]{\text{1 eq. } (CH_3)_2 \ SO_4} R-S-Me$$

when R is phenyl and tolyl, then the yield was 72 and 58%, respectively.

6.3 ARYLATION

6.3.1 C-ARYLATION

An efficient and ecofriendly microwave irradiated solvent-free benzoylation method was developed by Al-Masum et al. (2011). 50 mol% AlCl$_3$ was used as a Lewis acid catalyst at 130 °C for C-benzoylation and the reaction was completed in 10 min. The isolated yield was between 71–100%. N-benzoylation was also conducted in a catalyst-free environment at 130 °C in 10 min but the yield was between 80–100%.

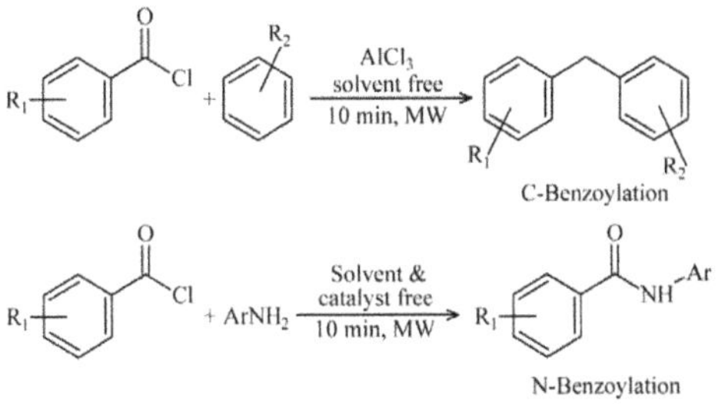

C-Benzoylation

N-Benzoylation

A highly efficient procedure for the synthesis of N-heteroaryl-4-(2-chloroethyl) piperazines and N-heteroaryl-4-(2-chloroethyl)piperidines was developed under microwave irradiation by Wang et al. (2009). Irradiation of electron deficient heteroaryl chlorides with 1,4-diazabicyclo[2.2.2] octane (DABCO) at 160 °C for 15 min led to N-heteroaryl–4-(2-chloroethyl)piperazines in good yields. In a similar manner, microwave irradiation of electron deficient heteroaryl chlorides with quinuclidine provided N-heteroaryl-4-(2-chloroethyl)piperidines in good to excellent yields. Extension of this method was demonstrated by the development of a one-pot, two-step microwave-assisted protocol for the synthesis of 4-(2-acetoxyefhyl)-substituted N-heteroarylpiperazines and N-heteroarylpiperidines.

6.3.2 N-ARYLATION

Copper-catalyzed N-arylation of 2-imidazolines has been carried out by Davis et al. (2013). The reaction provides compounds with advantageous lead-like characteristics in good yields with useful simplicity under inexpensive, ligand-free conditions. The cross coupling was successful with electron-rich and electron-poor aromatic iodides. Substrates having halides, esters, nitriles, and free hydroxyls are well tolerated, providing reactive handles for further functionalization. In addition, the regioselective N-arylation of 4-substituted imidazoline has also been reported.

Veiga et al. (2013) developed a rapid method for efficient palladium catalyzed N-arylation of polynitrogenated macrocycles. Its applicability for functionalization of protected azamacrocycles of various sizes with substituted aryl bromides of optional electronic properties has been established. The compatibility of the protocol with common N-protecting schemes as well as the impact of electronic versus steric factors was also discussed. This method provides moderate to excellent yields of N-arylated azamacrocycles (45–96%) using a commercially available catalytic system and easily available alkoxide or phenoxide base.

where R = tert-Butyl or 2,4,6-Tris(tert-butyl) phenyl.

A one-step, high-yielding and catalyst free method was reported for N-arylation of azoles and indoles from unactivated monofluorobenzenes by Diness and Fairlie

(2012) . This reaction tolerates a wide range of substituents and can also generate halogenated N-aryl products. The reaction can also be performed simultaneously with or subsequent to a copper- or palladium catalyzed cross-coupling reaction in the same pot.

Microwave-assisted direct arylation reactions were successfully employed in the synthesis of azafluoranthene alkaloids. Ponnala and Harding (2013) reported that direct arylation reactions on a diverse set of phenyltetrahydroisoquinolines produced the indeno[1,2,3-ij]isoquinoline nucleus, required towards a high-yielding azafluoranthene synthesis. The method was used as a key step in the efficient preparation of the natural products rufescine and triclisine. This synthetic approach may be generally applicable to the preparation of natural and un-natural azafluoranthene alkaloids as well as azafluoranthene-like isoquinoline alkaloids.

Dong et al. (2012) reported a highly efficient macrocyclization reaction via the palladium catalyzed C-H arylation of the side chains of tryptophan with halophenyl-containing amino acids. This method allows for direct access to 15- to 25-membered biaryl macrocycles in 40–75% yield in 30–45 min, at moderate concentration, with C-H arylation proceeding exclusively at the C-2 position of the tryptophan indole.

An efficient synthesis of the imidazo[1,2-b]pyrazole core has been developed and the first regioselective palladium catalyzed direct arylation of the C-3 position has been described by Grosse et al. (2012). Good to excellent yields were obtained for a extensive range of aryl partners with electron-rich and electron-poor substituents. This method allows rapid access to a large variety of imidazo[1,2-b] pyrazole products and could open the way to the design of new biologically active compounds.

where R = OCH$_3$, CH$_3$, CF$_3$.

Abdo et al. (2012) reported an improvement in an efficient regioselective direct C-H arylation of thieno[3,4-b]pyrazine (TP) and its 2,3-dimethyl derivative with bromoalkylthiophenes (BATs), under Heck experimental conditions using Pd(OAc)$_2$/Bu$_4$NBr as the catalytic system, giving rise to a variety of valuable aryl-substituted thienopyrazines. It was observed that the 2-position of the TP moiety is less reactive towards C-H arylation than the 5- and 7-positions. Moreover, the 3-position of the TP moiety showed almost no significant reactivity when all other positions were arylated. The C-H arylation of 2,3-dimethyl-TP with an excess amount of bromoalkylthiophenes (BAT) go on smoothly, affording the corresponding diarylated thienopyrazine derivatives in excellent yields, without any additional products.

Sharma et al. (2013) described the development of direct C-H arylation approaches through the application of focused microwave irradiation. The synergistic combination of microwave-assisted techniques with the rapidly evolving province of C-H arylation has opened new vistas in the proficient synthesis of a various assortment of biologically important (hetero) arenes.

$$Ar_1-X + Ar_2-H \xrightarrow{MW} Ar_1-Ar_2$$
$$Ar_1-H + Ar_2-H \xrightarrow{MW} Ar_1-Ar_2$$

A series of aryl pyrogallol[4]arenes were efficiently synthesized in excellent yields by Dawood and El-Deftar (2010). They observed the catalytic action of a benzimidazole-oxime Pd (II)-complex towards Suzuki and Heck C-C cross-coupling reactions of activated and deactivated aryl- and heteroaryl bromides under microwave irradiation (2 min) as well as thermal heating using water as a green solvent with 95% yields of products. The turnover frequency reached 420,000 h^{-1} under microwave condition.

The structures of aryl pyrogallol[4]arenes were confirmed by characterization of their acylated derivatives. Under microwave irradiation, alkylation reactions of aryl pyrogallol[4]arenes with some alkylating reagents such as n-butyl iodide, benzyl chloride, and ethyl α-chloroacetate were completed quickly to yield fully O-alkylated products. The ^{1}H NMR spectra and crystal structures showed that the acylated and alkylated aryl pyrogallol[4]arenes existed mainly in rctt(cis-trans-trans) configuration (Yan et al., 2007).

The use of sodium hydrogen carbonate under solvent-free conditions and microwave irradiation is best method for N-alkylating pyrazoles. The yields are good and the method is devoid of side reactions like quaternization, isomerization and hydrogen halide elimination. Solvent-free conditions are the only ones that allow the preparation of 1-substituted pyrazoles from secondary halides (Almena et al., 2009).

Zeolite Hβ is found to be an effective catalyst for the acylation of amines and alcohols with acetic acid under microwave irradiation. The process is environmentally safe and heterogeneous with excellent yields (Krishna Mohan et al., 2006).

$$R{-}HX \ + \ CH_3COOH \xrightarrow[\text{MW}]{\text{Zeolite}} R{-}XCOCH_3 \ + \ H_2O$$

where X = O or NH

Methylation of *cassava* starch was achieved by Hou et al. (2012). They used thiocarbamide as an active catalyst, incorporating dimethyl carbonate as a solvent and methylating reagent and applying microwave irradiation as energy resource. It can be performed efficiently to a high degree of substitution within 4 min.

6.4 DEALKYLATION

Park et al. (2013) reported a highly reliable dealkylation protocol of alkyl aryl ethers, whose alkyl groups are longer than methyl group. Various ethyl, n-propyl, and benzyl aryl ethers were successfully cleaved using an ionic liquid, 1-n-butyl-3-methylimidazolium bromide, [bmim][Br], under microwave irradiation. In spite of many characteristics such as less toxicity and lower cost of the alkylating agents and greater hydrophobicity of the products, longer alkyl ethers have been extensively less exploited than methyl ethers, probably due to more difficulty in the deprotection step. This method has the same advantages like mild conditions, short reaction time, and small use of the ionic liquids. The dealkylation protocol can greatly encourage the broader use of longer alkyl groups in the protection of phenolic groups.

The enhancement of the toluene steam dealkylation reaction is feasible by microwave irradiation (Litvishkov et al., 2012). It has been found that the most likely cause of the positive effect of microwave radiation on the reaction rate is an increase in the preexponential factor of the Arrhenius equation for the temperature dependence of the reaction rate. This effect is presumably due to an increase in the active surface area of the catalyst formed by the microwave-assisted thermal treatment.

$$C_6H_5CH_3 + 2\ H_2O \longrightarrow C_6H_6 + CO_2 + 3\ H_2$$

Guchhait and Madaan (2010) reported a novel microwave-assisted one-pot novel tandem de-tert-butylation of tert-butyl amine in an Ugi-type multicomponent reaction product. Tert-butyl isocyanide has been explored as a useful convertible isonitrile for the first time affording access to molecular diversity of pharmaceutically important polycyclic N-fused imidazo-heterocycles.

Delgado et al. (1991) reported oxidation of several Hantzsch 4-alkyl-1,4-dihydropyridines to the corresponding aromatic systems. An unexpected mixture of 4-alkyl pyridines and/or dealkylated pyridines was formed.

Replacement of hydrogen by an alkyl or aryl group can be carried out under microwave irradiation in a green chemical pathway using some green chemical reagent like DMC, DEC, etc. This reaction is also helpful in protection of a hydrogen containing groups and gives varieties of alkylation/arylation like N-, O-, S-, etc.

KEYWORDS

- **Arylation**
- **C-Alkylation**
- **Dealkylation**
- **N-Alkylation**
- **O-Alkylation**

REFERENCES

Abdo, N. I., El-Shehawy, A. A., El-Barbary, A. A., & Lee, J.-S. (2012). European Journal of Organic Chemistry, 28, 5540–5551.

Abenhdim, D., Loupy, A., Mahieu, C., & Semeria, D. (1994). Synthetic Communications, 24, 1809–1816.

Ainsworth, C. (1956). Journal of the American Chemical Society, 78, 1635–1636.

Al-Masum, M. A., Wai, M. C., & Dunnenberger, H. (2011). Synthetic Communications, 41, 2888–2898.

Almena, I., Díz-Barra, E., La Hoz, A. D, Ruiz, J., Sínchez-Migallón, A., & Elguero, J. (2009). Journal of Heterocyclic Chemistry, 35, 1263–1268.

Barbry, D., & Torchy S. (1996). Synthetic Communications, 26, 3919–3922.

Barcelo, G., Grenouillat, D., Senet, J., & Sennyey, G. (1991). Tetrahedron, 46, 1839–1848.

Basak, A., Nayak, M. K., & Chakraborti, A. K. (1998). Tetrahedron Letters, 39, 4883–4886.

Beam, C. F., Schady, D. A., Rose, K. L., Kelley Jr., W., Rakkhit, R., Hornsby C. D., & Studer-Martinez, S. L. (2000). Synthetic Communications, 30, 3391–3404.

Belov, P., Campanella, V. L., Smith, A. W., & Priefer, R. (2011). Tetrahedron Letters, 52, 2776–2779.

Bram, G., Loupy, A., Majdoub, M., Gutierrez, E., & Rui-Hitzky, E. (1990). Tetrahedron, 46, 5167–5176.

Buchi, G., & Mak, C. P. (1977). Journal of Organic Chemistry, 42, 1784−1786.

Caddick, S. (1995). Tetrahedron, 51, 10403–10432.

Davis, O. A., Hughes, M., & Bull, J. A. (2013). Journal of Organic Chemistry, 78, 3470–3475.

Dawood, K. M., & El-Deftar, M. M. (2010). Arkivoc, (ix), 319–330.

Delgado, F., Alvarez, C., Garcia, O., Penieres, G., & Marquez, C. (1991). Synthetic Communications, 21, 2137–2141.

Delledonne, D., Rivetti, F., & Romano, U. (1995). Journal of Organometallic Chemistry, 488, C15–C19.

Ding, J. Gu, H., Wen, J., & Lin, C. (1994). Synthetic Communications, 24, 301–303.

Diness, F., & Fairlie, D. P. (2012). Angewandte Chemie International Edition, 51, 8012–8016.

Dong, H., Limberakis, C., Liras, S., Price, D., & James, K. (2012). Chemical Communications, 48, 11644–11646.

Fache, F., Valot, F., Milenkovic, A., & Lamaire, M. (1996). Tetrahedron, 52, 9777–9784.

Fang, J. X., Li, S. M., & Shi, Y. N. (1998). Acta Scientiarum Naturalium Universitatis Nankaiensis, 31, 97.

Freifelder, M. (Ed.) (1971). Practical catalytic hydrogenation. New York: Wiley-Interscience, 333–389.

Giguere, R. J., & Herberich, B. (1991). Synthetic Communications, 21, 2197–2201.

Glasnov, T. N., Holbrey, J. D., Kappe, C. O., Seddon, K. R., & Yan, T. (2012). Green Chemistry, 14, 3071–3076.

Grabowska, H., Zawadzki, M., & Syper, L. (2008). Catalysis Letters, 121, 103–110.

Grosse, S., Pillard, C., Massip, S., Léger, J. M., Jarry, C., Bourg, S., Bernard, P., & Guillaumet, G. (2012). Chemistry - A European Journal, 18, 14943–14947.

Guchhait, S. K., & Madaan, C. (2010). Organic and Biomolecular Chemistry, 8, 3631–3634.

Guerrero, L. R., & Rivero, I. A. (2008). Arkivoc (xi), 295–306.

Heravi, M. M., Ahari, N. Z., Oskooie, H. A., & Ghassemzadeh, M. (2005). Phosphorus, Sulfur, and Silicon and the Related Elements, 180, 1701–1712.

Hou, C., Chen, Y., & Li, W. (2012). Carbohydrate Research, 355, 87–91

Jakopin, Z., & Dolenc, M. S. (2010). Synthetic Communications, 40, 2464–2474.

Jiang, Y. L., Hu, Y. Q., Feng, S. Q., Wu, J. S., Wu, Z. W., & Yuan, Y. C. (1996). Synthetic Communications, 26, 161–164.

Jinchang, D., Hengjie, G., Jinzhu, W., & Caizhen, L. (1994). Synthetic Communications, 24, 301–303.

Johnstone, R. A. W., & Rose, M. E. (1979). Tetrahedron, 35, 2169–2173.

Khadilkar, B. M., & Jaizinghani, H. G. (1999). Synthetic Communications, 29, 3693–3698.

Kim, J. S., Shon, O. J., Rim, J. A., Kim, S. K., & Yoon, J. J. (2002). Organic Chemistry, 67, 2348–2351.

Krishna Mohan, K. V. V., Narender, N., & Kulkarni, S. J. (2006). Green Chemistry, 8, 368–372.

Lee, W. S., Park K. H., & Yoon Y. J. (2000). Synthetic Communications, 30, 4241–4245.

Lidstrom, P., Tierney, J., Wathney, B., & Westman, J. (2001). Tetrahedron, 57, 9225–9283.

Litvishkov, Y. N., Tret'yakov, V. F., Talyshinskii, R. M., Efendiev, M., Guseinova, E. M., Shakunova, N. V., & Muradova, P. A. (2012). Petroleum Chemistry, 52, 186–188.

Liu, X., Chan, K., Chu, I. K., & Li, J. (2008). Carbohydrate Research, 343, 2870–2877.

Loupy, A., Pettit, A., Hamelin, J., Texier-Boullet, F., Jacquault, P., & Mathe, D. (1998). Synthesis, 9, 1213–1234.

Lourenco, M. A. O., Siegel, R., Mafara, L., & Ferreira, P. (2013). Dalton Transactions, 42, 5631–5634.

Malik, S., Nadir, U., & Pandey, P. (2008). Synthetic Communications, 38, 3074–3081.

Maras, N., Polanc, S., & Kocevar, M. (2010). Acta Chimica Slovenica, 57, 29–36.

Maras, N., Polanc, S., & Kocevar, M. (2008). Tetrahedron, 64, 11618–11624.

Marzaro, G., Guiotto A., & Chilin, A. (2009). Green Chemistry, 11, 774–776.

Matsuhashi, M., & Arata, K. (1991). Bulletin of the Chemical Society of Japan, 64, 2065–2076.

Montgomery, J. H. (1993). Agrochemicals desk reference: Environmental data; Lewis: Chelsea, MI, 34, 229–234.

Negwer, M. (Ed.) (1994). Organic-chemical drugs and their synonyms. Berlin: Akademie Verlag GmbH.

Nunomoto, S., Kawakami, Y., Yamashita, Y., Takeuchi, H., & Eguchi, S. (1990). Journal of the Chemical Society, Perkin Transactions 1, 111–114.

Onaka, M., Umezono, A., Kawai, M., & Izumi, Y. J. (1985). Chemical Communications, 17, 1202–1203.

Ottani, O., Cruz R., & Alves, R. (1998). Tetrahedron, 54, 13915–13928.

Park, S. K., Battsengel, O., & Chae, J. (2013). Bulletin of the Korean Chemical Society, 34, 174–178.

Perosa, A., Selva, M., Tundo, P., & Zordan, F. (2000). Synlett, 2, 272–274.

Ponnala, S., & Harding, W. W. (2013). European Journal of Organic Chemistry, 6, 1107–1115.

Rajabi, F., & Saidi, M. R. (2004). Synthetic Communication, 34, 4179–4188.

Rajput, A. P., & Gore, R. P. (2012). Der Pharma Chemica, 4, 2222–2227

Reineeke, M. G., Sebastian, J. F., Johnson, H.W. Jr., & Pyun, C. (1972). Journal of Organic Chemistry, 37, 3066 −3068.

Romera, J. L., Cid, J. M., & Trabanco, A. A. (2004). Tetrahedron Letters, 45, 8797–8800.

Roodbeen, R., Pedersen, S. L., Hossein, M., & Knud J. (2012). European Journal of Organic Chemistry, 36, 7106–7111.

Runhua, D., Yuliang, W., & Yaozhong, J. (1994a). Synthetic Communications, 24, 111–115.

Runhua, D., Yuliang, W., & Yaozhong, J. (1994b). Synthetic Communications, 24, 1917–1921.

Rusek, M. (1991). Studies in Surface Science and Catalysis, 59, 359–365.

Saidi, M. R., & Rajabi, F. (2003). Phosphorus, Sulfur and Silicon and the Related Elements, 178, 2343–2348.

Sharma, A., Vacchani, D., & Van Der Eycken, E. (2013). Chemistry—A European Journal, 19, 1158–1168.

Sharma, S., Ameta, S. C., & Sharma, V. K. (2010). Proceedings of the World Congress on Engineering and Computer Science, San Francisco, 2, 721–723.

Shi, Y. N., Fang, J. X., & Li, S. M. (1995). Chemical Journal of Chinese Universities, 16, 588.

Shieh, W. C., Dell, S., & Repic, O. (2001). Organic Letters, 3, 4279–4281.

Shieh, W., Dell, S., & Repic, O. (2002). Journal of Organic Chemistry, 67, 2188–2199.

Shimizu, I., & Lee, Y. (1998). Synthetic Letters, 1063–1064.

Singh, A. V., Nath, L. K., Guha, M., & Kumar, R. (2011). Pharmacology and Pharmacy, 2, 42–46

Singh, V., Tiwari, A., Tripathi, D. N., & Malviya, T. (2003). Tetrahedron Letters, 44, 7295–7297.

Stauffer, S. R., & Hartwig, J. F. (2003). Journal of the American Chemical Society, 125, 6977–6985.

Stefanidakis, G., & Gwyn, J. E. (1993). Chemical processing handbook, Boca Raton: CRC Press, 80–138.

Stouton, R. S., & Topham, A. (1953). Journal of the Chemical Society, 1889–1894.

Suwanprasop, S., Nhujak, T., Roengsumran, S., & Petsom, A. (2004). Industrial and Engineering Chemistry Research, 43, 4973–4978.

Tundo, P. (1991). Continuous flow methods in organic syntheses. Chichester (UK): Horwood Publication.

Varma, R. S. (1999). Green Chemistry, 1, 43–55.

Veiga, A. X., Arenz, S., & Erdélyi, M. (2013). Synthesis, 45, 777–784.

Villemin, D., & Alloum, A. B. (1990). Synthetic Communications, 20, 925–932.

Wang, H. J., Wang, Y., Csakai, A. J., Earley, W. G., & Herr, R. J. (2009). Journal of Combinatorial Chemistry, 11, 355–363.

Wathey, B., Tierney, J., Lidstrom, P., & Westman, J. (2002). Drug Discovery Today, 7, 373–380.

White, T. R., Renzelman, C. M., Arthur, C. R., Rezai, T., McEwen, C. M., Gelev, V. M. et al. (2011). Nature Chemical Biology, 7, 810–817.

Yan, C., Chen, W., Chen, J., Jiang, T., & Yao, Y. (2007). Tetrahedron, 63, 9614–9620.

Yuliang, W., & Yaozhong, J. (1992). Synthetic Communications, 22, 2287–2291.

Yulin, J., & Yuncheng, Y. (1994). Synthetic Communications, 24, 1045–1048.

Yuncheng, Y., & Yulin, J. (1992). Synthetic Communications, 22, 3109–3114.

Yuncheng, Y., Yulin, J., Jun. P., Xiaohui, Z., & Conggui, Y. (1993). Gazetta Chimica Italiana, 123, 519–520.

CHAPTER 7

ADDITION

ABHILASHA JAIN, PRIYA PARSOYA, and DIPTI SONI

CONTENTS

Addition reactions are simply those reactions, where two or more compounds are added without loss of any fragment (molecule, radical or ion). Addition reactions are truly atom economic reactions (100%) as these use all the atoms of reactants and no waste is produced. The field of addition reactions has not been properly explored as far as microwave-assisted reactions are concerned. These reactions can be basically classified in two categories depending on nature of the attacking reagent; electrophilic or nucleophilic.

7.1 NUCLEOPHILIC ADDITION

Some nucleophilic addition reactions have been carried out successfully under microwave irradiations.

Varma et al. (1997a) reported that the reactions of primary and secondary amines with aldehyde and ketones, are substantially accelerated by microwaves under solvent-free conditions to give imines and enamines, respectively. They have used montmorillonite K10 clay as the catalyst.

$$-\overset{|}{\underset{|}{C}}-\overset{|}{C}=O + HN-R_1 \xrightarrow[MW]{K\ 10\ Clay} -\overset{|}{\underset{|}{C}}-\overset{|}{C}=N-R_1 \text{ or } -\overset{|}{\underset{|}{C}}=C=\underset{\underset{R}{|}}{C}=N-R_1$$

Vass et al. (1999) used montmorillonite K10 clay in the microwave-assisted synthesis of N-sulfonylimines. They also used calcium carbonate as a benign reagent for reaction between aromatic aldehydes with sulfonamides.

Hydrazone synthesis from a mixture of benzophenone and hydrazine hydrate in toluene was also observed (Gadhwal et al., 1990). 95% Yield of the hydrazone was obtained within 25–30 min. These hydrazones were later reduced to corresponding hydrocarbons by Wolff-Kishner reduction.

$$\overset{Ph}{\underset{Ph}{\diagdown}}C=O + RNHNH_2 \longrightarrow \overset{Ph}{\underset{Ph}{\diagdown}}C=N-NHR$$

An efficient rapid, solvent-free and noncatalyzed method for the synthesis of 2-subsituted-2-oxazolines has been reported (Marreo-Terreo and Loupy, 1996). It involves two successive nucleophilic additions with the formation of an amide as an intermediate leading to cyclohydrated product.

$$\begin{array}{c} Ar \\ \diagdown \\ C=O + H_2NCH(CH_2OH)_3 \xrightarrow{-H_2O} \\ HO \diagup \end{array}$$

Hajipour et al. (1999) reported that the reactions of aliphatic alcohols with p-toluenesulfinic acid were accelerated by microwave irradiation under solvent-free conditions in the presence of silica gel. It afforded a high yield of alkyl p-toluene-sulfinates.

$$\underset{Ar}{\overset{O}{\underset{\|}{S}}}\!\!-\!OH + ROH \xrightarrow[MW]{Silica \ gel} \underset{Ar}{\overset{O}{\underset{\|}{S}}}\!\!-\!OR$$

Microwave-mediated Leuckart reductive amination of carbonyl compounds was carried out (Loupy et al., 1996). They observed a strong specific (nonthermal) activation effect of microwaves. The yields obtained were excellent (75–97%) within short reaction time compared to conventional harsh conditions.

$$\underset{R_1}{\overset{O}{\underset{\|}{\diagup}}}\!\!\diagdown_{R_2} \xrightarrow[MW, \ 60 \ W, \ 30 \ min]{HCO_2NH_4 \ or \ HCO_2H, \ HCONH_2} \underset{R_1}{\overset{NHCHO}{\diagup}}\!\!\diagdown_{R_2}$$

Addition of vinyl pyrazoles to ethyl N-trichloroethylidene carbonate was investigated by Carrillo et al. (1999). It was interesting to note that this reaction can be performed only under microwave irradiation where as under classical heating, decomposition or dimerization of pyrazole derivatives has been observed.

Gan et al. (2006) observed a rapid microwave-assisted solvent-free Henry reaction. The corresponding adducts (secondary alcohols) were obtained with moderate to good yields.

Conjugated nitroalkenes were also synthesized by the reaction between aromatic aldehydes and nitroalkenes in the presence of catalytic amount of ammonium acetate by microwave-assisted Henry reaction in solvent-free conditions (Varma et al., 1997b). The β-nitroalcohols were formed as intermediates.

Synthesis of α-keto amides by microwave-assisted reaction has been developed by Chen and Deshpande (2003) using an acyl chloride and isonitrile. This results in significantly greater yields (21–74%).

7.1.1 MICHAEL ADDITION

Microwave-assisted 1,4-Michael addition of primary and cyclic secondary amines to acrylic esters leading to several β-amino acid derivatives in high yield within short reaction time has been reported by Romanova et al. (1997).

where R = Me, Bu.

In the reactions of several amines with β-substituted acrylic acid esters from D(+) mannitol, 1,2-asymmetric induction (up to 76%) of diastereoisomeric excess was observed in solvent-free condition under microwave irradiation.

Imidazole has been condensed via a 1,4-Michael addition with ethyl acrylate using basic clays (Li⁺ and Cs⁺ montmorillonites) under solvent-free conditions with microwave irradiations by Martin-Aranda et al. (1997). When Li⁺ montmorillonites was used, the yield was 40% in 1 min and 72% in 5 min; while it was 0% and 27%, respectively under conventional heating.

Amore et al. (2006) have established a very simple and rapid methodology for the synthesis of N-aryl functionalized β-amino ester by microwave promoted Michael addition of anilines to α, β-unsaturated esters. Hydrolysis of this ester may give H-functionalized β-amino acid, easily.

An environmentally benign protocol has been developed for catalyst free benzylic C-H functionalization of methyl quinolines and Michael addition to β-nitro styrenes under microwave irradiation (Rao and Meshram, 2013).

It was observed that the microwave activation of the Michael addition between chiral α-alkoxy imine and methyl acrylate at 100 °C resulted into the corresponding Michael adduct with better yield compared to thermal rearrangement (Camara et al., 2003). The product has same regio- and stereoselectivity.

Microwave-assisted preparation of α- and β-substituted alanine derivatives by α-amido alkylation or Michael addition reaction in presence of silica supported Lewis acids as catalyst has been reported by de la Hoz et al. (2001).

It has been observed that microwave-assisted reaction of 2'-hydroxychalcones in the presence of DBU led to the formation of the unknown dimers by conjugate addition of the intermediate of the cyclic ketone to the starting enone (Patonay et al., 2001).

A novel protocol for the synthesis of highly substituted pyridines in a single synthesis step by the microwave-assisted Michael addition-cyclodehydration of ethyl β-aminocrotonate, an alkynone, has been developed by Bagley et al. (2002). This new one-pot Bohlmann-Rahtz procedure was conducted at 170°C in a self tunable microwave synthesizer giving high yield with total control of regiochemistry.

Truong and Vo-Thanh (2010) developed an efficient method for the synthesis of functionalized chiral ammonium imidazolium and pyridinium based ionic liquids derived from (1R, 2S) ephedrine by solvent-free asymmetric Michael addition under microwave irradiation with good yields in very short reaction time.

It has been found that MgAl hydrotalcite is the most efficient catalyst for the aza-Michael reaction (Mokhtar et al., 2012). This microwave-assisted reaction provides an ecofriendly alternative to traditional synthesis, where soluble bases are used. The reaction was completed in shorter time with superior yield under microwave irradiation.

It has been shown that the combination of water and microwave irradiation promotes the catalyst free nitro Michael addition of pyrroles and indoles with good yields (99%) (Rosa and Soriente, 2010).

The synthesis of a series of 6,6-bisbenzannulated spiroketals has been achieved by a novel microwave-assisted double intramolecular hetero-Michael addition approach, with good yield (Choi et al., 2009). In this synthesis, coupling of an aryl acetylene and an aryl aldehyde led to an alkymol via acetylide anion addition, which was followed by oxidation to give the desired ynone. Spirocyclization afforded bisbenzannulated spiroketals.

o-Aminochalcones were cyclized rapidly to tetrahydroquinolones in dry media using montmorillonite K10 clay by intramolecular Michael addition reaction under microwave irradiation with good yields (70–80%) (Varma and Saini, 1997).

Similar studies have also been carried out by them with o-hydroxy chalcones in dry media in silical gel.

where R_1 = H, OCH_3; R_2 = H.

A series of Michael adducts has been synthesized with high yields by solvent-free microwave promoted Michael addition of aza-nucleophiles to benzo[b]thiophen-2-yl-2-propenone (Pessoa-Mahana et al., 2009). In this method, aliphatic and aromatic amines act as Michael donors and on treatment with 1-(4,7-dimethoxybenzo[b]thiophen-2-yl)-2-propen-1-one led to formation of β-aminoketones and azaheterocyclic compounds. These aza-nucleophiles were impregnated on silica gel-manganese dioxide as solid inorganic support.

It was observed that solvent-free microwave-mediated facile Michael addition of active methylene compounds to α, β-unsaturated carbonyl compounds takes place on the surface of K_2CO_3 under microwave irradiation (Surya et al., 2005).

Kim et al. (2010) investigated Michael reaction under focused microwave irradiation between diethyl ethoxymethylenemalonate (EMME) and various amines, for example, diphenylamine, diisopolyamine, 4-nitrobenzenomine, etc., where solvent-free conditions and solid support alumina in presence of potassium carbonate catalyst. It gives moderate to high yield of diethylmalonate analogs having en-amine moieties.

It has been reported that ammonium chloride is an efficient catalyst for one pot, three component microwave-assisted Michael addition reactions of indole, benz-aldehyde and N,N- diethylbarbituric acid under solvent-free conditions, which re-sulted into the formation of 3- substituted indole with give yield (Anjum and Sultan, 2012).

Michaud et al. (1997) reported that nitromethane reacts via a diastereoselective double Michael addition with electrophilic alkenes in the presence of piperidine under solvent-free condition and focused microwave irradiation. It afforded func-tionalized cyclohexenes and there was no formation of cyclopropane.

7.2   ELECTROPHILIC ADDITION

Very little has been described about electrophilic addition reactions under micro-wave exposure. However, some studies have been carried out but mostly with sin-gle-walled carbon nanotubes (SWNTs).

Microwave induced electrophilic addition alkyl halides to single-walled carbon nanotubes (SWNTs) has been reported using Lewis acid as a catalyst and followed by hydrolysis (Xu et al., 2008). This reaction provides alkyl and hydroxyl groups attached on the surface of the nanotubes within few min. It was observed that iodo-alkanes show higher reactivity with SWNTs than chloro and bromoalkanes.

An electrophilic addition of chloroform to SWNTs was followed by hydrolysis and it resulted in the addition of hydroxyl groups to the surface of the nanotubes. Its esterification with propionyl chloride led to the corresponding ester derivatives (Tagmatarchis et al., 2002).

Microwave-assisted covalent functionalization of the external wall of C60-SW-CNT peapods, by in situ generated aryl diazonium salts was also reported (Karousis et al., 2010). The electrochemistry revealed that there are three reversible reductions of encapsulated C60. They have also reviewed latest trends in microwave-assisted functionalization of carbon nanostructured materials.

Since the mid 90's microwave-assisted synthesis has found its way into carbon nanostructures, offering significant advantages towards their functionalization. Economopoulos et al. (2011) have reviewed the latest trends in microwave-assisted functionalization of carbon nanostructured materials covering major breakthroughs achieved in the last few years, which is a novel, intriguing and nonconventional synthetic approach.

Nucleophilic addition reactions can conveniently proceed under microwave exposure where as it is slightly difficult to carry out electrophilic addition reactions. Addition reactions have their importance in synthetic organic chemistry being greener in nature as compared to other reactions except rearrangements. Microwaves have become a powerful ally in any synthetic procedure, shortening reaction times considerably and in some cases, facilitating the reaction by offering higher yields or fewer by products.

KEYWORDS

- **Electrophilic Addition**
- **Michael Addition**
- **Nucleophilic Addition**

REFERENCES

Amore, K. M., Leadbeater, N. E., Miller, T. A., & Schmink, J. R. (2006). Tetrahedron Letters, 47, 8583–8586.

Anjum, R., & Sultan, S. (2012). International Journal of Research in Pharmaceutical and Biomedical Sciences, 3, 1567–1569.

Bagley, M. C, Lunn R., & Xiong, X. (2002). Tetrahedron Letters, 43, 8331–8334.

Camara, C., Keller, L., & Dumas, F. (2003), Tetrahedron Asymmetry, 14, 3263–3266.

Carrillo, J. R., Diaz-Ortiz, A., de la Hoz, A., Gomez-Escalonilla, M. J., Moreno, A., & Prieto, P. (1999). Tetrahedron, 55, 9623–9630.

Chen, J. J., & Deshpande, S. V. (2003). Tetrahedron Letters, 44, 8873–8876.

Choi, P. J., Dominea, C. K., Rathwell, & Brimble, M. A. (2009). Tetrahedron Letters, 50, 3245–3248.

De La Hoz, A., Díaz-Ortiz, A., Gómez, M. V. Mayoral, J. A., Moreno, A., Sánchez-Migallón, A., & Vázquez, E. (2001). Tetrahedron, 57, 5421–5428.

Economopoulos, S. P., Karousis, N., Rotas, G., Pagona G. and Tagmatarchis, N. (2011). Current Organic Chemistry, 15, 1121–1132.

Gadhwal, S., Baruah, M., & Sandhu, J. S. (1990). Synlett, 10, 1573–1574.

Gan, C., Chen, X., Lai, G., & Wang, Z. (2006). Synlett, 3, 387–390.

Hajipour, A. R., Mallakpour, J. E., & Afrousheh, A. (1999). Tetrahedron, 55, 2311–2316.

Karousis, N., Economopoulos, S. P., Iizumi, Y., Okazaki, T., Liu, Z., Suenaga K., & Tagmatarchis, N. (2010). Chemical Communications, 46, 9110–9112.

Kim, K. W., Lee, H. J., Jo, J. I., & Kwon, T.W. (2010). Bulletin of Korean Chemical Society, 31, 1155–1159.

Loupy, A., Monteux, D., Petit, A., Aizpurua, J. M., Dominguez, E., & Palomo, C. (1996). Tetrahedron Letters, 37, 8177–8180.

Marrero-Terrero, A. L., & Loupy A. (1996). Synlett, 3, 245–246.

Martin-Arnada, R. M., Vincente-Rodriguez, M. A., Lopez-Pestane, J. M., Lopez-Perinado, A. J., Jerez, A., Lopez-Gonzalez, J. de D., & Banares-Munoz, M. A. (1997). Journal of Molecular Catalysis, A : Chemical, 124, 115–121.

Michaud, D., Ayoubi, S. A., Dozias, M.-J., Toupet, L., Texier-Boullet, F., & Hamelin, J. (1997). Chemical Communications, 1613–1614.

Mokhtar, M., Saleh, T. S. & Basahel, (2012) S. N. Journal of Molecular Catalysis A: Chemical, 353–354, 122–131.

Patonay, T., Varma, R. S., Vass, ALévai, A., & Dudás, J. (2001). Tetrahedron Letters, 42, 1403–1406.

Pessoa-Mahana, H., Gonalez, M., Gonalez, M., Pessoa-Mahana, D., Araya-Maturana, R., Ron, N, & Saitz, C. (2009). Arkivoc, (xi), 316–325.

Rao, N. N., & Meshram, H. M. (2013). Tetrahedron Letters, 54, 1315–1317.

Romanova, N. N., Gravis, A. G., Shaidullina, G. M., Leshcheva, I. F., & Bundel, Y. G. (1997). Mendeleev Communications, 7, 235–236.

Rosa M. De, & Soriente, A. (2010). Tetrahedron, 66, 2981–2986.

Surya, H., Rao, P., & Jothilingam, S. (2005). Journal of Chemical Society, 117, 323–328.

Tagmatarchis, N., Georgakilas, V., Prato M., & Shinohara, H. (2002). Chemical Communications, 38, 2010–2011.

Truong, T. K. T., & Vo-Thanh, G. (2010). Tetrahedron, 66, 5277–5282.

Varma, R. S., & Saini, R. K. (1997). Synlett, 7, 857–858.

Varma, R. S., Dahiya, R., & Kumar, S. (1997a). Tetrahedron Letters, 38, 2039–2042.

Varma, R. S., Dahiya, R., & Kumar, S. (1997b). Tetrahedron Letters, 38, 5131–5134.

Vass, A., Dudas, J., & Varma, R. S. (1999). Tetrahedron Letters, 40, 4951–4954.

Xu, T., Wang, X., Tian, R., Li, S., Wan, L., & Li, M. (2008). Applied Surface Science, 254, 2431–2435.

CHAPTER 8

CYCLOADDITION

ABHILASHA JAIN, K. L. AMETA, PINKI B. PUNJABI, and
SURESH C. AMETA

CONTENTS

A cycloaddition reaction is a pericyclic reaction in which two or more unsaturated compounds combine with the formation of a cyclic adduct. Cycloaddition reactions have been greatly benefited from the use of microwave irradiation not only by completing the reactions in few minutes, but avoiding many of the disadvantages of the conventional reaction conditions like solvent evaporation and poor yield. As MORE (Microwave organic reaction enhancement) chemistry provides cleaner reaction products as well as conditions for carrying out those reactions, which are unattainable under conventional heating conditions. Microwave-assisted cycloaddition reactions give rapid access to fused multicyclic and heterocyclic systems in a single step process.

Cycloaddition reactions are commonly divided in following categories;
(i)   [2+2] Cycloaddition
(ii) [3+2] Cycloaddition (1, 3-Dipolar addition)
(iii) [4+2] Cycloaddition
(iv) Other Cycloadditions

8.1   [2+2] CYCLOADDITION

The microwave-assisted [2+2] cycloaddition reactions of 2-amino-3-dimethylamino propenoates with acetylenecarboxylates lead to produce highly functionalized 1-amino-4-(dimethylamino) buta-1,3-dienes in high yield (Ursic et al., 2008). It has also been observed that single geometrical isomers were produced consistently in all these reactions.

where Z = COPh, COMe, Cbz; R_1 = COOMe, COOEt, COOt-Bu; R_2 = COOMe, COOEt, COO-t-Bu, CF_3, H.

A novel protocol has been developed by Benzensek et al. (2010) for the synthesis of highly functionalized (2E,3E)-dimethyl-2-[(dimethylamino)]-3-substituted succinates by microwave-assisted [2+2] cycloaddition of (E)-3-dimethylamino-1-heteroaryl-prop-2-en-1-ones to dimethyl acetylene dicarboxylates.

An efficient method has been investigated by Johnstone et al. (2010) for the synthesis of a range of cyclobutene diesters by ruthenium-catalyzed microwave-assisted [2+2] cycloaddition, where better yields of products have been achieved only in 2 min.

Many [2+2] cycloaddition reactions are synthetically important as they provide rapid and efficient synthesis of strained cyclobutane rings and their derivatives. The reactions are typically performed under thermal or photochemical activation or with the Lewis acids. However, there are some reports of the application of microwave irradiation to perform [2+2] cycloaddition reactions.

Brummond and Chen (2005) investigated the microwave-assisted intramolecular [2+2] cycloaddition reaction of alkynyl allenes to afford bicyclomethylenecyclobutenes under microwave irradiation in toluene at 110 °C for 15 min. However, only starting material was obtained. After screening several conditions, they finally arrived at the application of a 3 M solution of the ionic liquid 1-ethyl-3-methylimidazolium hexafluorophosphate in toluene under microwave irradiation at 250 °C for 15 min, to give the compounds in good yields of 54–83%.

where R_1 = Me, Aryl, TMS, Alkyl; R_2 = H, Me.

Ovaska and Kyne (2008) have reported that a series of 1-allenyl-2-propargyl substituted cyclopentanol derivatives undergo facile intramolecular microwave-assisted [2+2] cycloaddition affording strained tricyclic 5–6–4 ring systems (70–90% yield), as present in the natural product (–) sterpurene. Surprisingly, only those allenic systems were found to be reactive, which bear the allenyl and propargylic moiety in a trans relationship.

where R = TMS, TBS, Ph, Et, H.

Wipf et al. (2005) reported microwave irradiated rapid O,N-acylation-cyclo-dehydration cascade reaction of oximes and acid chlorides to give oxazoles. The microwave irradiation allowed considerable acceleration of the rate of oxazole formation and increased the yield of this synthetically attractive heterocycle formation process. The starting oximes were readily obtained from commercially available ketones in yields exceeding 90%, while the yields of isolated oxazoles ranged from 23% to 62%.

The [2+2] cycloaddition reaction of 9-substituted anthracenes with levoglucose-none has been reported (Sarrotti et al., 2006). The corresponding cycloadducts were obtained in only 10 min in 60–82% yield with a ceiling temperature of 95 °C and 50 W maximum power for 5 h.

where R = H, Me, Ac, Ph, TBS.

Nicolaou et al. (2007) have carried out the synthesis of the key artochamin intermediate via a cascade sequence involving a microwave-assisted [3,3] sig-matropic rearrangement of the appropriately functionalized stilbene, which was followed by a [2+2] cycloaddition of the generated intermediate at 180°C in the presence of a catalytic amount of Ph_3PO in o-xylene in 20 min. The yield of 55% was obtained.

8.2   1,3-DIPOLAR ADDITION

A novel method has been developed for the synthesis of indole based constrained mimetic scaffolds by microwave-assisted regio- and stereo-selective 1,3-dipolar cycloadditions. Here, C-(3-indolyl)-N-phenylnitrone reacts with number of olefinic dipolarophiles to afford isoxazolidine in superior yields (Bhella et al., 2009).

Recently, Oukani et al. (2013) also investigated the 1,3-dipolar cycloaddition reaction of chiral carbohydrate derived α,β-unsaturated esters and nitrones, which were derived from threitol under microwave irradiation for the synthesis of long chain sugars with nine contiguous chiral centers.

It has been observed that Mg-Al-hydrotalcite is an efficient catalyst for microwave-assisted region-selective 1,3-dipolar cycloaddition of nitrillimines with the enaminone derivatives to produce pyrazole derivatives (Saleh et al., 2013).

A novel protocol has been reported by Chiacchio et al. (2007) for the synthesis of polyhedral oligomeric silsesquioxane (POSS) macromers by 1,3-dipolar cycloaddition reactions of vinyl- and styryl-POSS with N-methyl-C-ethoxycarbonylnitrone and ethoxycarbonyl nitrite oxide. These reactions are promoted by microwave irradiation.

Microwave-assisted copper catalyzed 1,3-dipolar addition of alkynes has been studied by Potewar et al. (2013), where 1-(1,'2,3,3,'4,4,'6-hepta-O-acetyl-6'-deoxy-sucros-6'-yl)-4-substitued 1,2,3-triazoles were synthesized in excellent yields (83–93%) in short duration.

where X = OAc, N$_3$ or ___ ; R = Alkyl, Aryl, Acyl

A series of 3- and 4- pyridyl substituted pyrroles was prepared from N-acylated amino acid under microwave irradiation (Harju et al., 2009). In this synthesis, dehydration of the acylated amino acids gave cyclic intermediates (munchnones or azlactones), which were further treated in situ with alkynes in 1,3-dipolar cycloadditions.

It has been shown that microwave irradiation induces the 1,3-dipolar cycloaddition of cyclobutane epoxides with norbornenes to produce various [n] polynorbornane scaffolds.

[3] Polynorbarnane

A rapid protocol has been developed for the synthesis of number of 4-aza-2,'3'-dideoxynucleosides in high yields by microwave-assisted 1,3-cycloaddition of vinyl nucleobases with nitrones in solvent-free condition (Bortolini et al., 2008).

where B (Base) = Thymine, Uracil, Cyctosine, 5-Fluorocytosine, Adenine or 2-(N-trityl) guanine; R = Bu or Me.

The 1,3-dipolar cycloaddition of azido-2'-deoxyribose with terminal alkynes in presence of Cu (I) catalyst was dramatically enhanced under solvent-free condition and microwave irradiation, where 4-substituted-1,2,3-triazalyl nucleosides was formed as product rapidly and efficiently with high yield (91–98%) (Guezgueg et al., 2006).

The regio- and stereoselective synthesis of novel tetraspiro-bispyrrolidine and bisoxindolopyrrolidine derivatives has also been achieved by 1,3-dipolar cycloaddition reaction in one pot three component reaction under solvent-free microwave conditions (Rajesh and Raghunathan, 2010).

Andrade et al. (2008) reported an expeditious microwave-assisted neat synthesis of α-phenyl-tert-butylnitrone (PBN) and other alkyl & aryl nitrones. Further, a rapid synthesis of isoxazolidines has been accomplished by 1,3-dipolar cycloaddition of these nitrones to ethyl-transcrotonate.

where R_1 = Me, t-Bu; R_2 = Aryl.

An efficient approach to the synthesis of a novel polyhydroxylated indolizidine derivatives containing an amino group has been developed. The key step of the synthesis involved microwave-assisted 1,3-dipolar cycloaddition of azasugar nitrone and methacrylate for installing a potential amino group and ester group with an extended chain (Li et al., 2009).

An expedient method has been developed by Mabrour et al. (2007) for the synthesis of novel isoxazolines and isoxazoles of N-substituted saccharin derivatives by 1,3-dipolar cycloaddition of n-allyl or propargyl n-substituted saccharin with arylnitrite oxide under microwave irradiation, where high yield of products (91–95%) has been achieved without alteration of the selectivity.

91-95% yield

Fordyce et al. (2010) have enveloped an improved method for the synthesis of isothiazoles and 1,2,4-thiadiazoles. The 1,3-dipolar cycloaddition reactions of nitrile sulfides, generated by microwave-assisted decarboxylation of 1,3,3-oxathiazal-2-ones have been studied. The adducts 1,2,4-thiadiazole-5-carboxylates were synthesized by cycloaddition of the nitrite sulfides to ethyl cyanoformate.

Using microwave irradiation under solvent-free condition, hexahydrochromeno [4,3-b] pyrroles have been synthesized by intramolecular 1,3-dipolar cycloaddition reactions by Pospisil and Potacek (2007). The yields range between 16–84% in 15–40 min microwave irradiation.

Formation of spiro-pyrrolidines/pyrrolizidine has been achieved under microwave exposure by using the alkene unit of Baylis-Hillman adduct of ninhydrin with sarcosine/proline and various activated ketones through 1,3- dipolar cycloaddition reactions (Ramesh et al., 2007).

Rijkers et al. (2005) synthesized multivalent dendrimeric peptides via a microwave-assisted CuAAC reaction between azido peptides and dendrimeric alkynes. They carried out the reaction in the presence of $CuSO_4$, sodium ascorbate and Cu-wire at room temperature for 16 h, showing smooth formation of monovalent cycloadduct while the divalent product was sluggishly formed in 43% yield. A tremendous improvement was achieved by carrying out the reaction under microwave irradiation at a ceiling temperature of 100 °C yielding the divalent compound in 63% yield in 10 min.

Criado et al. (2013) have investigated the functionalization of single wall carbon nanotubes (SWCNT) by cycloaddition reaction with arynes under microwave irradiation. Various techniques were employed to monitor the efficiency of the functionalization. The use of microwave irradiation is crucial for the rapid chemical functionalization of single walled carbon nanotubes (Wang et al., 2005). The noticeable rate enhancement and reduction in the number of steps in reaction procedure have been observed. It comprised of two model reactions namely, amidation and 1,3-dipolar cycloaddition of SWNTs. Here, 1,3-dipolar cycloaddition was carried out in 15 min under microwave conditions compared to conventional methods, where synthesis required 5 days.

A significant rate enhancement by microwave irradiation has been observed, when o-dichlorobenzene solution of 2,5-disubstituted thienosultine was refluxed with [60] fullerene for 2–24 h. In this reaction, nonKekule biradical intermediates were formed by cheletropic extrusion of SO_2 from sultines (Chi et al., 2004). These intermediates were trapped by [60] fullerene leading to formation of corresponding cycloadducts (Cruz et al., 1997a). It was shown that the 1,3-dipolar cycloaddition reactions of [60] fullerene are substantially accelerated by microwave irradiation to afford high yields of respective cycloadducts (Cruz et al., 1997b).

A number of suitable 1,3-dipoles like nitrones, isonitriles, etc. can be used. Although it lacks the remarkable regioselectivities of the CuAAC reactions, even then these examples are of prime interest to the synthetic chemists due to their value and potential for the generation of interesting heterocyclic moieties and fused-ring systems with high to excellent stereoselectivities. Another important factor in the usefulness of these nonazide 1,3-dipoles is in synthesizing structurally novel entities, which resemble the cores or skeletons of some naturally products with interesting biological properties.

Veverkova and Toma (2005) observed an increase in the reaction rate of 1,3-dipolar cycloaddition reactions of 9-azidoacridines with various acetylenes using microwave irradiation. Complicated mixture of decomposition products was obtained containing low amount of product in 3 min.

where R_1 = COOCH$_3$, CH$_2$OCOC$_2$H$_5$, COOC$_2$H$_5$, Ph, CH$_2$Br, Si(CH$_3$)$_3$; R_2 = COOCH$_3$, H, CH$_3$, Ph.

Zhang et al. (2006) have reported an advance fluorous synthesis of hydantoin-, piperazinedione-, and benzodiazepinedione-fused tricyclic and tetracyclic ring systems using a sequence of microwave-assisted, fluorous multicomponent reactions (F-MCRs) and fluorous solid-phase extractions (F-SPEs). They used microwave-assisted one pot, three-component [3+2] cycloaddition of azomethine ylides with flourous tagged amino esters to synthesize the novel triaza-tricyclic and tetracyclic ring systems with four totally controlled stereocenters. The reaction between a suitable fluorous aminoester, an aldehyde and an N-alkylmaleimide was carried out under microwave irradiation at 130 °C for 20 min in DMF, to generate the fluorous bicyclic intermediates in good yields of 70–95% with remarkable >90% purity levels of each target molecule after the F-SPE.

where R_1 = p-OMe p-Cl, m-Cl, p-Br; R_2 = Me, Et, t-Bu, Bn, Cy; R_3 = H, Me, t-Bu, Bn.

Another interesting example of a microwave-assisted [3+2] cycloaddition was reported by Meng et al. (2007), where a small library of polycyclic pyrrolidines was designed. They explored a microwave-assisted intramolecular [3+2] cycloaddition of azomethine ylides derived from N-allyl-2-carboxaldehyde benzimidazoles with secondary R-amino esters in a solution phase to diastereoselectivity generate the target fused pyrrolidine heterocycles. N-Allyl-benzimidazole-2-carboxaldehydes were synthesized from the corresponding suitably functionalized o-diaminoarenes in three steps following a well established condensation-allylation-deprotection sequence. The condensations of the aldehydes with secondary amino esters and the subsequent 1,3-dipolar cycloaddition of the S-shaped ylides were carried out under microwave-assisted conditions in xylene at 130°C for 20 min. It was observed that the azomethine ylide cycloadditions performed under classical heating conditions required longer reaction times with poor yield and diastereoselectivity, while the microwave-assisted protocol proved to be superior.

S-Shaped ylide intermediate

where R_1 = H, Me, Cl, -(CH)$_4$-; R_2 = Me, Bn, Ph.

Chakraborty et al. (2012) studied microwave induced 1,3-dipolar cycloaddition reactions of dihydropyran derived nitrone with various activated alkenes. They found that new isoxazolidine derivatives were obtained with moderate selectivity.

Chiacchio et al. (2007) have also demonstrated the synthesis of methylene-isoxazolidine nucleoside analogs by microwave-assisted nitrone cycloaddition. The insertion of a methylene isoxazolidine spacer unit between the nucleobase and the hydroxyl-methyl group in the N, O nucleoside analogs could control the conformational mobility of the system. The cycloaddition between a proper N-methylated nitrone and a suitably functionalized allene, (often thymallene) in CCl$_4$ or EtOH at 70°C for 10 min generated the methylene-isoxazolidine nucleoside analogs with good yields (45–72%) and purity. The reaction rate was enhanced, when the reactions were run under microwave irradiation, in comparison to conventional reaction conditions.

where R$_1$ = COOEt and R$_2$ =

Recently, Arigela et al. (2013) reported a microwave-assisted three component domino synthesis of indolodiazepinotriazoles involving N-1 alkylation of 2-alkynylindoles with epichlorohydrin, ring opening of the epoxide with sodium azide and intramolecular Huisgen azide/internal alkyne 1,3-dipolar cycloaddition domino sequence. Various 2-alkyl indoles (aromatic/aliphatic) with epichlorohydrin and sodium azide have been used affording annulated tetracyclic indolodiazepinotriazoles with satisfactory yields.

Microwave-assisted [3+2] cycloaddition reactions were also explored for the synthesis of the core structure of Pinnaic acid (Yang and Caprio, 2007). It is a spiro-bicyclic alkaloid and its derivatives can be an interesting leads in antiinflammatory drugs. The key spirocyclic nitrone intermediate was synthesized by the m-CPBA-mediated oxidative cleavage of the corresponding isoxazole, which can otherwise be synthesized in 5 steps by established protocols. It was synthesized by the microwave-assisted [3+2] cycloaddition of the nitrone with the suitable alkene in toluene at 165°C for 1 h, whereas the same reaction under conventional heating conditions furnished quite poor yields. The intermediate was then converted to the desired core structure of pinnaic acid in three steps using conventionally known synthetic procedure.

Pinnaic acid

where R$_1$ = Bn; R$_2$ = H or TBDPS.

Chiacchio et al. (2005) investigated microwave-assisted, Zn(OTf)$_2$-promoted [3+2] cyloaddition between C-(2-thiazolyl)nitrones and allylic alcohols for the synthesis of an enantiomerically pure isoxazolidinyl analog of the C-nucleoside tiazofurin. This reaction is 100% exo-cis selective and show 80% disterofacial selectivity. This compound exhibits potent antitumor activity against several human cancers.

where X = OEt or NH$_2$.

A novel protocol for the synthesis of new dispiro-pyrrolo/pyrrolizidino ring systems has been achieved by the [3+2] cycloaddition of azomethine with dipolarophile 9-arylidine-fluorene (Jayashankaran et al., 2004). They have reported superior yield and high rate of reaction by this solvent-free microwave-assisted approach.

Majumdar et al. (2012) have developed an efficient ecofriendly microwave-assisted one pot approach for the synthesis of triazolobenzothiadiazepine-1,1-dioxide derivatives by the reaction of 2-azido-N substituted benzenesulfonamides and propargyl bromide on basic alumina through [3+2] azide-alkyne cycloaddition reaction. The yield of the product was 83–95% and that too in 10 min.

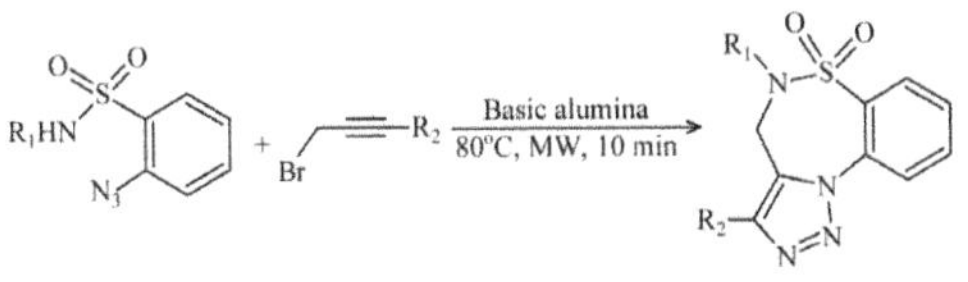

It has been observed that by using microwave technology, sterically hindered 2,4-disubstituted 3-(5-tetrazolyl) pyridines can be synthesized from the corresponding nicotinonitriles (Lukyanov et al., 2006).

Three component and two component microwave-assisted [3+2] cycloadditions of various azides to benzyne, 3-methoxybenzyne and 4,5-difluorobenzyne have been carried out for the synthesis of benzo [d] [1,2,3] triazoles (Ankati and Biehl, 2009). An azide has been prepared in situ in a three component reaction, where as in a two component system, freshly prepared azide was added to the reaction vessel.

Three component reaction:

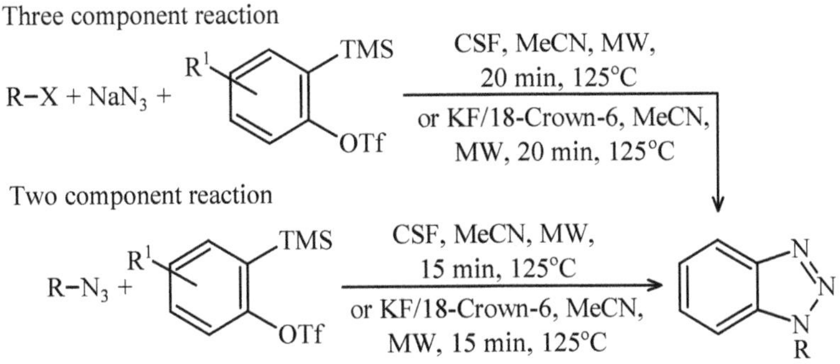

An efficient synthesis of dispiro-oxindolopyrrolidines and pyrrolizidines were reported through [3+2] cycloaddition reaction of azomethine ylides with [E]-2-oxoindolino-3-ylidene acetophenones under microwave irradiation in presence of $ZrOCl_2.8H_2O$ (Babu et al., 2007) while Wilson et al. (2001) have developed a novel method for the rapid synthesis of a library of substituted prolines by microwave-assisted [3+2] cycloaddition reaction. In this process, α-aminoesters and aldehydes generates imines under microwave irradiation resulting into the [3+2] cycloadducts.

It has been shown that microwave irradiation induced a noticeable acceleration of the [3+2] dipolar cycloaddition reaction of pyridazinium ylides to activated alkenes and alkynes (Butnariu and Mangalogiu, 2009). A rapid three components, [3+2] cycloaddition/annulation domino protocol for the regio- and diastereoselective series of cage penta and hexacyclic compounds has been developed by Kumar et al. (2011) This synthesis involved the generation of two heterocyclic rings and five contiguous stereocenter.

It has been proved by many researches that microwave irradiation has a profound influence on the various cycloaddition reactions and it has been considered as an ecofriendly approach for the synthesis of plethora heterocyclic and alicyclic compounds (Appukkuttan et al., 2010; Bougrin and Benhida, 2012).

Katritzky and Singh (2002) reported 1,3-dipolar cycloaddition of organic azides to acetylenic amides under solvent-free microwave irradiation producing the corresponding N-substituted C-carbamoyl-1,2,3-triazoles in good to excellent yields in 30 min.

8.3   [4+2] CYCLOADDITION

[4+2] Cycloadditions certainly represent one of the most important area in synthetic organic chemistry. Diels-Alder cycloaddition is an important step in a wide variety of natural products skeleton constructions. Since this is a known thermal reaction, the influence of microwave irradiation should prove to be significant. The Diels–Alder reaction of (hetero) dienes with various dienophiles generating a versatile array of useful heterocyclic systems and fused-ring systems, often plays a role in the synthesis of key intermediates of complex natural products and biologically active molecules. The [4+2] cycloaddition reactions furnishes high degrees of chemo-, regio- and stereoselectivities, and generate upto four stereo centers in one step. These reactions gained extreme importance for the synthesis of highly substituted and stereoselective polycyclic systems in a easy workup.

Gomez et al. (2009) reported that the cycloaddition reactions of nitropyrroles in solvent-free conditions under microwave irradiation give 27–71% yields of the aromatic indoles followed by elimination of the nitro group and subsequent aromatization.

Zhang et al. (2007) described an efficient [4+2] cycloaddition/rearrangement method that plays a crucial role in the synthesis of (-)-strychnine, a strychnos alkaloid possessing highly toxic properties including disruption of nerve-cell signaling. They applied MgI_2-catalyzed microwave-assisted intramolecular [4+2] cycloaddition-rearrangement cascade, generating the key aza-tetracycle from the furanylindole intermediate in a remarkable yield (95%) in 30 min.

Kranjc and Kocevar (2008) have proposed an interesting microwave-assisted [4+2] cycloaddition between suitably functionalized alkynes and 2H-pyran-2-ones, followed by a microwave-assisted cyclization for the synthesis of the indole ring, where the positions on the aromatic ring can be preselected. The cycloaddition reactions were run in toluene at 150°C in 90–180 min and the acid mediated cyclization of the formed intermediates into the corresponding indole analogs were performed at 120°C in ethanol. This cycloaddition reaction requires 15 to 138 days under high-pressure conventional heating.

where R_1 = Me, CH_2CO_2Et; R_2 = COMe, COEt, CO_2Me, CO_2Et; R_3 = Me, Ph, Bn.

Singh et al. (2008) studied a microwave-assisted [4+2] cycloaddition reaction for the synthesis of unsymmetrically substituted 1,4-dihydropyridines. 1,4-Dihydropyridine analogs such as Nifedipine and Nimodipine are well known for their potent biological activities. Microwave-assisted [4+2] cycloaddition of 1,4-diaryl-1-aza-1,3-butadienes with allenic esters at 100°C for 5–17 min gave cycloadducts in excellent yields (83–96%), which after a tandem 1,3-H-shift furnished unsymmetrically substituted 1,4-dihydropyridines.

where R_1 = H, Me, OMe, Cl, CN; R_2 = Me, Et.

A microwave-assisted, diversity-oriented, inverse electron-demand [4+2] cyclo-addition of functionalized 1,2,4-triazines with alkyne dienophiles was carried out (Hajbi et al., 2008). 2,3-Dihydrofuro[2,3-b]pyridines and 3,4-dihydro-2H-pyrano-[2,3-b]pyridines on functionalization at the 3 and 4 positions have close structural similarity with well-known potent serotoninergic ligands like quinolines, substitut-ed pyridines, and chromanes.

where R = 2-Thienyl, 2-Pyridinyl, 4-Me-Ph, 4-OMe-Ph.

The [4+2] cycloaddition reactions of some triazole substituted 3,5-dichloro-2(1H)-pyrazinones have been studied by Kaval et al. (2006). They developed the required intermediates from the corresponding 3,5-dichloropyrazinones, follow-ing either a nucleophilic substitution or a Sonogashira reaction to install the alkyne handle on the scaffold. The various triazole derivatives linked to the pyrazinones were synthesized through a microwave-assisted CuAAC reaction. Then [4+2] cy-cloaddition reactions were carried out under microwave irradiations with dimethyl acetylenedicarboxylate (DMAD) at 180–200 °C for 10–20 min, and the correspond-ing highly decorated pyridine and pyridine scaffolds were isolated in good yields.

where R_1 = p-MeO-Bn, CH_2COOEt, Ph; R_2 = H, Me; R_3 = 4-MeO, 4-Me, 4-Cl, 4-NO_2, 4-NMe, 4-$CONH_2$.

Choshi et al. (2008) have developed a novel method for the total synthesis of the 2-azaanthraquinone alkaloid in nine steps, out of which, one key step involves a microwave-assisted [4+2] cycloaddition reaction for the constriction of a 2-azaan-thraquinone frame work.

Safaei-Ghomi et al. (2004) reported microwave enhanced Diels-Alder reactions of furan, 2,5-dimetylfuran, 1,3-cyclohexadiene and anthracene with dimethyl acet-ylenedicarboxylate (DMAD) and diethyl acetylenedicarboxylate (DEAD) to give [4+2]-cycloadducts in high yields in domestic microwave oven. Aluminum (III) chloride and dichloromethane were used in combination with microwave irradiation for increasing the reaction rate.

8.3.1 DIELS-ALDER ADDITION

In Diels-Alder reaction, a diene reacts with a dienophile to give an addition product.

Microwave-assisted Diels-Alder cycloaddition of 2-fluoro-3-methony-1,3-buta-diene gives 70–90% yield within 5–25 min compared to thermal reaction, which gives only 20–65% in 30 min to 3 days (Timothy et al., 2007).

A facile microwave-assisted Diels-Alder reaction of vinylboronates has been in-vestigated by Sarotti et al. (2010). This is an example of microwave-assisted Diels-Alder reaction of boron substituted dienophiles. The yield of the addition product was 75–100%.

A tetracyclic core of marine diterpenoid was obtained by a Diels-Alder reaction of 1,3,3-trimethyl-2-vinyl-1-cyclohexene with chromones under microwave irradiation with and without $TiCl_4$ (Rajesh et al., 2010).

These diterpenoids are related to Puupehenone and Kampanels.

Zheng et al. (2013) observed Diels-Alder cycloaddition between Danishefsky's diene and derivatives of ethyl α-(hydroxymethyl) acrylate, where the hydroxy group was protected, giving high yield with faster rate of reaction. These adducts are required for the synthesis of a biotin conjugate of monocyclic cyanoenone with high antiinflammatory activity.

Bansal et al. (2003) reported the use of microwave technique for the synthesis of azaphospholes by the Diels-Alder reaction of 1,3-azaphospholo(5,1-a)isoquinolines.

Microwave-assisted solid phase Diels-Alder cycloaddition reaction of 2(1H) pyrazinones with dienophiles has been investigated by Kaval et al. (2003).

In this synthesis, a traceless linking concept was applied and the products (pyridines) were separated from the by products (pyridinones), which remained on the solid support, where as pyridine was released in the solution. It was observed that the rate of reaction was accelerated from hours or days to minutes, when carried out under high temperature microwave conditions.

The synthesis of isoquinolinone carboxylates by Diels-Alder reaction was carried out by using arecoline (or its isomer methyl tetrahydropyridine carboxylate) with Danishefsky's diene under thermal and microwave conditions. It was found that with microwave technique, higher yields of the adduct were achieved as well as a new α, β-unsaturated pyridyl ketone was also formed (Jankowski et al., 2001).

Garrigues et al. (1996) reported the microwave-assisted Diels-Alder reaction between anthracene and azadienes supported on graphite, while Diaz-Ortiz et al. (2000) studied the solvent-free microwave-assisted Diels-Alder cycloaddition reaction, where a 1,2,3-triazole ring serve as a diene towards DMAD.

An improvement in Diels-Alder irreversible cycloaddition of 1,3-cyclohexadiene, 3-carbomethoxy-2-pyrone and 2-methoxythiophene with acetylenic compound was observed (Loupy et al., 2004) while a combined effect of microwave irradiation and solid supports in ionic liquids on the Diels-Alder reaction of 1,3-cyclopentadiene and different nucleophiles was investigated by Lopez et al. (2007).

Diels-Alder reaction of tetrakis (pentafluorophenyl)porphyrin with pentacene and naphthacene gave mono-adduct product within 1 min under microwave irradiation with 83% and 23% yields, respectively whereas only 22% and no reaction was observed in classical heating. Bis-adducts was also formed in microwave heating, which were not obtained by classical method (Silva et al., 2005).

Microwave-assisted tandem Witting intramolecular Diels-Alder cycloaddition of ester tethered 1,3,9-decatrienes to give 3,4,4a,7,8,8a-hexahydroisochromen-1-ones was reported with high yields (53–89%) (Wu et al., 2011). Three consecutive carbon-carbon bonds in the end product were formed rapidly during this tandem process under the microwave irradiation efficiently.

Wu et al. (2006) achieved 90–99% yield in the intramolecular Diels-Alder cycloaddition of different esters tethered 1,3,8-nonatrienes under controlled microwave heating.

Microwave-assisted intramolecular dehydrogenative Diels-Alder reactions have been carried out for the synthesis of the functionalized naphthalene by Benedetti et al. (2012), where styrenyl derivatives generated a variety of functionalized cyclopenta[b]naphthalenes.

It has been shown that by using the inverse electron demand, Diels-Alder reaction under sealed tube microwave irradiation, polysubstituted 2,3-dihydrofuro[2,3-b] pyridines can be synthesized effectively within shorter reaction time (Hajbi et al., 2007).

where R = H, Aryl, Heteroaryl.

Microwave-mediated intramolecular Diels-Alder cyclization of biodihydroxyl-
ated benzoic acid derivatives has been investigated by Mithovilovic et al. (2004).

The multicomponent microwave-assisted organocatalytic Knoevenagel/hetero-
Diels-Alder reaction for the synthesis of 2,3-dihydropyran [2,3-c] pyrazoles (a novel
protocol) has been developed by Radi et al. (2009). These compounds have potential
antituberculorsis activity.

Montmorillonite K10 clay-mediated Knoevenagel/hetero Diels-Alder reaction
for synthesis of polycyclic pyrano[2,3,4-kl] xanthene derivative, was carried out
under microwave irradiation, which gave the products in stereoselective manner
with high yields (Ramesh and Raghunathan, 2009).

An efficient one pot three component aza-Diels-Alder reaction for the synthesis of some complex spiroquinoline derivatives has been carried out under microwave irradiation and solvent-free condition (Bhuyan et al., 2012).

A rapid synthesis of 1,4- dihydropyridine was achieved by microwave-assisted aza-Diels-Alder reaction (Lee and Kim, 2011).

An efficient protocol for the catalyst free synthesis of dihydropyrido[4,3-d] pyrimidine by aza-Diels-Alder has been reported using microwave promoted solvent-free reaction.

Xing et al. (2006) reported the acid-mediated three-component aza-Diels-Alder reaction of 2-aminophenols under controlled microwave heating. These reactions were carried out in the presence of catalytic amount of CF_3COOH in CH_3CN at 60°C for 15 min affording highly functionalized 8-hydroxy,1,2,3,4-tetra-hydroquiniolines in significantly reduced reaction time.

Microwave-assisted synthesis of tetrahydroisoquinolin-pyrrolo-pyridinones by triple process was investigated by Islas-Jacome et al. (2011). Ugi-3CR-aza-Diels-Alder reaction, S-oxidation and Pummerer reactions were carried out for this synthesis.

A two-step synthesis of annulated dihydropyrano[3,4-c]chroheme derivatives has been developed via Knoevenagel condensation followed by a microwave-assisted intramolecular hetero-Diels-Alder reaction in presence of 20 mol% CuI in MeCN (Jha et al., 2011). The yield was 60–71% under microwave irradiation.

An efficient synthesis of pyrrlo[2,3-d]pyrimidine annulated pyrano[5,6-c]coumarin/[6,5-c]chromone derivatives was achieved by intramolecular hetero-Diels-Alder reaction by microwave irradiation under solvent-free and solid supported conditions (Ramesh and Raghunathan, 2008).

A novel protocol for the synthesis of 2-amino substituted-2-perfluoroalkyl-3,6-dihydro-2H-thiopyrans by hetero-Diels-Alder reactions of fluorinated thioamides under microwave heating was developed by Mikhailichenko et al. (2010).

where $R_1 = R_2 = H$, p-Tolyl, Morpholino; $R_3 = CF_3$, $n-C_3F_7$, $-(CF_2)_4H$.

A rapid synthesis of a series of acetylene-tethered pyrimidines was achieved by intramolecular hetero-Diels-Alder addition under microwave irradiation. The substrates gave moderate to high yields (60–90%) of the fused bicyclic pyridines within shorter reaction time. This method has been applied successfully to the synthesis of both; fused lactones and lactam (Shao, 2005).

where $X = -O$, $-NH$, $-NMe$; $Y = CH_2$, $-C(CH_3)_2$, $-CH_2CH_2$.

Bandini et al. (2010) developed a method for the synthesis of optically active conhydrines by hetero-Diels-Alder cycloaddition reaction mediated by microwaves.

R — ᵐᵐH (-)- -Conhydrine
R = ◀H (-)- -Conhydrine

where R = β-D-Ribofuranose.

Jimenez-Alonso et al. (2008) synthesized some novel pyrano embelin derivatives through domino Knoevenagel-hetero Diels–Alder reactions of embelin with paraformaldehyde and electron rich alkenes. The synthetic approach is highly efficient, when microwave irradiations and EtOH are used at 120 °C for 20 min. It gave the best result with the ratio of embelin, ethyl vinyl ether and paraformaldehyde as 1:3:8.

8.4 OTHER CYCLOADDITIONS

A facile route for the synthesis of dimers via [3+3] dipolar cycloaddition under microwave irradiation has been reported. A tetrahydropyrazino-diphthalezine structure was assigned for this dimer. This [3+3] dipolar cycloaddition reaction of phthalazinium ylides occurs in a highly cis selective manner (Zbancioc et al., 2005).

The synthesis of azafulvenium has been carried out by microwave irradiation of 2,2-dioxa-1H, 3H-pyrrolo-[1,2-c]thiazoles. [8π+2π] Cycloaddition of this 1,7-dipole intermediate led to the formation of a range of pyrole derivatives (Soares et al., 2008).

An expedient method has been reported by Angel et al. (2006) for the synthesis of pyrazolo [3,4–6] pyridines by cycloaddition of pyrazolylimines with aromatic and aliphatic nitroalkenes under microwave irradiation while a microwave enhanced [6+4] cycloaddition reaction between 6-aminofulvene and pyrones followed by CO_2 extrusion produces azulene-indoles with antineoplastic activity (Hong et al., 2001).

Shanmugasundaram et al. (2007) carried out microwave-assisted iridium catalyzed [2+2+2] cycloaddition reaction of resin bound dipropargylamine with alkynes, where isoindoline derivatives were produced in high purity with moderate yields.

It has been reported that Cu on porous glass is a recyclable catalyst for the azide-alkyne cycloaddition in water under microwave irradiation (Jacob et al., 2013). This microwave-assisted cycloaddition of benzyl azide with phenyl acetylene in water gave 1,4-tviazole almost exclusively after 10 or 20 min at 120° or 100°C, respectively.

Synthesis of tetrazolo[5,1-a]pyridines and 4,5-disubstituted-2H-1,2,3-triazoles has been accomplished under microwave irradiation in high yields by cycloaddition of 2-alkynylbenzonitriles with sodium azide (Tsai et al., 2009).

where R = Alkyl, Aryl, Heterocyclic

Parmar et al. (2012) reported an efficient microwave-assisted one pot method for the synthesis of aryldiazenylchromeno[4,3-b]pyrrolidines through intramolecular azomethine ylide cycloaddition reaction.

where R = Me, Et, Bn; R_1 = Me, Et, n-Pr, n-Bu.

Synthesis of triazoles and acetyl group migration involving microwave-assisted 1,3-dipolar cycloaddition reactions of vinylic glycosides with aryl azides has been studied by Marta et al. (2009). Three different carbohydrate derivatives were reacted with p-methoxybenzylazide and benzylazide, under MW irradiation in the absence of solvent. It was observed that the carbohydrate derivative plays two important roles in the cycloaddition reaction. It acts as an electron donating group in a way that controls the regioselectivity of the reaction, but also as a good leaving group, avoiding decomposition of the intermediate triazoline with formation of the corresponding triazole.

Kale et al. (2012) reported a rapid and green approach to achieve 1,4-disubstituted triazoles via three component reactions of alkynes and in situ generated azides in the presence of copper apatite (25% weight) as a heterogeneous and recyclable catalyst in aqueous media without using any base. The yield was same (99%) in both the methods, but it needs 1.5–6 h under conventional heating while it requires only 5–20 min in presence of microwave radiations.

Copper catalyzed azide-alkyne cycloaddition (CuAAC reaction) is the well-known Huisgen [3+2] cycloaddition reaction of an azide with a terminal alkyne. The CuAAC gives a mild efficient reaction, which requires no protection groups, and no purification in many cases. Appukuttan et al. (2004) reported a one pot, three-component synthesis of various 1,4-substituted-1,2,3-triazoles using the corresponding alkyl halides, sodium azide and alkynes, under microwave irradiation condition for 10 to 15 min at 70–125°C in t-BuOH and water (1 : 1). Complete regioselective formation of the 1,4-regioisomer was achieved although the reactions were performed at relatively high temperature. The yield was 81–93% with 100% regioselectivity.

$$R_1 \diagup\!\!\diagdown X \;+\; NaN_3 \;+\; \equiv\!\!-R_2 \;\xrightarrow[\substack{\text{t-BuOH/H}_2\text{O (1:1),} \\ \text{10-15 min}}]{\substack{\text{Cu (0), CuSO}_4, \\ \text{MW, 70-125}^{\circ}\text{C}}}\; $$

where X = Cl, Br, I; R_1 = Bn, Subs. Bn, Me; R_2 = Ph, HO-$(CH_2)_3$-, COOEt, TMS.

An interesting case of dendrimers synthesis has also been reported by Yoon et al. (2007). The selective monofunctionalization of dendrimers with biologically specific targeting and recognition moieties should allow transportation of the dendrimer to the biological targets of interest.

Cycloaddition of various types like [2+2], [3+2], [4+2], etc. have been successfully achieved under these conditions with good yields and that too in limited time. Such cycloaddition reactions are helpful in the synthesis of homocyclic and heterocyclic compounds, which are useful in preparing compounds of desired structures.

KEYWORDS

- [2+2] Cycloaddition
- [4+2] Cycloaddition
- 1,3-Dipolar Addition
- Diels-Alder Addition

REFERENCES

Andrade, M. M., & Barros, M. T. (2009). Arkivoc, (xi), 299–306.

Andrade, M. M., Barros, M. T., & Pinto, R. C. (2008). Tetrahedron, 64, 10521–10530.

Angel, D.-O., Cozar, A., Prieto, P., Hoz, A., & Moreno, A. (2006). Tetrahedron Letters, 47, 8761–8764.

Ankati, H., & Biehl, E. (2009). Tetrahedron Letters, 50, 4677–4682.

Appukkuttan, P., Dehaen, W., Fokin, V. V., & Eycken, E. V. (2004). Organic Letters, 6, 4223–4225.

Appukkuttan, P., Mehta, V. P. & Eycken, E. V. (2010). Chem. Soc. Rev., 39, 1467–1477.

Arigela, R. K., Sharma, S. K., Kumar, B., & Kundu, B. (2013). Beilstein Journal of Organic Chemistry, 9, 401–405.

Babu, A. R. Suresh & Raghunathan, R. (2007). Tetrahedron Letters, 48, 305–308.

Bandini, E., Corda G., D'Aurizio A., & Panunzio M. (2010). Tetrahedron Letters, 51, 933–934.

Bansal, R. K., Dandia, A., Gupta, N., & Jain, D. (2003) Heteroatom Chemistry, 14, 560–563.

Benedetti, E., Kocsis, L. S., & Brummond, K. M. (2012). Journal of the American Chemical Society, 134, 12418–12421.

Bezenšek, J., Koleša, T., Grošelj, U., Wagger, J., Stare, K., Meden, A., Svete, J., & Stanovnik, B. (2010). Tetrahedron Letters, 51, 3392–3397.

Bhella, S. S., Pannu, A. P. S., Elango, M, Kapoor, A., Hundal, M. Singh, & Ishar, M., P. S. (2009). Tetrahedron, 65, 5928–5935.

Bhuyan, D., Sarma, R., & Prajapati, D. (2012). Tetrahedron Letters, 53, 6460–6463.

Bortolini, O., Agostino, M. D., Nino, A. De, Maiuolo, L., Nardi, M., & G. Sindona (2008). Tetrahedron, 64, 8078–8081.

Bougrin, K., & Benhida, R. (2012). Microwave-assisted cycloaddition reactions, in de la Hoz, A., & Loupy, A. (Eds). Microwaves in organic synthesis, 1, Weinheim: Wiley-VCH.

Brummond, K. M., & Chen, D. (2005). Organic Letters, 7, 3473–3475.

Butnariu, R. M., & Mangalagiu, I. I. (2009). Bioorganic and Medicinal Chemistry, 17, 2823–2829.

Chakraborty, B., Sharma, P. K., Rai, N., & Sharma, C. D. (2012). Journal of Chemical Sciences, 124, 679–685.

Chiacchio, U., Corsaro, A., Innazzo, D., Piperno, A., Romeo, G., Romeo, R., Saita, M. G., & Rescifina, A. (2007). European Journal of Organic Chemistry, 28, 4758–4764.

Chiacchio, U., Rescifina, A., Saita, M. G., Iannazzo, D., Romeo, G.,Mates, J. A., Tejero, T., & Merino, P. (2005). Journal of Organic Chemistry, 70, 8991–9001.

Chi, C. C., Pai, I. F., & Chung, W. S. (2004). Tetrahedran, 60, 10869–10876.

Choshi, T., Kumemura, T., Nobuhiro, J., & Hibino, S. (2008). Tetrahedron Letters, 49, 3725–3728.

Criado, A., Vizuete M., Gómez-Escalonilla, M. José, García-Rodriguez, S., Fierro, Jose Luis G., Cobas, A., Peña, D., Guitián, E., & Langa, F. (2013). Carbon, 63, 140–148.

Cruz, P. de la, Hoz, A. de la Langa, F., Illescas, B., & Martin, N., (1997a). Tetrahedron, 53, 2599–2608.

Cruz, P., de la Hoz, A. de la Langa, F., Illescas, B., Martin, N., & Seoane, C. (1997b). Synthetic Metals, 86, 2283–2284.

Díaz-Ortiz, A., Carrillo, J. R., Cossío, F. P., Gómez-Escalonilla, M. J., Hoz, A. De La., Moreno, A., & Prieto, P. (2000). Tetrahedron, 56, 1569–1577.

Economopoulos, S. P., Kaousis, N., Rotas, G., Pagoua, G., & Tagmatarchis, N. (2011). Current Organic Chemistry, 15, 1121–1132.

Fordyce, Euan A. F., Morrison, Angus J., Sharp, Robert D. & Paton, R. Michael (2010). Tetrahedron, 66, 7192–7197.

Garrigues, B., Laporte, C., Laurent, R., Laporterie, A., & Dubac, J. (1996). European Journal of Organic Chemistry, (5), 739–741.

Gómez, M. V., Ana. Aranda I, Moreno, A., Fernando, P. Cossío, Abel de Cózar, Díaz-Ortiz Áde la Hoz., & A., Prieto, P. (2009). Tetrahedron, 65, 5328–5336.

Guezguez, R., Bougrin, K., Akri, K. El, & Benhida, R. (2006). Tetrahedron Letters, 47, 4807–4811.

Hajbi, Y., Suzenet, F., Khouili, M., Lazar, S., & Guillaumet, G. (2007). Tetrahedron, 63, 8286–8297.

Hajbi, Y., Suzunet, F., Khouili, M., Lazar, S., & Guillaumet, G. (2009). Synthesis, 1, 92–96.

Harju, K., Manevski, N., & Yli-Kauhaluoma, J. (2009). Tetrahedron, 65, 9702–9706.

Hong, B.-C., Jiang, & Y.-F., Kumar, E. S. (2001). Bioorganic and Medicinal Chemistry Letters, 11, 1981–1984.

Islas-Jácome, A., González-Zamora, E., & Gámez-Montaño, R. (2011). Tetrahedron Letters, 52, 5245–5248.

Jacob, K., Stolle, A., Ondruschka, B., Jandt, K. D., & Thomas F. K. (2013). Applied Catalysis A: General, 451, 94–100.

Jankowski, C. K., LeClair, G., Bélanger, M. R. J., R. J. Paré J., & Van Calsteren, M.-R. (2001). Canadian Journal of Chemistry, 79, 1906–1909.

Jayashankaran, J., Durga, R., Manian, R. S., & Raghunathan, R. (2004). Tetrahedron Letters, 45, 7303–7305.

Jimenez-Alonso, S., Chavez, H., Estevez-Braun, A., Ravelo, A. G., Feresin, G., & Tapia, A. (2008). Tetrahedron, 64, 8938–8942.

Johnstone, Mark D., Lowe, Adam, J., Henderson, Luke, C., Pfeffer, & Frederick, M. (2010), Tetrahedron Letters, 51, 5889–5891.

Kale, S., Kahandal, S., Disale, S., & Jayaram, R. (2012). Current Chemistry Letters, 1, 69–80.

Kaousis, N., Economopoulos, S. P., Iizumi, Y., Okazaki, T., Liu, Z., Suenaga, K., & Tagmatarchis, N. (2010). Chemical Communications, 9110–9112.

Katritzky, A. R., & Singh, S. K. (2002). Journal of Organic Chemistry, 67, 9077–9079.

Kaval, N., Eycken, J. V., Caroen, J., Dehaen, W., Strohmeier, G. A., Kappe, C. O., & Eycken, E. V. (2003). Journal of Combinatorial Chemistry, 5, 560–568.

Kaval, N., Appukuttan, P., & Eycken, E. V. (2006). Topic in Heterocyclic Chemistry, 1, 267–304.

Kranjc, K., & Kocevar, M. (2008). Tetrahedron, 64, 45–52.

Kumar, R. S., Osman, H., Perumal, S., Menéndez, J. C., Ali, M. A., R. Ismail, & T. S. Choon (2011). Tetrahedron, 67, 3132–3139.

Lee, Y. A., & Kim, S. C. (2011). Journal of Industrial and Engineering Chemistry, 17, 401–403.

Li, X., Zhu, Z., Duan, K., Chen, H., Li, Z., Zhe, Li, & Zhang, P. (2009). Tetrahedron, 65, 2322–2328

López, I., Silvero, G., Arévalo., M. J., Babiano, R., Palacios, J. C., & Bravo, J. L. (2007). Tetrahedron, 63, 2901–2906.

Loupy A., Maurel, F., & Sabatié-Gogová, A. (2004). Tetrahedron, 60, 1683–1691.

Lukyanov, S. M., Bliznets, Igor, V., Sergey, V. Shorshnev, Aleksandrov Grigory G., Stepanov, Aleksandr E., & Vasil'ev, Andrei A. (2006). Tetrahedron, 62, 1849–1863.

Jha, M., Guy, S., & Chou, T.-Y. (2011). Tetrahedron Letters, 52, 4337–4341.

Mabrour M., Bougrin K., Benhida R., Loupy A., & Soufiaoui M. (2007). Tetrahedron Letters, 48, 443–447.

Mabrour, M., Bougrin, K., & Benhida, A., Loupy, A., & Soufiaoui, M. (2007). Tetrahedron Letters, 48, 443–447.

Majumdar, K. C., Ganai, S., & Sinha, B. (2012). Tetrahedron, 68, 7806–7811.

Mihovilovic, M. D., Leisch, H. G., & Mereiter, K. (2004). Tetrahedron Letters, 45, 7087–7090.

Meng, L., Fettinger, J. C., & Kurth, M. J. (2007). Organic Letters, 9, 5055–5058.

Mikhailichenko, Sergey S., Bouillon, Jean-P., Besson T., Yuri, G. Shermolovich (2010). Tetrahedron Letters, 51, 990–993.

Nicolaou, K. C., Lister, T., Denton, R. M., & Gelin, C. F. (2007). Angewandte Chemie International Edition, 46, 7501–7505.

Oukani, H., Pellegrini-Moïse, N, Jackowski, O., Chrétien, F., & Chapleur, Y. (2013). Carbohydrate Research.

Ovaska, T. V., & Kyne, R. E. (2008). Tetrahedron Letters, 49, 376–378.

Parmar, N. J., Pansuriya, B. R., Barad, H. A., Kant, R., & Gupta, V. K. (2012). Bioorganic and Medicinal Chemistry Letters, 22, 4075–4079.

Potewar, T. M., Petrova, K. T., & Barros, M. T. (2013). Carbohydrate Research, 379, 60–67.

Pospíšil, J., & Potáček, M. (2007). Tetrahedron, 63, 337–346.

Pospíšil, J., & Potáček, M. (2007). Tetrahedron, 63, 337–346.

Radi, Marco, B., Vincenzo, B., Beatrice, C., Daniele, P., Mafalda, & Botta, M., (2009). Tetrahedron Letters, 50, 6572–6575.

Rajesh, M., Kamble, M., & Ramana, M. V. (2010). Canadian Journal of Chemistry, 88, 1233–1239.

Rajesh, R., & Raghunathan, R. (2010). Tetrahedron Letters, 51, 5845–5848.

Ramesh, E., & Raghunathan, R. (2008). Tetrahedron Letters, 49, 1812–1817.

Ramesh, E., & Raghunathan, R. (2009). Synthetic Communications, 39, 613–625.

Ramesh, E., Kathiresan, M., & Raghunathan, R. (2007). Tetrahedron Letters, 48, 1835–1839.

Rijkers, Dirk T. S., Wilma van Esse, G., Merkx, R., Brouwer, Arwin J., Jacobs, Hans J. F., Pieters, Roland, J., & Rob, M. J. L. (2005). Chemical Communications, 4581–4583.

Safaei-Ghomi, J., Tajbakhsh, M., & Kazemi-Kani, Z. (2004). Acta Chimica Solvenica, 51, 545–550.

Saleh, T. S., Narasimharao, K., Ahmed, N. S., Basahel, S. N., Al-Thabaiti, S. A., & Mokhtar, M. (2013). Journal of Molecular Catalysis A: Chemical, 367, 12–22.

Sarotti, A. M., Joullie, M. M., Spanevello, R. A., & Suarez, A. G. (2006). Organic Letters, 8, 5561–5564.

Sarotti, A. M., Spanevello, R. A., & Suárez A. G. (2011). Organic Chemistry in Argentina, 7, 31–37.

Sarotti, A. M., Pisano, P. L., & Pellegrinet, S. C. (2010). Organic & Biomolecular Chemistry, 8, 5069–5073.

Shanmugasundaram, M., Aguirre, A. L., Leyva, M., Quan, B., & Martinez, L., E. (2007). Tetrahedron Letters, 48, 7698–7701.

Shao, B. (2005). Tetrahedron Letters, 46, 3423–3427.

Silva, A. M. G., Tomé, A. C., Maria, G. P. M. S., Neves, A. S., José Cavaleiro, C., & Kappe, O. (2005). Tetrahedron Letters, 46, 4723–4726.

Singh, L., Ishar, M. P. S., Elango, M., Subramanian, V., Gupta, V., & Kanwal, P. (2008). Journal of Organic Chemistry, 73, 2224–2233.

Soares, Maria I. L., & Teresa, M. V. D. Pinho e Melo (2008). Tetrahedron Letters, 49, 4889–4893.

Tagmatarchis, N., Georgakilas, V., Prato, M., & Shinohara, H. (2002). Chemical Communications, 2010–2011.

Timothy, B., Patrick, Gorrell, K., & Rogers, J. (2007). Journal of Fluorine Chemistry, 128, 710–713.

Tsai, C.-W., Yang S.-C., Liu, Y.-M, Wu, M.-J. (2009). Tetrahedron, 65, 8367–8372.

Ursic, U., Groselj, A., Meden, J., Svete, J., & Stanovnik, B. (2008). Tetrahedron Letters, 49, 3775–3778.

Veverkova, E., & Toma, S. (2005). Chemical Papers, 59, 350–353.

Wang, Y., Iqbal, Z., & Mitra, S. (2005). Carbon, 43, 1015–1020.

Wilson, N. S., Sarko, C. R., & Roth, G. P. (2001). Tetrahedron Letters, 42, 8939–8941.

Wipf, P., Fletcher, J. M., & Scarone, L. (2005). Tetrahedron Letters, 46, 5463–5466.

Wu, Jinlong, Sun, L., & Wei-Min, D. (2006). Tetrahedron, 62, 8360–8372.

Wu, Jinlong, Xiuqing, J., Jingjing, Xu., & Wei-Min Dai (2011). Tetrahedron, 67, 179–192.

Xing, X., Wu, J., & Wei-Min Dai. (2006). Tetrahedron, 62, 11200–11206.

Yang, S. H., & Caprio, V. (2007). Synlett, 1219–1222.

Yoon, K., Goyal, A., & Weck, M. (2007). Organic Letters, 9, 2051–2054.

Zhang, H., Boonsombat, J., & Padwa, A. (2007). Organic Letters, 9, 279–282.

Zhang, W., Lu, Y., Chen, C. H. T., Curran, D. P., & Geib, S. (2006). European Journal of Organic Chemistry, 2055–2059.

Zbancioc, G., Caprosu, M., Moldoveanu, C., & Mangalagiu, I. (2005). Arkivoc, (x), 189–198.

Zheng, S., Chowdhury, A., Ojima, I., & Honda, T. (2013). Tetrahedron, 69, 2052–2055.

CHAPTER 9

ELIMINATION

CHETNA AMETA, KUMUDINI BHANAT, ARPIT KUMAR PATHAK, and PINKI B. PUNJABI

CONTENTS

Elimination reaction involves the removal of two substituents, from a pair of adjacent atoms in a molecule, without being replaced by other atoms or groups. As a consequence of the removal of atoms or groups from the adjacent atoms of molecule, unsaturation is introduced. The most common multiple bonds formed are those of alkene, alkyne and their heteroatom variations such as carbonyl and cyano groups. The removal of the groups from a molecule takes place in either one or two-steps. In most of the organic elimination reactions, at least one hydrogen is lost to form the double bond and as a result, the unsaturation of the molecule increases. It is also possible that a molecule undergoes reductive elimination, by which the valence of an atom in the molecule decreases by two. Elimination may be considered the reverse of an addition reaction.

There is an acute need of new organic molecules for a variety of uses for mankind like pharmaceutical, agrochemicals, food, textile, cosmetics, petrochemicals, etc. Traditional methods of organic synthesis are slow enough to satisfy the demand of these compounds. The field of microwave-assisted chemistry has been developed to meet the increasing requirement of new compounds for drug discovery and within this field, speed is of the core importance. The efficiency of microwave chemistry in reducing reaction times dramatically (reduced from days and hours to minutes and seconds) has recently been proved in several different fields of organic chemistry. The time saved by using focused microwaves is potentially important in traditional organic synthesis but could be of even greater importance in high-speed combinatorial and medicinal chemistry.

9.1 SYNTHESIS OF ALKENES AND ALKYNES

The use of microwaves for synthesis of olefins via elimination has become increasingly interesting in recent years because olefins are the basic moiety of many olefinic natural products, for example, the insect hormones, pheromone, etc. It is also used in nonenzymic biogenetic-like cyclizations to form polycyclic compounds.

These elimination reactions have been known for many decades and have been shown to occur under both; acidic and basic conditions.

The cis-diphenoxy-2-butene was used to perform a base-promoted elimination reaction to yield the phenoxydiene. Conventionally, this reaction takes about 2.5 h at 0–25 °C to give 78% yield (Roversi et al., 2002), when the reaction was performed at - 60 °C, substantially below the conventional temperature using the subambient microwave apparatus and gives 81% yield only in 5 min (Barmhardt, 2004). This example highlighted the use of instantaneous temperatures to increase the rate of the reaction.

$$PhO\text{—}\!\!\!\!=\!\!\!\!\text{—}OPh \quad \xrightarrow[\text{THF}]{\text{n-BuLi}} \quad /\!\!/\text{—}\!\!\!\!\text{—}OPh$$

(Z)-1-Bromo-1-alkenes were efficiently prepared in high yields stereoselectively by microwave irradiation of the corresponding anti-2,3-dibromoalkanoic acids in DMF using Et$_3$N as a base. Microwave-assisted one-pot syntheses of terminal alkynes and enynes from 2,3-dibromoalkanoic acids were also developed (Kuang et al., 2005a).

$$
\underset{\substack{\text{Br}}}{R}\!\!\!\!\overset{\substack{\text{Br}}}{\diagup}\!\!\!\!CO_2H
\quad \xrightarrow[\substack{\text{DMF, MW (200 W)}\\\text{Open vessel, 0.5-1 min}}]{1.05 \text{ eq. NEt}_3}
\quad \left[R\diagdown\!\!\overset{}{\diagup}_{Br} \right]
\quad \xrightarrow[\substack{\text{MW (200 W)}\\\text{Open vessel, 1 min}}]{2 \text{ eq. DBU}}
\quad R\text{—}\!\!\equiv
$$

where R = Phenyl, p-Tolyl and –C$_6$H$_4$COOEt, the yield is 88 (0.5 min), 93 (1.0 min) and 99% (1.0 min), respectively.

(E)-β-Arylvinyl bromides were readily prepared in a short reaction time (1–2 min) by microwave irradiation of the corresponding 3-arylpropenoic acids in the presence of N-bromosuccinimide and a catalytic amount of lithium acetate (Kuang et al., 2005b).

$$
R\overset{H}{\underset{N}{\diagdown}}\!\!\diagup\!\!\!=\!\!O \quad + \quad O\!\!=\!\!\!\overset{\overset{X}{\underset{|}{N}}}{\diagup\!\!\diagdown}\!\!\!=\!\!O
\quad \xrightarrow[\text{CH}_2\text{Cl}_2\text{, r.t., 5 min}]{5 \text{ mol \% NEt}_3}
\quad \text{Ar}\diagup\!\!\diagdown\!\!\diagup X
$$

where X = Cl, Br, I.

In this reaction, if the products are phenyl alkynes or substituted phenyl alkynes, it was observed that the yield continuously increases.

Microwave-assisted reaction of (Z)-arylvinyl bromides involving an elimination and homocoupling in the presence of 1,8-diazabicyclo[5.4.0]undec-7-ene (DBU) and CuI in DMF affords a variety of symmetrical 1,3-diynes in excellent yields. This method provides an alternative to the conventional homocoupling methods for the synthesis of symmetrical 1,3-diynes (Zhang et al., 2013).

9.2　SYNTHESIS OF ISONITRILES

A simple and efficient method has been reported for the synthesis of both; aliphatic and aromatic isonitriles in high yields. In this method, aliphatic and aromatic formamides have been transformed to their corresponding isonitriles by using 1.3–3.0 equivalents of an inexpensive dehydration agent such as 2,4,6-trichloro[1,3,5]triazine (TCT) at 50–100°C within 3–10 min under controlled microwave exposure (Porcheddu et al., 2005).

A microwave promoted, concerted, facile and one-pot method for the synthesis of nitriles from aldehydes via elimination of tert-butanesulfenic acid from tert-butanesulfinyl protected imines has been presented by Tanuwidjaja et al. (2007). The mild reaction conditions exhibit broad functional group compatibility and generate products in good yields (69–87%). This method provides a rapid and mild procedure for the synthesis of various functionalized nitriles.

Highly diastereoselective intra and intermolecular self condensation reactions of N-tert-butanesulfinyl aldimines have been developed and applied to the rapid, asymmetric synthesis of trans-2-aminocyclopentanecarboxylic acid and the drug candidate SC-53116. Key to both these syntheses is a novel microwave-assisted reaction in which N-sulfinyl aldimines are cleanly converted into nitriles in high yielding concerted elimination processes (Schenkel and Ellman, 2004).

9.3   SYNTHESIS OF IMIDAZOLE DERIVATIVES

A simple reaction of *o*-phenylenediamine with ethoxymethylenemalononitrile at room temperature gave 2-[(2-aminophenylamino) methylene]malononitrile. Then an intramolecular cyclization of 2-[(2-aminophenylamino) methylene]malononitrile under microwave conditions generated the benzimidazole ring in quantitative yield only in 7 min by elimination reaction of malononitrile (Faria, 2013).

Hernandez et al. (2010) reported a microwave-assisted deacylation of (20R)-20-acetyl-23,24-dinorcholanic lactones by hydrazine hydrate in high yields. The elimination of the 20-acetyl group proceeded with retention of configuration, while other deacylation methods give a mixture of diastereoisomers.

A variety of 3-vinyl-substituted imidazo[1,5-*a*]indole derivatives were synthesized by intramolecular Pd catalyzed cyclization of the indole-2-carboxylic acid allenamides through either a domino carbopalladation/exo-cyclization process or a novel hydroamination reaction that proceeds smoothly under microwave irradiation. Both the observed pathways involve a π-allyl-palladium (II) complex arising from insertion of the allene group into a palladium (II) species, the latter being formed in situ by the intervention of an aryl iodide or of the N−H group. In both these cases, the role of nucleophile is covered by the indole nitrogen (Beccalli et al., 2010).

Carbonamination   Hydroamination

9.4   DIELS-ALDER ELIMINATION

Various electron-deficient 2H-pyran-2-ones and ethyl vinyl ether undergo Diels-Alder elimination reaction under microwave exposure. This microwave-accelerated sequence of a cycloaddition followed by a retro-Diels-Alder reaction (the elimination of CO_2) and a second step elimination of ethanol yields substituted aniline derivatives (Kranjc and Kokevar, 2008). This reaction is significantly accelerated by the application of DABCO (1,4-diazabicyclo[2.2.2]octane) as a suitable base.

A microwave promoted aza-Diels–Alder reaction between 6-[2-(dimethylami-no)vinyl]-1,3-dimethyluracil and aldimines has been developed for the synthesis of dihydropyrido[4,3-d]pyrimidines. Urea was effectively employed as an environmentally benign source of ammonia in the absence of any catalyst or solvent. The key step in the reaction is in situ generation and trapping of the reactive aldimine formed from urea and aldehyde by the diene system of the uracil. The reaction is clean in nature and excellent yields were obtained in few minutes (Sarma et al., 2012).

A set of Diels-Alder reactions of fused pyran-2-ones with ethyl vinyl ether (an appropriate synthetic equivalent of acetylene) gave fused carbocyclic systems. DABCO was used as a catalyst for the elimination of ethanol under microwave irradiation (Juranovic et al., 2012). The Diels-Alder cycloaddition reaction in 3-nitro-1-(p-toluenesulfonyl)indole with dienes under microwave irradiation in solvent-free conditions gave carbazole derivatives after elimination of the nitro group and in situ aromatization (Victoria et al., 2009).

9.5 CYCLOADDITION–ELIMINATION

The multifunctionalized 2-azadiene system could easily undergo inter and intramolecular cycloaddition-elimination reactions to afford a large variety of interesting and highly decorated heterocyclic scaffolds like pyridines, pyridinones, α- and β-carbolines, (benzo)fluoro- or (benzo)pyrano pyridines, etc. The reactions were carried out under microwave irradiation conditions with dimethyl acetylenedicarboxylate (DMAD) at 180–200°C for 10–20 min and the corresponding highly decorated pyridine and pyridine scaffolds were isolated in good yields (Appukkuttan et al., 2010).

where R₁ = CH₂COOEt, Ph; R₂ = H, Me; R₃ = 4-MeO, 4-Me, 4-Cl, 4-NO₂.

Microwave-assisted heating in the presence of amines provides a simple and efficient methodology for regioselective alkylation of exocyclic nitrogen of cyclic amidines, which results in N-alkylated 3,4-dihydropyrazino[2,1-b]quinazolin-6-ones in good yields. The reaction occurs by addition-elimination process involving a first attack of the amine on the electrophilic carbon of the amidine function (Pereira et al., 2007).

Same reaction has also been carried out with methyl iodide in place of amine and in the presence of sodium hydride to yield N-methyl-3,4-dihydropyrazino[2,1-b] quinazolin-6-ones. The novel pyrimido[4,5-d]pyrimidine derivatives of biological significance from electron rich 6-[(dimethylamino)methylene]amino uracil undergoes [4+2] cycloaddition reactions with various in situ generated glyoxylate imine and imine oxides, after elimination of dimethylamine from the (1:1) cycloadducts and oxidative aromatization. This method provides a suitable method for the direct synthesis of pyrimido[4,5-d]pyrimidines in excellent yields, when reaction was carried out in the solid state and under microwave irradiations (Prajapati et al., 2006).

[4+2] Cycloaddition reactions of electron rich 6-[(dimethylamino)methylene] aminouracil with various electron deficient substrates gives pyrimido[4,5-d]pyrimidines and pyrido[2,3-d]pyrimidines. This reaction takes place in two steps, firstly formation of cycloadducts and then elimination of dimethylamine from the cycloadducts and oxidative aromatization (Gohain et al., 2004). The reaction gives excellent yields, when carried out under microwave irradiation and solvent-free conditions.

The reaction of aldoxime with dimedone and ammonium acetate in glycol under microwave irradiation has been carried out, where elimination and cyclization take place and acridine derivatives were obtained with excellent yields (80–95%) within a short reaction time (4–8 min) (Tu et al., 2004).

Addition-elimination reactions of 3-phenylimino-2-indolinones with 2-hydrazino-3-(p-chlorphenyl)-1,8-naphthyridine in methanol containing a few drops of glacial acetic acid under microwave irradiation have been described by Mogilaiah et al. (2003).

where R = H, 5-CH$_3$, 7-CH$_3$, 5-Cl, 5-Br.

Microwaves also provide an easy access to 2-alkyl and 2-aryl-4-quinolones by the addition-elimination reaction of 5-methylthioalkylidene isopropylidene malonates with arylamine, followed by cyclization (Huang et al., 2000). A condensation, addition, cyclization and elimination reactions of aromatic aldehydes, 5,5-dimethyl-1,3-cyclohexandione and isopropylidene malonate afforded a series of 4-aryl-7,7-dimethyl-2, 5-dioxo-1,2,3,4,5,6,7,8-octahydroquinoline and 4-aryl-7,7-dimethyl-5-oxo-3,4,5,6,7,8-hexahydrocoumarin under microwaves only in 3–5 min (Tu et al., 2002). Basic silica gel (NaOH/SiO$_2$) in solvent-free conditions and microwave exposure acts as a very efficient medium for the β-elimination (dehydrosulfenylation) of sulfoxides (Moghaddam and Jamshidi, 2001).

9.6 MICHAEL ADDITION-ELIMINATION

Michael reaction of 3-(2'-nitrovinyl)indole with 3-unsubstituted indoles furnished unsymmetrical bis(indolyl)nitroethanes in 7–12 min under microwave irradiation while at room temperature, it takes about 8–14 h.

In contrast, the p-TsOH catalyzed reaction of the nitrovinylindole with 3-unsubstituted and two 3-substituted indoles in solution by conventional method furnished both; unsymmetrical and symmetrical bis(indolyl)nitroethanes (BINEs). The symmetrical bis(indolyl)nitroethanes results from novel tandem Michael addition-elimination-Michael addition reactions (Chakrabarty et al., 2004).

where R/R$_1$/R$_2$ = H, Alkyl, CH$_2$CH$_2$NHAc; X = H, Br.

9.7 HOFMANN ELIMINATION

In two-phase systems, absorption of microwave radiation by dielectric heating and ionic conduction enables individual phases to be heated at different rates, potentially affording sizable temperature differences. Raner et al. (1995) performed a Hoffmann elimination using a two-phase water-chloroform system. The reaction carried out in water at 105 °C led to polymerization of the final product. However, the reaction proceeds adequately under microwave irradiation in a two-phase water-chloroform system. The temperatures of the aqueous and organic phases were 110 and 50 °C, respectively, due to differences in the dielectric properties of the solvents. This difference avoids the decomposition of the final product. Comparable conditions would be difficult to obtain by traditional heating methods. Reactions on the starting salts progressed rapidly in the more strongly microwave absorbing aqueous phase. The thermally unstable monomeric products were simultaneously extracted and diluted into the cooler and denser, organic phase.

9.8 COPE ELIMINATION

Photo-mediated asymmetric synthesis of (2)-cuparene has been studied by Grainger and Patel (2003). In this procedure, generation of a benzylic quaternary stereocenter *via* the photo-mediated cyclization of a chiral (aminobutyl) styrene followed by a microwave-assisted Cope elimination led to a total synthesis of the sesquiterpene (2)-cuparene.

9.9 COREY–WINTER ELIMINATION

The Corey-Winter olefin synthesis is a series of chemical reactions for conversion of 1,2-diols into olefins. A change of the conformation prior to ring closure was necessary to achieve it. Clerc et al. (2010) imagined that this could be achieved through simple modification of the RCM (ring-closing metathesis) by replacing the central double bond by a spatially different five-membered ring system, arising from dihydroxylation and subsequent acetonide formation. Re-installment of the double bond could then be achieved by Corey–Winter elimination.

9.10   ELIMINATION OF SIMPLE ENTITIES

9.10.1   ETHYL ACETATE

The ecofriendly solvent-free reactions of o-aminothiophenol and aromatic or aliphatic β-keto esters with microwave irradiation produced 2-substituted benzothiazoles in excellent yields. The formation of the 2-substituted benzothiazoles probably involves the nucleophilic addition of the thiol group to the keto group of the β-keto ester followed by elimination of ethyl acetate from the resulting adduct. Then it undergoes an intramolecular addition of the o-amino group to the carbonyl group to give an adduct, from which water is eliminated to afford the 2- substituted benzothiazoles (Kamila et al., 2005).

9.10.2   NITROGEN

Synthesis of mono isoxazolinyl derivatives are reported from the reaction of chloroheterocycles with some isoxazolones under microwave irradiation and solvent-free conditions. The main advantages of this scheme are:
  (i)   Elimination of the nitrogen gas (N_2),
  (ii) Solvent-free conditions,
  (iii) Microwave irradiation,
  (iv) Avoiding the use of silica gel for purification of the products, and
  (v)   Higher and shorter reaction times (Ebrahimlo, 2012).

  Substituted bicyclic and spirocyclic N-nitroso-4,5-dihydropyrazoles eliminate nitrogen under microwave exposure (3–8 min) in solvent (chlorobenzene-DMF and chlorobenzene-AcOH) or in solvent-free conditions (on SiO_2) to afford the corresponding 4,5-dihydroisoxazole derivatives (Stepakov et al., 2007). This methodology represents an improvement on thermolysis and allows 4,5-dihydroisoxazole systems to be obtained in good yields.

9.10.3　AMMONIA

The one pot microwave-assisted reaction of 1,2-phenylenediamine and 2-amino-benzylamine to yield 5-thia-4b,10-diaza-indeno[2,1- a]indene-5,5-dioxide and 10H-11-thia-5,10a-diaza-benzo[b]fluorene-11,11- dioxide has been reported by Cole et al. (2009).

In this reaction, sulfonamide acts as an intermediate, which is generated through reaction of the appropriate diamine with 2-cyanobenzenesulfonyl chloride. This sulfonamide provides corresponding tetracycle by a multistep intramolecular transformation and subsequent elimination of ammonia.

9.10.4　HYDROGEN SULFIDE

Microwave-assisted reaction of 1-benzoyl-3-benzylthiourea and benzoyl-ethyl-thiocarbamate with benzylamine in dry media conditions using KF-Al$_2$O$_3$ provides 1-benzoyl-3-benzylguanidine and 1-benzoyl-3-benzyl-O-ethylisourea in good yields (68 and 76%, respectively) (Marquez et al., 2006). Strong nucleophilic benzylamines promoted the elimination of sulfur as H$_2$S by attack on the thiocarbonyl group in both; thiourea and thiocarbamates to afford guanidines and isourea, respectively. Very important nonpurely thermal MW specific effects were found responsible for stabilization by coulombic interactions between materials and waves.

where X = NH$_2$, OC$_2$H$_5$.

9.10.5　AMINOALKOXY GROUP

Terminal chelation controlled Heck vinylations of electron-rich amino-functionalized vinyl ethers were performed with high regioselectivity to furnish the corresponding 1-alkoxy-1,3-butadienes. Controlled microwave heating effectively accelerated these palladium catalyzed reactions and full conversion could be achieved within 30 min. Subsequent Diels–Alder reactions with dimethyl acetylenedicarbox-ylate under microwave irradiation resulted in partly aromatized bi- and tricyclic compounds by elimination of the aminoalkoxy group. Thus, the selected dimeth-ylamino auxiliary, a preferred metal presenting functionality, controlled the regio-chemistry in the palladium-catalyzed vinylation as well as easily displaced in the aromatization process (Stadler et al., 2004).

9.11   ELIMINATION OVER SOLID SUPPORTS

The use of alumina and bentonite as heterogeneous catalysts for various organic reactions is well documented (Caddick et al., 1995; Gedye et al., 1988; Gutierrez et al., 1989; Smith, 1992). 1,2-Dibromocompounds with KF supported on alumina, or bentonite/Et$_3$N under microwave irradiation leads to an elimination reaction, which produces functionalized alkenes in good yield. Elimination from dibromide carried out over KF-alumina under microwave irradiation led to an isomeric mixture of α-bromoalkenes, and in some cases, debromination occurs to give alkenes (Saoudi et al., 1998).

Trans-4-(4-fluorophenyl)-3-chloromethyl-1-methylpiperidine was subjected to elimination reaction on alumina or KF-alumina under solvent-free conditions and microwave irradiation. A solvent-free supported microwave methodology for dehy-drochlorination of compound, trans-4-(4-fluorophenyl)-3-chloromethyl-1-methylpi-peridine, provided significantly better yields in comparison with conventional heat-ing in the presence of organic base and solvent (Navratilova et al., 2004).

9.12   ELIMINATION IN IONIC LIQUIDS

The research and application of green chemistry principles have led to the develop-ment of cleaner processes. Use of ionic liquids for organic synthesis is one of such green technology. During the present century, an ever-growing number of studies have been carried out using ionic liquids (ILs) as solvent, catalyst, or templates to develop more environmentally friendly and efficient chemical transformation for their use in both; academic and industry. Ionic liquids are ionic, salt-like materials that are liquid normally below 100°C. ILs typically comprises an organic cation and an anion and are liquids in nature.

Ranu and Jana (2005) observed that ionic liquid, [pmim][BF$_4$], can be used as an effective catalyst as well as solvent for the stereoselective debromination of a broad range of vicinal dibromides to the corresponding (E)-alkenes in a domestic micro-wave oven. The [pmim][BF$_4$] remained intact and could be reused several times.

R_1, R_2 = Alkyl, Aryl, CN, CO$_2$Me, CO$_2$Et, COPh, NO$_2$.

Another application of [pmim][BF$_4$] for the debromination reaction has been reported by Ranu et al. (2007). α-Bromoketones were selectively debrominated to either monobromo or debromoketones under microwave irradiation. The utilization of [pmim][BF$_4$] was also found appropriate for the dehalogenation of α-haloketones & esters and vic-bromoacetals.

where R$_1$ = Ar, OEt; R$_2$ = COPh, COMe, Me, H.

Microwave radiations are the powerful tool for the elimination reaction of the poly(phenylene vinylene)-polyelectrolyte precursor polymer (PPV precursor) into the fully conjugated polymer. The extent of the elimination process was determined from the amount of residual sulfur in the film after the irradiation process. The microwave induced elimination process was also attempted using PPV precursor films that have been impregnated with anionic compounds containing iron. The molecular mechanisms for the observations are discussed and comparison with the elimination reaction induced by other energy sources has been made (Torres-Filho et al., 1994).

Green chemistry involves design of chemical synthesis to prevent pollution and thereby, solve the environmental problems. The microwave chemistry is a current approach in green chemistry. Microwave-mediated reactions occur more rapidly, safely and in environment-friendly manner with high yields. Such reactions reduce the amount of waste products and increase the pure required products. Thus, one can conclude that microwave-mediated synthesis is a green chemical technology because microwave not only accelerate chemical processes but also improve yield, selectivity, reduces pollution and enable reaction to occur in solvent-free conditions.

Elimination reactions are commonly used for synthesizing unsaturated products and these unsaturated products containing C=C, C≡C, C=N, etc. These unsaturated units have their own importance in synthetic organic chemistry as electron rich entities. Microwave provides a facile route to prepare these species.

Furthermore, many new applications of microwave are being developed like region-selective and chemoselective synthesis, polymer synthesis, ceramic products, intercalation products, organometallic and coordination compounds, macro-molecules, radiopharmaceuticals, etc.

KEYWORDS

- Cope Elimination
- Corey-Winter Elimination
- Cycloaddition-Elimination
- Diels-Alder Elimination
- Hofmann Elimination
- Michael Addition-Elimination

REFERENCES

Appukkuttan, P., Mehta, V. P., & Eycken, E. V. (2010). Chemical Society Reviews, 39, 1467–1477.

Barnhardt, S. K., Presented at the 227th ACS National Meeting (2004) Anaheim, C. A.

Beccalli, E. M., Bernasconi, A., Borsini, E., Broggini, G., Rigamonti, M., & Zecchi, G. (2010). *Journal Organic Chemistry*, 75, 6923–6932.

Caddick, S. (1995). Tetrahedron, 51, 10403–10432.

Chakrabarty, M., Basak, R., Ghosh, N., & Harigaya, Y. (2004). Tetrahedron, 60, 1941–1949.

Clerc, J., Schellenberg, B., Groll, M., Bachmann, A. S., Huber, R., Dudler, R., & Kaiser, M. (2010). European Journal of Organic Chemistry, 3991–4003.

Cole, A. G., Kultgen, S. G., & Henderson, I. (2009). Synthetic Communications, 39, 3607–3610.

Ebrahimlo, A. R. M. (2012). South African Journal of Chemistry, 65, 104–107.

Faria, J. (2013). Synlett, 2, 264–265.

Gedye, R. N., Smith, F. E., & Westaway, K. C. (1988). Canadian Journal of Chemistry, 66, 17–27.

Gohain, M., Prajapati, D., Gogoi, B. J., & Sandhu, J. S. (2004). Synlett, 7, 1179–1182.

Grainger, R. S. & Patel, A. (2003). Chemical Communications, 1072–1073.

Gutierrez, E., Loupy, A., Bram, G., & Ruiz-Hitzky, E. (1989). Tetrahedron Letters, 30, 245–248.

Hernandez, L., Ma, G., Sandoval, R. J., Meza, R. S., Montiel, S. S., Fernandez, H. M. A., & Bernes, S. (2010). Steroids, 75, 240–244.

Huang, Z.-Z., Wu, L.-L., & Huang, X., (2000). Chinese Journal of Organic Chemistry, 20, 88–90.

Juranovic, A., Kranjc, K., Polanc, S., Perdih, F., Kocevar, M. (2012). Monatshefte fur Chemie, 143, 771–777.

Kamila, S., Zhang, H., & Biehl, E. R. (2005). Heterocycles, 65, 2119–2126.

Kranjc, K., & Kocevar, M. (2008). Synlett, 17, 2613–2616.

Kuang, C., Yang, Q., Senboku, H., & Tokuda, M. (2005b). Synthesis, 1319–1325.

Kuang, C., Yang, Q., Senboku, H., & Tokuda, M. (2005a). Tetrahedron, 61, 4043–4052.

Marquez, H., Loupy, A., Calderon, O., & Perez, E. R. (2006). Tetrahedron, 62, 2616–2621.

Moghaddam, F. M., & Jamshidi, H. (2001). Sulfur Letters, 24, 269–273.

Mogilaiah, K., Kavitha, S., & Babu, H. R. (2003). Indian Journal of Chemistry, 42 B, 1750–1752.

Navratilova, H., Kriz, Z., and Potacek, M. (2004). Synthetic Communications. 34, 2101–2115.

Pereira, M. D. F., Thiery, V., & Besson, T. (2007). Tetrahedron Letters, 48, 7657–7659.

Porcheddu, A., Giacomelli, G., & Salaris, M. (2005). Journal of Organic Chemistry, 70, 2361–2363.

Prajapati, D., Gohain, M., & Thakur, A. J. (2006). Bioorganic and Medicinal Chemistry Letters, 16, 3537–3540.

Raner, K. D., Strauss, C. R., Trainor, R. W., & Thorn, J. S. (1995). Journal of Organic Chemistry, 60, 2456–2460.

Ranu, B., & Jana, R. (2005). Journal of Organic Chemistry, 70, 8621–8624.

Ranu, B. C., Chattopadhyay, K., & Jana, R. (2007). Tetrahedron Letters, 63, 155–159.

Roversi, E., Monnat, F., & Vogel, P. (2002). Helvetica Chimica Acta, 85, 733–760.

Saoudi A., Hamelin, J., & Benhaouaa, H. (1998). Tetrahedron Letters, 39, 4035–4038.

Sarma, R., Sarmah, M. M., and Prajapati, D. (2012). Journal of Organic Chemistry, 77, 2018–2023.

Schenkel, L. B., & Ellman, J. A. (2004). Organic Letters, 6, 3621–3624.

Smith, K., Ed., (1992). Solid Supports and Catalysts in Organic Synthesis, New York: Horwood.

Stadler, A., Von Schenck, H., Vallin, K. S. A., Larhed, M., & Hallberg, A. (2004). Advanced Synthesis and Catalysis, 346, 1773–1781.

Stepakov, A. V., Galkin, I. A., Kostikov, R. R., Starova, G. L., Starikova, Z. A., & Molchanov, A. P. (2007). Synlett, 8, 1235–1238.

Tanuwidjaja, J., Peltier, H., & Lewis, J. (2007). Synthesis, 3385–3389.

Torres-Filho, Afranio, Lenz, & Robert W. (1994). Journal of Applied Polymer Science, 52, 377–386.

Tu, S., Miao, C., Gao, Y., Fang, F., Zhuang, Q., Feng, Y., & Shi, D. (2004). Synlett, 2, 255–258.

Tu, S.-J., Deng, X., Zhou, J.-F., Sun, B.-W., & Feng, J.-C. (2002). Kao Teng Hsueh Hsiao Hua Heush Hsueh Pao/ Chemical Journal of Chinese Universities, 23, 225.

Victoria, G. M., Aranda, A. I., Moreno, A., Cossio, F. P., de Cozar, A., Diaz, O. A., De la, H. A., & Prieto, P. (2009). Tetrahedron, 65, 5328–5336.

Zhang, W. S., Xu, W. J., Zhang, F., & Qu, G. R. (2013). Chinese Chemical Letters, 24, 407–410.

CHAPTER 10

CONDENSATION

SANYOGITA SHARMA, ABHILASHA JAIN, and RAKSHIT AMETA

CONTENTS

Condensation is a class of organic reactions, where two molecules combine, usually in the presence of a catalyst, with elimination of water or some other simple molecule. The combination of two identical molecules is known as self-condensation. Aldehydes, ketones, esters, alkynes (acetylenes) and amines are among several organic compounds that combine with each other and except for amines, among themselves to form larger molecules, many of which are useful intermediate compounds in organic syntheses. Catalysts commonly used in condensation reactions include acids, bases, cyanide ion, and complex metal ions.

In general, in most of the reactions, traditional heat transfer equipments such as oil baths, sand baths and heating jackets are used for heating. These techniques are slow, more time consuming and have temperature gradients. Sometime overheating in reaction can lead to substrate, reagent and product decomposition. These local heating techniques required more fuels and power for generating heat. Industries apply these techniques for heat generation and the major requirements are fossils fuels, the use of which is very harmful for our ecosystem and environment. Therefore, there is an urgent need for some alternate heating technique, which can overcome these kinds of problems.

The microwave energy is introduced for completing this demand for more efficient and ecofriendly methods of performing reactions in short times and enhanced reactions rates. Since energy transfer occurs in less than a nanosecond (10^{-9} s), the molecules are unable to completely relax ($\sim 10^{-5}$ s) or reach equilibrium. This creates a state of nonequilibrium that results in a high instantaneous temperature (T_i) of the molecules and is a function of microwave power input. The instantaneous temperature is not directly measurable, but it is much greater than the measured temperature T_B ($T_i \gg T_B$). Thus, the greater is the intensity of microwave power being administered to a chemical reaction, the higher and more consistent will be T_i. Precedence exists, where the concept of instantaneous temperatures has been used to explain reactions occurring at a lower bulk temperature than expected, while using microwave irradiation (De Pomerai, et al., 2003). Microwave heating is playing an important role in treatment of domestic and hazardous industrial & nuclear waste also.

Microwave heating can be advantageously used for waste management in areas where human exposure can cause health problems. Different materials possess different properties, which influence the interaction with microwaves. These materials can be classified in three categories. (i) Electrical conductors, for example, metals, graphite, etc., that reflect microwaves. (ii) Good insulators, for example, quartz glass, porcelain, ceramics, etc. that let microwaves penetrate through it without absorption and (iii) Polar materials that can absorb microwaves (Bogdal and Prociak, 2007).

10.1   ALDOL CONDENSATION

In an aldol condensation, an enol or an enolate ion reacts with a carbonyl compound to form a β-hydroxyaldehyde or β-hydroxyketone, followed by dehydration to give a conjugated enone.

Aldol condensations are important in organic synthesis, as these provide a good approach to form carbon–carbon bonds. It involves the nucleophilic addition of a ketone enolate to an aldehyde to form a β-hydroxyketone, or aldol (aldehyde + alcohol), a structural unit found in many naturally occurring molecules and pharmaceuticals.

Marjani et al. (2009) have developed a preparative method for the synthesis of hydroxycyclopentenones in high yields by the cross-aldol reactions of benzil with various ketones. When the reactions were performed in various solvents (KOH/EtOH) under classical heating for a long time, even then they produced the products in relatively low yields.

where G = Br, H; R = H, CH$_3$, C$_6$H$_5$; R$_1$ = H, CH$_3$; R$_2$ = H, CH$_3$

The powdered KOH, benzil and ketone were ground together with a mortar and pestle. When irradiated in a microwave oven, the mixture burnt due to the tendency of KOH to absorb the microwave energy. On the other hand, a mixture of KOH in ethanol, benzil and ketone was irradiated by microwave. It was observed that ethanol evaporated and the reaction was not completed. Therefore, the mixture of benzil, ketones and a solution of KOH in ethanol was irradiated in a water bath inside the microwave oven.

The $HgSO_4$ catalyzed hydrative cyclization of 1,6-heptadiynes was reported by Zhang et al. (2011). This reaction proceeded smoothly under the mild conditions for differently 4-sustituted 1,6-diynic substrates giving corresponding 3-methyl-2-cyclohexenones with high to excellent yields. The microwave-assisted aldol condensation of cyclohexenones under the catalysis of $BiCl_3$ afforded 3-styryl-cyclohexenones.

(90% yield)   (91% yield)

10.2 BAYER CONDENSATION

A facile synthesis of an array of triarylmethanes by Bayer condensation of different arylaldehydes carrying activated and deactivated groups and N,N-dimethylaniline using $ZrOCl_2$ as a catalyst under solvent-free microwave irradiation condition was carried out by Reddy et al. (2009). The catalytic activity of $ZrOCl_2$ was compared with Lewis acid as a catalyst and it was concluded that $ZrOCl_2$ is an excellent catalyst for this synthetic route. This green protocol gives a good yield (70–96%).

10.3 CLAISEN CONDENSATION

It is a carbon–carbon bond forming reaction that occurs between two esters or one ester and another carbonyl compound in the presence of a strong base, resulting in a β-keto ester or a β-diketone (Carey, 2006).

Ester   β-keto ester   Alcohol

10.4 CLAISEN-SCHIMDT CONDENSATION

Mohan et al. (2010) have synthesized benzimidazole by reaction between anthranilic acid and o-phenylenediamine. The acetylated product of benzimidazole undergoes Claisen-Schmidt condensation with aryl aldehyde to produce corresponding chalcones.

Mathew et al. (2011) synthesized (2E)-1-(1H-benzimidazol-2-yl)-3-phenylprop-2-en-1-ones. o-Phenylenediamine reacts with lactic acid to give 2(α-hydroxyethyl) benzimidazole, which on oxidation in presence of potassium dichromate produced 2-benzimidazoles. The chalcones were prepared with Claisen-Schmidt condensation by reacting 2-acetylbenzimidazole with appropriate aldehydes in the presence of a base.

10.5 CYCLOCONDENSATION

A microwave-assisted method has been developed by Ighilahriz et al. (2008) for the synthesis of 4(3H)-quinazolinones by condensation of anthranilic acid, orthoesters (or formic acid) and substituted anilines, using Keggin-type heteropolyacids ($H_3PW_{12}O_{40} \cdot 13H_2O$, $H_4SiW_{12}O_{40} \cdot 13H_2O$, $H_4SiMo_{12}O_{40} \cdot 13H_2O$ or $H_3PMo_{12}O_{40} \cdot 13H_2O$) as catalysts. They found that the use of $H_3PW_{12}O_{40} \cdot 13H_2O$ acid coupled with microwave irradiation allows a solvent-free rapid (~13 min) and high-yielding reaction.

The following mechanism has been proposed for the reactions.

N-Methylaniline was reacted with two moles of diethyl malonate in a cyclo-condensation reaction with elimination of four moles of ethanol to give the product pyranoquinoline on a 0.2 molar scale (Razzaq and Kappe, 2007). On this scale, the distillation took only 82 min as compared to 3–5 h by conventional method. When the reaction was performed in a sealed-vessel microwave system, only <5% product formation was achieved.

A novel one pot synthesis of pyrano[2,3-d]pyrimidines via the cyclocondensation of barbituric acids, benzaldehyde, and aliphatic nitriles under microwave irradiation in the solid state has been reported (Devi et al., 2003). The synthesis of pyrido[2,3-d]pyrimidines has also been performed under identical conditions via condensation of 6-aminouracils or 6-hydroxyaminouracils, benzaldehyde, and aliphatic nitriles.

A one-pot synthesis of highly functionalized tetrahydrobenzo pyrans has also been reported by Devi and Bhuyan (2004) via sodium bromide catalyzed cyclocondensation of aryl aldehydes, aliphatic nitriles, and dimedone under solvent-free microwave irradiation conditions. It was observed that the activity of cyanoacetamide is poor as compared to malononitrile and ethyl cyanoacetate.

Yan et al. (2007) reported that polysubstituted annulated pyridines can be synthesized in high yields by four-component, one-pot cyclocondensation reactions of N-phenacyl pyridinium bromide, aromatic aldehydes, acetophenones or cyclic ketones in the presence of ammonium acetate and acetic acid, assisted by microwave irradiation. Cyclic ketones with two α-CH$_2$ groups yield annulated pyridines with additional α-benzylidene groups, which are derived in situ from double aldol condensation of cyclic ketones with two moles of aromatic aldehydes.

Gududuru et al. (2004) have reported a convenient microwave method for the synthesis of 4-thiazolidinones via the one pot condensation of a primary amine, an aldehyde, and mercaptoacetic acid. The reactions were conducted using an environmentally benign solvent ethanol in open vessels at atmospheric pressure in the presence of a base and molecular sieves. Amine components bearing a chiral center have also participated in the formation of diastereomeric products.

$$R_1-NH_2.HCl + R_2-CHO + HS\!-\!\!\overset{O}{\underset{}{\big\|}}\!\!-OH \xrightarrow[\text{MW}]{\text{EtOH}} R_1N\underset{R_2}{\overset{O}{\diagup}}S$$

A green and inexpensive preparation method for synthesis of 2-arylbenzothiazoles has been reported by Lee et al. (2012). A series of 2-arylbenzothiazoles was synthesized from the condensation reaction of aryl aldehyde with 2-aminothiophenol in the presence of L-proline under solvent-free and microwave irradiation conditions.

A series of reaction for the synthesis of 2-aryl-1-arylmethyl-1H-1, 3-benzimidazoles from ortho phenylenediamine and aromatic aldehydes in the presence of montmorillonite K-10 under microwave irradiation in the absence of solvent with good yields has been reported (Perumal et al., 2004).

$$\text{(diamine)} + 2\ ArCHO \xrightarrow[\text{MW}]{\text{K-10 Clay}} \text{(benzimidazole)}-Ar$$

Kalirajan et al. (2010) synthesized a series of pyrazole derivatives of benzimidazole. Condensation of substituted o-phenylenediamine with lactic acid under microwave irradiation and further oxidation of the product with potassium dichromate gave intermediate 2-acetyl benzimidazole, which was then reacted with aromatic aldehydes. Finally the product was cyclized with hydrazine to form final product.

Pasha and Nizam (2012) reported highly selective synthesis of 2-aryl-benzimidazoles by the condensation of o-phenylenediamine with various acraldehyde. This greener reaction was catalyzed by Amberlite IR- 120 and proceeds efficiently in the absence of any organic solvent under microwave irradiation within 3–8 min.

2-Aryl-benzimidazoles were also synthesized by the condensation of o-phen-ylenediamine with various arylaldehydes (Forouzani and Bosra, 2012). These eco-friendly protocols were catalyzed by 1,3-dibromo-5,5-dimethylhydantoin (DBH), and proceeds efficiently in the absence of any organic solvent under thermal condition, but with microwave irradiation in high yields.

Santos et al. (2006) reported an alternative method for the synthesis of phthalimide derivatives by exploiting the condensation of phthalic anhydride with amino groups under microwave radiation. The results showed that phthalimide derivatives were obtained in shorter reaction times (5–10 min) and higher yields (60–89%) than with conventional heating.

Microwave-mediated one-step condensation of malonamide with o-diamino-aromatics under solvent-free conditions generated bis(heteroaryl-2-yl)methanes in good yields without using any catalyst (Duan et al., 2006).

Rana et al. (2011) synthesized {6-methyl-4- (substituted phenyl)-2-thioxo-1, 2, 3, 4-tetrahydropyrimidin-5-carboxylic acid ethyl ester by condensation of acetoace-tic ester, thiourea and substituted aromatic aldehydes using piperidine as catalyst with stirring for 4 h. While these were also obtained under microwave heating in five min using ethanol as a solvent and HCl as a catalyst.

A simple, solventless and efficient method for the synthesis of dihydropyrimidinone derivatives was developed through the rapid condensation of various aromatic aldehydes, ketones and urea using efficient and readily obtainable MnO_2-CNT nanocomposites as catalyst under microwave irradiation (Safari and Gandomi-Ravandi, 2013). This simple procedure is efficient and can be applied to the synthesis of a wide variety of dihydropyrimidinones in good to excellent yields.

Saxena et al. (2005) have reported a quick method for the condensation reaction of an aldehyde, ethyl acetoacetate, and urea or thiourea to synthesize substituted 3,4-dihydropyrimidin-2(1H)-ones using iodine-alumina as the catalyst under microwave irradiation and solvent-free conditions. This method required 1 min with a variety of aromatic, substituted aromatic, and heterocyclic aldehydes; however, no aliphatic aldehyde has been used.

10.6   FRIEDLANDER CONDENSATION

The Friedlander synthesis is a chemical reaction of 2-aminobenzaldehydes with ketones to form quinoline derivatives. This reaction has been catalyzed by trifluoroacetic acid (Shaabani et al., 2007) toluenesulfonic acid (Jia et al., 2006), iodine (Wu et al., 2006) and Lewis acids (Varala et al., 2006).

Agrawal and Joshipura (2005) carried out an environmental friendly Friedlander reaction catalyzed by acid as well as without catalyst under solvent-free microwave irradiation with increase in yields and decrease in the reaction time period. 2-Aminobenzophenones, ketones and 2–3 drops of conc. H_2SO_4 were mixed and irradiated in a domestic microwave oven at an output of 70 W for 3.5 to 13 min to obtain 4-arylquinolines in excellent yields (70–99%).

However, when a mixture of 2-aminobenzophenones and ethyl acetoacetate was irradiated at higher output (700 W) for 5–7 min without a catalyst, 3-acetyl-4-aryl-2-oxoquinolines were obtained in excellent yields (80–91%).

10.7 ISAY CONDENSATION

In this type of reaction, certain diaminopyrimidines were transformed into pterins by condensation with a 1,2-dicarbonyl compound, such as 2,3-butanedione.

Goswami and Adak (2002) have developed a new method for the synthesis of 6-substituted pterins, pterin sugar derivatives and 2-substituted quinoxalines under microwave conditions. In addition to its simple reaction conditions, this procedure has the advantages of very short reaction times, simple experimental and work-up procedures and most importantly, its regiospecificity for the C-6 position of the pterin, which makes it useful for the synthesis of pterin and also quinoxaline heterocycles.

10.8   KNOEVENAGEL CONDENSATION

Knoevenagel condensation is a classic C-C bond formation reaction. It occurs between aldehydes or ketones and active methylene compounds with ammonia or another amine as a catalyst in organic solvents. The Knoevenagel reaction is considered to be a modification of the aldol reaction. The main difference between these approaches is the higher acidity of the active methylene hydrogen as compared to a carbonyl hydrogen.

This type of condensation is applicable to the synthesis of unsaturated acids, which are used as precursors for perfumes, flavonoids and as building blocks of many heterocycles. Gupta and Wakhloo (2007) have studied Knoevenagel condensation between carbonyl compounds and active methylene compounds, such as malonic acid, using tetrabutylammonium bromide (TBAB) and potassium carbonate in water forming unsaturated acids in excellent yield and purity under microwave irradiation.

Bhuiyan et al. (2012) reported that arylidene-malononitrile can be prepared by Knoevenagel condensation reaction of malononitrile with corresponding aromatic aldehydes in presence of ammonium acetate (NH_4OAc), using microwave irradiation under solvent-free condition. The reaction proceeds in a clean manner with shorter reaction time, mild reaction condition, ecofriendly and with excellent yields as compared to conventional methods. A variety of functional groups such as nitro, chloro, amino and ether survived under the reaction conditions.

where Ar = 3-Nitrophenyl, 4-Hydroxyphenyl, 2-Methylphenyl, 3-Methylphenyl, 3-Methoxyphenyl, 4-Methoxyphenyl, 4-N,N'-Dimethylphenyl, 4-Hydroxy-3-me-thoxyphenyl 2,4-Dimethoxyphenyl, 1-Naphthyl, 9-Anthracyl, 2-(4-N,N'-Dimethyl) ethenyl.

Aminopropylated functionalized hexagonal mesoporous silicas (HMS) and SBA-15 materials with different amino-loadings (5–30 wt. % NH$_2$) were synthesized by Pineda et al. (2013). These play important role as catalyst in the microwave-assisted Knoevenagel condensation of cyclohexanone and ethyl cyanoacetate as well as in the Michael reaction between 2-cyclohexen-1-one and nitromethane. The low loaded HMS-5%NH$_2$ and higher loaded SBA-15–20% NH$_2$ were found to give the best activities in the reactions. High activities and selectivities to the condensation product could be achieved in short times of microwave irradiation for both these base-catalyzed processes.

Gracia et al. (2009) found the Na-SBA-1 is most active material in the Knoeve-nagel condensation of benzaldehyde and benzylacetone with ethyl cyanoacetate, both under conventional heating as well as microwave irradiation conditions.

Ethyl 2-cyano-3-methyl-5-phenyl-2-pentenoate

Hydrolysis and decarboxylation

Citronitrile

3-Cinnamoyl-4-hydroxy-6-methyl-2-pyrone was synthesized via Knoevenagel condensation of dehydroacetic acid with benzaldehyde derivatives (Baziz et al., 2008). The reaction rate was dramatically enhanced by a specific microwave effect, when the reaction was performed under solvent-free conditions, in a microwave re-actor. Excellent isolated yields (upto 90%) were obtained within short reaction times (typically 2 to 10 min). They concluded that Knoevenagel condensation assisted by solvent-free microwave procedure was better than conventional procedure in terms of energy saving, rapidity, yield and cleanliness.

(5Z)-2-Alkylthio-5-arylmethylene-1-methyl-1,5-dihydro-4H-imidazol-4-ones have been synthesized by Bourahla et al. (2011). The first step involves a solvent-free Knoevenagel condensation under microwave irradiation that produced (5Z)-5-arylmethylene-2-thioimidazolidin-4-ones in good yields and high purity, which was followed by a chemoselective S-alkylation with retention of double bond con-figuration.

Biradar and Sasidhar (2011) reported a new series of novel indole analogs via Knoevenagel condensation and evaluated their in vitro antioxidant and cytotoxin activities using three tumor cell lines. 2,5-Disubstituted indole-3-carboxaldehydes undergo NH_4OAc catalyzed Knoevenagel condensation with barbiturates, thiazolidine-2,4-dione and 3-methyl-1H-pyrazol-5(4H)-one in solvent-free microwave-assisted reaction.

Shindalkar et al. (2006) developed an ecofriendly and rapid methodology for Knoevenagel condensation of 4-oxo-(4H)-1-benzopyran-3-carbaldehyde with Meldrum's acid using acidic alumina under solvent-free microwave irradiation in excellent yield (90–95%).

Villemin et al. (1998) reported that 3(2)-naphthofuranone was condensed efficiently and rapidly in the presence of Al_2O_3-KF with aromatic aldehydes into Z-arylidene naphthofuranones without solvent under focused microwave irradiation. The method is very simple, safe and convenient.

Knoevenagel condensation of aromatic aldehydes with active methylene compounds under solvent-free, microwave irradiation conditions gives arylidene derivatives (Mogilaiah, et al., 2010a). This convenient process was catalyzed by NH-$_2$SO$_3$NH$_4$ and the yield is excellent and purity is high.

$$Ar-CHO + \overset{CN}{\underset{R}{\diagdown}} \xrightarrow[MW]{NH_2SO_3NH_4} \overset{Ar}{\underset{H}{\diagup}} = \overset{CN}{\underset{R}{\diagdown}}$$

10.9 PECHMANN CONDENSATION

Sinhamahapatra et al. (2011) reported that the microwave-assisted synthesis is the most appropriate method for synthesis of coumarins as it provides improved yield in very less time. Mesoporous zirconium phosphate (m-ZrP) is used as a solid acid catalyst for the synthesis of coumarins via Pechmann condensation reaction. Among the substituted phenols, m-amino phenol is more reactive and 100% yield was obtained in very short time at low temperature due to the presence of ring activating amine group in meta position. The m-ZrP is also active towards phenols for the synthesis of 4-methyl coumarin, where only 57% yield was obtained in conventional heating method.

10.10 UGI CONDENSATION REACTION

The Ugi reaction is a multicomponent reaction in organic chemistry involving a ketone or aldehyde, an amine, isocyanide and carboxylic acid to form a bis-amide.

$$R_1-NH_2 + R_5-N\overset{+}{\equiv}C^- + \overset{O}{\underset{R_2}{\overset{\|}{C}}}R_3 + \overset{O}{\underset{R_4}{\overset{\|}{C}}}OH$$

The Ugi reaction is exothermic in nature and usually completed within minutes of adding the isocyanide. High concentration (0.5 M–2.0 M) of reactants gave the highest yields. Polar aprotic solvents, like DMF work well; however, methanol and ethanol have also been used successfully. This uncatalyzed reaction has an inherent high atom economy as only a molecule of water is lost and chemical yield are high in general.

Three-component Ugi condensation was reported by Ireland et al. (2003). The condensation of 2-aminopyridine, an aldehyde, and isocyanide successfully leads to fused 3-aminoimidazoles in 10 min under microwave irradiation.

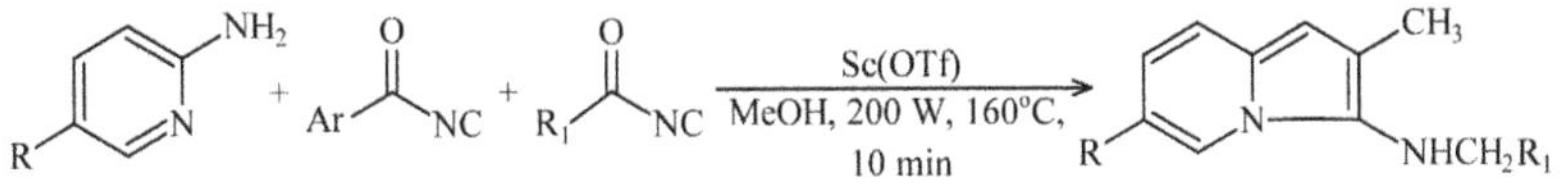

where R = CH$_3$, Br; R$_1$ = Ph, CO$_2$Et; Ar = 2 Pyridinyl, 2-Naphthyl.

Zhang and Tempest (2004) reported that the construction of heterocyclic compounds has been improved through the incorporation of microwave and fluorous technologies.

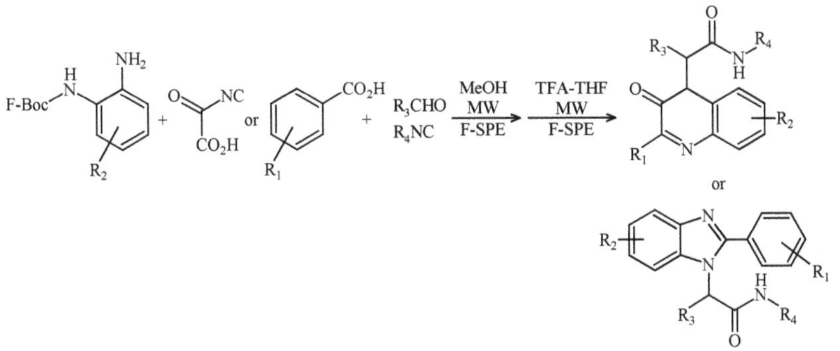

In the synthesis of substituted quinoxalinones and benzimidazoles, a fluorous-Boc protected diamine has been used for the Ugi reactions.

1. *Quinoxalinone synthesis*

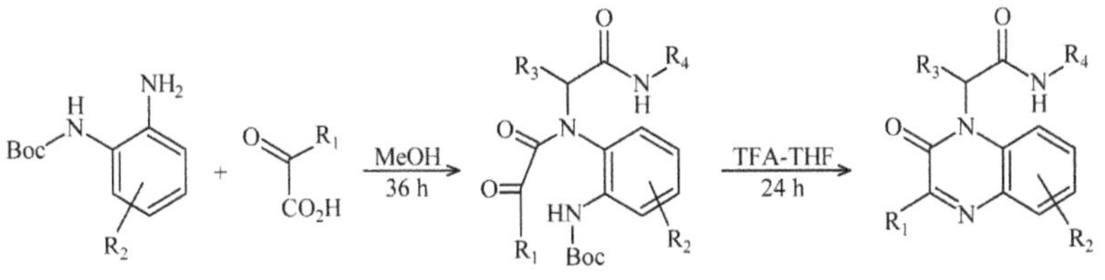

2. *Benzimidazole synthesis*

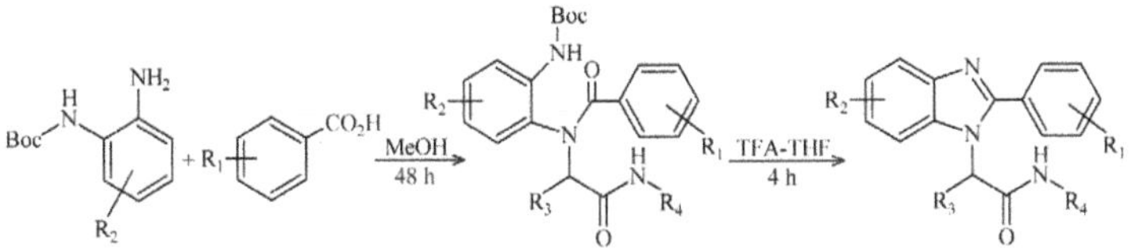

Both the Ugi and the postcondensation reactions proceed rapidly under microwave irradiation and the reaction mixtures were purified by solid-phase extraction.

A rapid and efficient optimization of a microwave-assisted procedure for the Ugi three- component condensation of levulinic acid with an amine and an isonitrile to afford lactam derivatives was developed (Tye and Whittaker, 2004). This reaction is completed in shorter reaction times (30 min) compared to the conventional procedure (48 h) using methanol as a solvent; however, ethanol was used as a solvent in the case of ethyl isocyanoacetate substrate to avoid problems with transesterification.

10.11 MISCELLANEOUS

Shih et al. (2007) reported that 3-aryl-4-formylsydnones react with symmetrical 1,2-dicarbonyl compounds, such as benzil, 4,4'-dimethoxybenzil, 4,4'-difluorobenzil and di-2-thienylethanedione in glacial acetic acid, using ammonium acetate as the ammonia source yielding 4,5-diaryl-2-sydnonyl-substituted imidazoles under conventional heating. In a similar treatment, 4,5-diaryl-2-sydnonyl-1-substituted imidazoles can be prepared by the one-pot condensation of 3-(4-ethoxyphenyl)-4-formylsydnone, benzil derivatives, ammonium acetate, and primary amines. However, such reactions, which take 1–3 days at high temperature by conventional method, are completed successfully within a few minutes under microwave irradiation.

An efficient one-pot procedure for the synthesis of ionic liquids based on nitrogen-containing heterocycles, imidazolium or pyridinium salts by following the conditions of green process has been developed (Aupoix et al., 2010). Imidazolium salts and DBU have been found to catalyze efficiently the benzoin condensation giving good yields within very short reaction time using solvent-free microwave activation conditions.

2-Amino-2-chromenes are the main constituents of many natural products like pigments, cosmetics and potential agro-chemicals (Mekheimer and Sadek, 2009). These are generally prepared either under conventional heating, benign reagents or very recently, a three-component condensation in polyethylene glycol-water in the presence of nanosized magnesium oxide. Most of these methods require prolonged reaction time, reagents in stoichiometric amounts and generate moderate yields of products. When a mixture of benzaldehyde, malononitrile and α-naphthol in 10 mL of ethanol and piperidine as a catalyst was refluxed under microwave irradiation for 5–8 min at 80 °C, 2-amino-2-chromene derivatives were obtained in almost quantitative yield (85–94%). All reactions proceeded smoothly to afford the corresponding 2-amino-2-chromene derivatives in high yield, but of course with a little bit decrease, when substituent is an electron-withdrawing group.

Shaker et al. (2005) developed a three-component condensation of ninhydrin, malononitrile, and some nucleophilic reagents in the presence of piperidine under microwave irradiation without solvent, which afforded the corresponding spiro-fused pyran derivatives.

A microwave-assisted, one-pot synthesis of dihydropyridopyrimidine derivatives has been reported by Tu et al. (2005a) via condensation of 2,6-diaminopyrimidin-4-one, aldehydes, and acyclic 1,3-dicarbonyl compounds in glycol without the use of a catalyst. The main advantages were shorter reaction times, environmentally benign methodology and it can also be applied to aliphatic as well as aromatic aldehydes.

The same group has also disclosed an efficient synthesis of a series of pyrimi-doquinoline derivatives through a one pot condensation of 2,6-diaminopyrimidin-4-one, an aldehyde, and a cyclic 1,3-dicarbonyl compound in glycol under microwave irradiation without a catalyst (Tu et al., 2005b).

Labrini et al. (2005) have reported a rapid synthesis of 2,5-disubstituted-1,3,4-thiadiazoles via the one pot condensation reaction of aromatic aldehydes, hydrazine, and sulfur in ethanol under microwave irradiation. Microwave-assisted synthesis of 2-aminothiophene-3-carboxylic derivatives (Gewald synthesis) has been reported by Hu et al. (2006) under microwave conditions using a functional ionic liquid as a soluble support.

Microwave-assisted synthesis of 4(3H)-quinazolinone derivatives has been developed by Dabiri et al. (2004) via a one pot condensation reaction of isatoic anhydride, primary amines, and ortho esters in the presence of catalytic amounts of p-toluenesulfonic acid.

A one pot synthesis of series of some γ-spiroiminolactone derivatives has been reported by Azizian et al. (2004) via a microwave-assisted condensation reaction of isocyanides, dialkyl acetylenedicarboxylates and indenoquinoxilin-11-ones in DMF or in a solvent less operation on montmorillonite K10 support.

A fast and facile microwave enhanced N-alkenyl condensation between lactams and various aldehydes such as n-propanal, isopropanal, n-butanal, n-hexanal, n-oc-tanal and phenylacetaldehyde in moderate (55%) to high yields (86%) was reported (Kim et al., 2006).

where R = H, CO$_2$Et; R$_1$ = CH$_3$, CH$_2$CH$_3$, (CH$_2$)$_2$CH$_3$, (CH$_2$)$_5$CH$_3$, Ph; R$_2$ = H or CH$_3$ or R = H; R$_1$ = H, CH$_2$CH$_3$; R$_2$ = H.

Schweitzer and Inazu (2001) have reported the condensation of enolate of tert-butyl bromoacetate with 1,5-lactones giving directly exocyclic epoxides, which were subsequently transformed into desired compounds. Cationic reduction of epoxides provided C-glycosyl compounds. Upon esterification of the latter as trifluormethanesulfonate, reaction with primary amines furnished the corresponding C-glycosylamine esters.

Mladenovic et al. (2009) reported that 4-hydroxy-chromene-2-one derivatives were easily prepared through condensation reactions with microwave heating. All tested compounds are, in principle, very good potential microorganism growth inhibitors, but their structural diversity resulted in great differences in their inhibitory potential.

Kidwai and Rastogi (2005) have reported a solvent-free synthetic route for the synthesis of tetrahydroacridinones through microwave-assisted condensation reaction of an aldehyde, dimedone and a primary aromatic amine. The best microwave method was found to be the "neat reaction" technology without the use of any solvent or inorganic solid supports. The reactions were completed within 5 min with 82–87% yields, when compared to the conventional solution phase method with longer reaction times (hours) and comparatively lower yields.

Moligaiah et al. (2010b) carried out the synthesis of 1-[3-(3-fluorophenyl) [1,8] napthyridn-2-yl-]-3-(2-oxo-2H-chromenyl)-1H-4-pyrazolecarbaldehydes   from 3-[2-(3-(3-fluorophenyl) [1,8]- napthyridin-2yl)ethanhydrazonyl]-2H-2-chromenones (hydrazones).

Under   similar   conditions,   3-(2-oxo-2H-chromenyl)-1-[3-(4-methoxyphe-nyl) [1,8] napthyridin-2-yl]-1H-4-pyrazolecarbaldehyde was also been synthesized   by   condensation   of   3-[2-(3-(4-methoxyphenyl)[1,8]-napthyridin-2-yl) ethanhydrazonyl]-2H-2-chromenones in the presence of Vilsmeer-Haack reagent (DMF-POCl₃) using silica gel as solid support under microwave irradiation (Mogligaiah et al., 2009).

Elimination of a smaller molecule from two or more molecules gives condensed product. These products may be used in the synthesis of larger molecules, macromolecules or even polymerized products. Microwave irradiation helps in the formation of these products with a variety of properties in limited time and with excellent yields.

KEYWORDS

- **Aldol Condensation**
- **Claisen Condensation**
- **Claisen-Schmidt Condensation**
- **Cyclocondensation**
- **Friedlander Condensation**
- **Isay Condensation**
- **Knoevenagel Condensation**
- **Pechmann Reaction**
- **Ugi Condensation**

REFERENCES

Agrawal, Y. K., & Joshipura, H. M. (2005). Indian Journal of Chemistry, 44B, 1649–1652.

Aupoix, A., Pe´got, B., & Vo-Thanh, G. (2010). Tetrahedron, 66, 1352–1356.

Azizian, J., Karimi, A. R., Mohammadi, A. A., & Mohammadizadeh, M. R. (2004). Heterocycles, 63, 2225–2229.

Baziz, N. A., Rachedi, Y., Chemat F., & Hamdi, M. (2008). Asian Journal of Chemistry, 20, 2610–2622.

Bhuiyan, M. M. H., Hossain, M. I. M., Alam, A., & Mahmud, M.M. (2012). Chemistry Journal, 2, 30–36.

Biradar, J. S., & Sasidhar, B. S. (2011). European Journal of Medicinal Chemistry, 46, 6112–6118.

Bogdal, D., & Prociak, A. (First Ed.) (2007). Microwave-Enhanced Polymer Chemistry and Technology. New Jersey: Blackwell.

Bourahla, K., Paquin, L., Lozach, O., Meijer, L., Carreaux, F., & Bazureau, J. P. (2011). Molecules, 16, 7377–7390.

Carey, F. A., (6th Ed.) (2006). Organic Chemistry, New York: McGraw-Hill.

Dabiri, M., Salehi, P., Khajavi, M. S., & Mohammadi, A. A. (2004). Heterocycles, 63, 1417–1421.

De Pomerai, D. I., Smith, B., Dawe, A., & North, K. (2003). FEBS Letters, 543, 93–97.

Devi, I., & Bhuyan, P. J. (2004). Tetrahedron Letters, 45, 8625–8627.

Devi, I., Kumar, B. S. D., & Bhuyan, P. J. (2003). Tetrahedron Letters, 44, 8307–8310.

Duan, G. Y., Sun, Y.-W., Liu, J.-Z., Zhao, G.-L., Zhang, D.-L., & Wang, J.-W. (2006). Journal of the Chinese Chemical Society, 53, 455–458.

Forouzani, M., & Bosra, H. G. (2012). E-Journal of Chemistry, 9, 1064–1069.

Goswami, S., & Adak, A. K. (2002). Tetrahedron Letters, 43, 8371–8373.

Gracia, M. D., Jurado, M. J., Luque, R., Campelo, J. M., Luna, D., Marinas, J. M., & Romero, A. A. (2009). Microporous and Mesoporous Materials, 118, 87–92.

Gududuru, V., Nguyen, V., Dalton, J. T., & Miller, D. D. (2004). Synlett, 13, 2357–2358.

Gupta, M., & Wakhloo, B. P. (2007). Arkivoc, (i), 94–98.

Hu, Y., Wei, P., Huang, H., Han, S-Q., & Ouyang, P-K. (2006). Heterocycles, 68, 375–380.

Ighilahriz, K., Boutemeur, B., Chami, F., Rabia, C., Hamdi, M., & Hamdi, S. M. (2008). Molecules, 13, 779–789.

Ireland, S. M., Tye, H., & Whittaker, M. (2003). Tetrahedron Letters, 44, 4369–4371.

Jia, C.-S., Zhang, Z., Tu, S.-J., & Wang, G.-W. (2006). Organic and Biomolecular Chemistry, 4, 104–110.

Kalirajan, R., Rathore, L., Jubie, S., Gowramma, B., Gomathy, S., Sankar, S., & Elango, K. (2010). Indian Journal of Pharma, 44, 358–359.

Kidwai, M., & Rastogi, S. (2005). Heteroatom Chemistry, 16, 138–141.

Kim, K. W., Ahn, H. S., Lee, H. J., Song, S. J., Kim, C. G., & Kwon, T. W. (2006). Bulletin of Korean Chemical Society, 27, 286–290.

Lebrini, M., Bentiss, F., & Lagrenee, M. (2005). Journal of Heterocyclic Chemistry, 42, 991–994.

Lee, A. S.-Y., Chung, C.-H., Chang, Y.-T., & Chen, P.-L. (2012). Journal of Applied Sciences & Engineering, 15, 311–315.

Marjani, K., Asgari, M., Ashouri, A., Mahdavinia, G. H., & Ahangar, H. A. (2009). Chinese Chemical Letters, 20, 401–403.

Mathew, B., Unnikirishnan, G., Shafeer, V.P., Musthafa, M. C., Femina, P. (2011). Der Pharma Chemica, 3, 627–631.

Mekheimer, R. A., & Sadek, K. U. (2009). Chinese Chemical Letters, 20, 271–274.

Mladenović, M., Vuković, N., Nićiforović, N., Sukdolak, S., & Solujić, S. (2009). Molecules, 14, 1495–1512.

Mohan, S. B., Behera, T. P., Kumar, R. B. V. V. (2010). International Journal of ChemTech Research, 2, 1634–1637.

Mogilaiah, K., Babu, H. S., Vidya, K., & Kumar, K. S. (2010a). Indian Journal of Chemistry, 49B, 390–393.

Mogilaiah, K., Jagadeeshwar, K., & Prasad, R. S., (2009). Indian Journal of Chemistry 48B, 1466–1469.

Mogilaiah, K., Prasad, R. S., & Kumar, K. S. (2010b). Indian Journal of Chemistry, 49B, 1417–1421.

Pasha, M. A., & Nizam, A. (2012). Journal of Saudi Chemical Society, 16, 237–240.

Perumal, S., Mariappan, S., & Selvaraj, S. (2004). Arkivoc, (viii), 46–51.

Pineda, A., Balu, A. M., Campelo, J. M., Romero, A. A., & Luque, R. (2013). Catalysis Communications, 33, 1–6.

Rana, K., Kaur, B., Chaudhary, G., Kumar, S., & Goyal, S. (2011). International Journal of Pharmaceutical Sciences and Drug Research, 3, 226–229.

Razzaq, T., & Kappe, C. O. (2007). Tetrahedron Letters, 48, 2513–2517.

Reddy, C. S., Nagaraj, A., Srinivas, A., & Reddy, G. P. (2009). Indian Journal of Chemistry 48B, 248–254.

Safari, J., & Gandomi-Ravandi, S. (2013). Journal of Molecular Catalysis A: Chemical, 373, 72–77.

Santos, J. L., Lima, L. M., & Chung, M. C., (2006). Journal of Basic and Applied Pharmaceutical Sciences, 27, 163–167.

Saxena, I., Borah, D. C., & Sarma, J. C. (2005). Tetrahedron Letters, 46, 1159–1160.

Schweizer, F., & Inazu, T. (2001). Organic Letters, 3, 4115–4118.

Shaabani, A., Soleimani, E., & Badri, Z. (2007). Synthetic Communications, 37, 629–635.

Shaker, R. M., Mahmoud, A. F., & Abdel-Latif, F. F. (2005). Journal of the Chinese Chemical Society, 52, 563–567.

Shih, M.-H., Tsai, C.-H., Wang, Y.-C., Shieh, M.-Y., Lin, G.-L., & Wei, C.-Y. (2007). Tetrahedron, 63, 2990–2999.

Shindalkar, S. S., Madje, B. R., & Shingare, M. S. (2006). Indian Journal of Chemistry, 45B, 2571–2573.

Sinhamahapatra, A., Sutradhar, N., Pahari, S., Bajaj, H. C., & Panda, A. B. (2011). Applied Catalysis A: General, 394, 93–100.

Tu, S., Fang, F., Li, T., Zhu, S., & Zhang, X. (2005b). Journal of Heterocyclic Chemistry, 42, 707–710.

Tu, S., Zhang, J., Xiang, Z., Fang, F., & Li, T. (2005a). Arkivoc, (xiv), 76–81.

Tye, H., & Whittaker, M. (2004). Organic and Biomolecular Chemistry, 2, 813–815.

Varala, R., Enugala, R., & Adapa, S. R. (2006). Synthesis, 3825–3830.

Villemin, D., Martin, B., & Bar, N. (1998). Molecules, 3, 88–93.

Wu, J., Xia, H.-G., & Gao, K. (2006). Organic and Biomolecular Chemistry, 4, 126–129.

Yan, C-G., Ca, X-M., Wang, Q-F., Wang, T. Y., & Zheng, M. (2007). Organic and Biomolecular Chemistry, 5, 945–951.

Zhang, W., & Tempest, P. (2004). Tetrahedron Letters, 45, 6757–6760.

Zhanga, C., Wanga, B-S., Chena, S-F., Zhanga, S-Q., & Cui, D-M. (2011). Journal of Organometallic Chemistry, 696, 165–169.

CHAPTER 11

REARRANGEMENT

PARAS TAK, NIRMALA JANGID and PINKI B. PUNJABI

CONTENTS

11.1 INTRODUCTION

Rearrangement is a class of organic reactions, where the carbon skeleton of a molecule is rearranged to give a structural isomer of the original molecule. Often a substituent moves from one atom to another atom in the same molecule, that is, the mechanism of rearrangement is mainly intramolecular, although intermolecular rearrangements also take place. A rearrangement is not well represented by simple and discrete electron transfers. In pericyclic reactions, explanation by orbital interactions gives a better picture than simple discrete electron transfers. Three key rearrangement reactions are 1,2-rearrangements, pericyclic reactions and olefin metathesis.

Several rearrangement reactions have been reported using microwave irradiation. Some reactions are performed in solution phase while many others are performed on graphite or mineral support surfaces often doped with Lewis acids. Notable examples are benzil–benzilic acid rearrangement (Yu et al., 1999), Beckmann rearrangement on K10 clay (Bosch et al., 1995), Fries rearrangement (Kad et al., 1999), thia-Fries rearrangement of arylsulfonates using aluminum and zinc chloride on silica gel (Moghaddam and Dakamin, 2000), etc.

11.2 ARBUZOV REARRANGEMENT

Microwave-assisted Arbuzov rearrangement under solvent-free condition was found to be an efficient method for the preparation of dialkyl alkylphosphonates of alkyl halides (Kaboudin and Balakrishna, 2001). This method is an easy, rapid, and high-yielding reaction for the Arbuzov rearrangement.

$$R_1-Cl \ + \ P(OR_2)_3 \xrightarrow[\text{MW, 5-10 min}]{\text{Al}_2\text{O}_3 \text{ (n)}} R_1-\overset{\overset{\displaystyle O}{\|}}{P}(OR_2)_2 + R_2-Cl$$

The yield of 90% was obtained, when R_1 = Methyl or Ethyl and R_2 = PhCOCH$_2$-CH$_2$- in 5 min, but it was reduced to 80% in 10 min, when R_2 = PhCH$_2$-.

Benzyl chloride, carbon tetrachloride, allyl chloride, and also 3-chloro-1-phenyl-1-propanone with triethyl phosphate and trimethyl phosphite in the presence of neutral alumina afforded the desired products in excellent yields with microwave irradiation. The t-butyl chloride failed in this condition and no Arbuzov rearrangement product was detected. The other types of alumina (acidic and basic) were not as effective as neutral alumina and usually give relatively low yields of the corresponding products. Simple work-up, low consumption of solvent, fast reaction rates, mild reaction condition, and good yields of the reaction make this method an attractive and useful contribution to the present methodologies.

11.3   AZA-CLAISEN REARRANGEMENT

Generally, the replacement of the oxygen offers two advantages. The vinyl double bond of the sigmatropic framework can be built up with a high E selectivity, since a bulky C1 substituent and the chain branched nitrogen will adopt a maximal distance around the enamine moiety. Furthermore, only two valences of the nitrogen are occupied by allyl and vinyl substituents of the rearrangement system. The major problem of the aza-Claisen rearrangement is the extremely high temperature, excluding the presence of a variety of functional groups, upon running the reaction. Hill and Gilman (1967) reported that uncatalyzed simple allyl vinyl amines undergo the 3,3-sigmatropic conversion at about 250°C, while the somewhat more activated aromatic analogs require 200–210°C.

A significant acceleration of aza-Claisen rearrangements was observed using microwave irradiation (Takano et al., 1984). This eliminated problems like requirement of high temperatures, extended reaction times, and decomposition of the starting materials and products also. The Fischer indole synthesis represents a special type of aza-Claisen rearrangement incorporating two N atoms in the 3- and 4-positions of the rearrangement system.

Microwave-assisted aza-Claisen rearrangements of N-allylanilines proceeded in very short reaction times in the presence of Zn^{2+}-montmorillonite as a catalyst (Nubbemeyer, 2005).

Bermner and Organ (2008) reported [3,3]-aza-Claisen rearrangement under microwave irradiation to obtain good yield of pyrroles within 30 min as compared to one pot, domino aldehyde/amine condensation and imine-allene cyclization while Aza-Claisen rearrangement of the unexplored N-thiophenoxyacetyl-α-vinyl piperidine substrate and the oxone-induced transannulation via microwave accelerated, chiral transfer using the achiral phenylsulfide auxiliary has been described by Sim et al. (2012).

Aza-Claisen rearrangement of allylic imidates and thiocyanates in presence of o-xylene under microwave irradiation resulted in corresponding amides and isothiocyanates (Gonda et al., 2007). The reaction rate was enhanced 8–30 and 24–80 times, respectively.

where X = F, Cl

11.4 BAKER-VENKATARAMAN REARRANGEMENT

Baker-Venkataraman rearrangement of 2'6'-diaroyloxyacetophenones to 3-aroyl-5-hydroxyflavones under microwave irradiation was observed by Pinto et al. (2007). It was over in very short reaction time. These reactions afforded 5-hydroxyflavones as in classical heating conditions.

Abdel Ghani et al. (2008) used a modified Baker-Venkataraman rearrangement to synthesize 11 types of flavonoid derivatives with subsequent microwave-assisted closure of the heterocyclic ring while Kumar and Makrandi (2012) reported the Baker-Venkataraman rearrangement of 2-hydroxyacetophenone in aqueous medium using aroyl chlorides/cinnamoyl chloride, potassium carbonate and under microwave irradiation in 30–40 sec/40–45 sec in one step. The intermediate 2-aoryloxyacetophenone/2-cinnamoyloxyacetophenone was not isolated and hydroxydibenzoylmethanes/2-hydroxybenzoyleinnamylmethanes directly. It is a green synthesis because it avoids the need of organic solvent at any stage of the reaction.

1-(2-Hydroxyphenyl)-5-phenylpent-4-ene-1, 3-diones was prepared by Baker-Venkataraman rearrangement under microwave irradiation (Goel, 2012). This rearrangement was carried out by heating 2-cinnamoyloxyacetophenones with pulverized potassium hydroxide in pyridine medium in the presence of triton-B adsorbed on fly ash. It was completed in 30–60 sec with 93–97% yield under microwave exposure while it gives 65–75% yield in 60–75 min.

Agrawal et al. (2013) observed Baker-Venkataraman rearrangement of 1-(2-hydroxy-phenyl)-propan-1-one (derivatives of o-hydorxypropiophenone) with benzoyl chloride under microwave radiation in a microwave synthesizer at 40% in 40–50 sec. 3-Methyl-2-phenyl-chromen-4-one has been synthesized by this rearrangement.

11.5   BECKMANN REARRANGEMENT

Phosphoric acid is an efficient and mild reagent for conversion of ketoximes to amide or lactam by Beckmann rearrangement (Banerjee and Mitra, 2005). The reactions are carried out in solvent-free condition under microwave irradiation. It gives good to excellent yields (73–82%).

A new and efficient microwave induced bismuth trichloride catalyzed Beckmann rearrangement of oximes in the solid state has been achieved by Thakur et al. (2007). Beckmann rearrangement of aryl ketoximes catalyzed by In(OTf)$_3$ gave amides in ionic liquid under microwave irradiation. These aryl ketoximes were converted to corresponding amides in good yields within very short times (10–270 sec). The catalyst and the ionic liquid were easily recovered and reused (Sugamoto et al., 2011).

Microwave radiation promoted liquid-phase Beckmann rearrangement of cyclohexanone oxime was also carried out by Cheng et al. (2011) for preparing ε-caprolactam over P_2O_5 catalyst.

The effects of the heating method, catalyst (H_3PO_4, $ZnCl_2$ and P_2O_5), reactor, power & time of microwave radiation, mass capacity of P_2O_5 solvent (cyclohexanone, water and N,N-dimethylformamide) on cyclohexanone oxime rearrangement reaction were also investigated. The yield of 95.75% was obtained, when the reaction was carried out under the following conditions : N,N-dimethyl formamide (10 mL), P_2O_5 (0.19 g), cyclohexanone oxime (1.153 g) microwave radiation power (280 W) and 5.0 min irradiation. This method has many advantages like simple operation, rapid reaction, high yield and potential industrial application prospects.

Silica supported dichlorophosphate has been found to be an efficient, recoverable and reusable catalyst for Beckmann rearrangement of a variety of ketoximes and dehydration of various aldoximes in tetrahydrofuran under microwave irradiation (Li and Lu, 2008). This protocol is also advantageous because of high conversion, high selectivity, short reaction time, no environmental pollution and simple work-up procedure.

Microwave-assisted Beckmann rearrangement of ketoximes was investigated using stannous (II) chloride in ionic liquid as an efficient catalyst (Niralwad et al., 2011). They carried out conversion of keto halides to lactams by means of sequential azidation and intramolecular Schmidt reaction in a combined flow format.

where X = Cl, Br; TFA = Trifluoroacetic acid.

Potassium dihydrogen phosphate (KH_2PO_4) was found to be a new and efficient medium for the Beckmann rearrangement in solvent-free conditions under microwave irradiation (Niralwad et al., 2013). It was observed that the yield of the products increases with reduction in time as the power of microwaves was increased. The yield was 93% in 25 min for power 600 watts.

Bosch et al. (1995) have also achieved the Beckmann rearrangement of ketoximes with montmorillonite K10 clay in dry media in good yields.

Bagheri and Karimkoshteh (2013) used nano silica-H_2SO_4 as an efficient and mild catalyst for the Beckmann rearrangement of ketoximes to amides. The reactions were carried out in solvent-free conditions under microwave irradiation (600 W) within 50–120 sec in good yields.

where R = Aryl, Alkyl.

Varma et al. (1999) reported that the role of the surface is critical since the same reagent supported on a clay surface delivers predominantly the Beckmann rearrangement products. They also observed a facile deoximation protocol with sodium periodate impregnated moist silica that is applicable exclusively to ketoximes (Varma et al., 1997).

The reactions were incomplete even after 4 h by conventional heating method (110°C in an oil-bath). Recently, amides have also been achieved by treatment of ketones with hydroxylamine hydrochloride under microwave irradiation in the presence of silica gel supported $NaHSO_4$ catalyst in good yields (Das et al., 2000).

Microwave-assisted solvent-free reactions were employed to synthesize amides from ketoximes in the presence of phosphoric acid without using any hazardous solvent (Kantharaju et al., 2005). Advantages of this method are environmental benign nature, economical procedure, short reaction time and high yields (74–82%).

11.6 BENZIL–BENZILIC ACID REARRANGEMENT

Normally, the benzyl derivatives with electron donating group takes a long time for benzil-benzilic acid rearrangement, for example, p-methoxy derivative undergoes this rearrangement and given 32% yield of the rearranged product and that too in 6 h (Toda et al., 1990).

$$\underset{Ar-C-C-Ar}{\overset{O\ \ O}{\underset{||\ \ ||}{}}} \xrightarrow{MW} \underset{Ar-C=C-Ar}{\overset{OH\ \ COOH}{\underset{|\ \ \ \ \ |}{}}}$$

where Ar = Phenyl, p-Methylphenyl, p-Methoxyphenyl, o-Chlorophenyl. It gave yields ranging between 86–98% under microwave irradiation for 45–65 sec.

A new procedure for carrying out the benzil-benzilic acid rearrangement in the solid state has been developed by Yu et al. (1999).

Benzil

5,5-Diphenylhydantoin
(Dilatin)

11.7   CLAISEN REARRANGEMENT

Solvent-free Claisen rearrangement of bis(4-allyloxyphenyl) sulfone under microwave irradiation gave high yields of bis(3-allyl-4-hydroxyphenyl) sulfone in 5 min (Yamamoto et al., 2003). It takes 2–30 h under conventional heating. It is used as a color developer for a heat- or pressure sensitive recording in industry.

Similarly, 30-allyl-20-hydroxy-acetophenone was obtained in quantitative yield from 20-allyloxy-acetophenone (Bennett et al., 2004).

Microwave-assisted double decarboxylative Claisen rearrangement of bis(allyl) 2-tosylmalonates to substituted 1,6-heptadienes was studied. These heptadienes may be alkylated, and then converted into pyridines by ozonolysis followed by reaction with ammonia generated in situ under microwave conditions (Craig et al., 2008).

Microwave irradiation strongly enhances the rate of Claisen rearrangement. This helps in solving the problem of the long-term heating under conventional conditions. The yield of rearranged products in classical method is about 85% in 6 h where as it was increased to 92% in 6 min by using N,N-dimethylformamide as solvent under microwave irradiation.

It was observed that the aromatic Claisen rearrangement allows the regioselective isoprenylation at the *para* position of flavonoids under microwave irradiation (Daskiewicz et al., 2001). Microwave irradiation successfully afforded the key step of the synthesis of R,R-dialkyl amino acids derived from benzocycloheptene. Transformation involving a double ortho ester Claisen rearrangement from 2-butyne-1, 4-diol was also carried out (Kotha et al. 2001). A microwave induced Claisen rearrangement of the propargylic enol ether was the key step in the synthesis of the skeleton present in the triterpenoid azadirachtin (Durand-Reville, et al., 2002). Sampath Kumar et al. (2000) reported that microwave irradiation can be used to carry out Claisen rearrangement in the solid phase of O-allyl aryl ethers (derived from salicylic acids anchored to a Merrifield resin), which afforded the corresponding trisubstituted aromatic systems in 4–6 min in high yields where as 10–16 h are required under conventional heating conditions.

Microwave-assisted decarboxylative Claisen rearrangement reactions of substituted acetate derivatives of 3-(hydroxyalkyl)indoles give dearomatised products (Camp et al., 2011). The reactivity of the resultant compounds was evaluated.

It was observed that the rate of catalytic enantioselective Claisen rearrangement was drastically increased under microwave irradiation (Nushiro et al., 2013) without any loss of the enantioselectivity. It was revealed that enantioselectivity decreased as the internal reaction temperature was increased. Therefore, this reaction acceleration would not be caused by only a simple thermal effect.

Two monoterpenols, perillyl alcohol and nerol, have been converted into their γ,δ-unsaturated ester derivatives following a modified process of microwave-assisted ortho ester Claisen rearrangement (Mehl et al., 2010). The yields obtained (>90%) are better than those previously obtained. This process needs less reaction time (5 min), smaller amount of reagent, and no solvent.

The ortho ester Claisen rearrangement was carried out in sealed tubes, without solvent, in a high-powered, focused microwave oven and with magnetic stirring (Herrero et al., 2008). In this reaction, the microwave increases reaction rates with thermal effect. In fact, using microwave technology, products were heated directly instead of convection and conduction, and one could work at a controlled temperature (190°C). A Biotage focused microwave with its powerful 400 W magnetron delivers precise heating control (measured by IR). Heating at this temperature was not possible with an oil bath because the temperature was limited by the boiling point of the reagent (138°C).

Claisen rearrangement of 4-methyl-7-[4-(substituted-but-2-ynyloxy]-coumarins under the microwave irradiation was reported by Valizadeh and Shockravi (2006). A good yield of 4'-substituted methyl-4,5'-dimethylangelicins was obtained in short reaction times.

Srikrishna and Nagaraju (1992) observed the three-step ortho ester Claisen rearrangement of allyl and propynyl alcohols in dry condition under microwave irradiation. The reaction was over in 10–15 min.

The Claisen rearrangement of aldehydes into substituted-2-oxohex-5-enoic acids in a tandem three step, one pot method, water, K_2CO_3 and under microwave irradiation (50 W) was reported (Quesada and Taylor, 2005). The reaction was competed in 10 min and good to excellent yield (55–93%) was obtained.

11.8 CURTIUS REARRANGEMENT

A new entry to the total synthesis of isocryptolepine (cryptosanguinolentine), isolated from *Cryptolepis sanguinolenta*, was achieved by constructing a tetracyclic ring system through a microwave-assisted tandem Curtius rearrangement and electrocyclic reaction of an aza 6π-electron system (Hayashi et al., 2012). The tetracyclic lactam was further converted to isocryptolepine in a four-step sequence. The acid is simply reacted with DPPA in toluene under microwave irradiation to deliver the desired fused lactam in 97% yield.

An efficient and highly versatile microwave-assisted Paal-Knorr condensation of various 1,4-diketones gave furans, pyrroles and thiophenes in good yields. In addition, transformations of the methoxycarbonyl moiety, such as Curtius rearrangement, hydrolysis to carboxylic acid, or the conversion into amine by reaction with a primary amine in the presence of Me_3Al, are described by Minetto et al. (2005).

where R = Methyl, Ethyl, t-Butyl, Phenyl; R_1 = Propyl, Phenyl, Benzyl; R_2 = Isobutyl, Benzyl.

2-Benzofuran isocyanate was prepared by Curtius rearrangement from 2-benzo-furoyl azide with acid hydrazide under microwave irradiation (Wang et al., 2006). A series of asymmetric semicarbazides was synthesized by reactions of 2-benzofuran isocyanate. This method gives high yield under mild conditions compared to the conventional method.

Patil et al. (2003) carried out Curtius rearrangement of Fmoc-amino acid azides in toluene to isocyanates under microwave irradiation for 60 sec. The resulting iso-cyanates were found to be analytically pure with 91–96% yield. The utility of these isocyanates is as building blocks in the synthesis of urea peptides. The coupling of isocyanates directly with N,O-bis(trimethylsilyl) derivatives of amino acids resulted in urea peptide acids with good yield in high purity.

Chai et al. (2013) reported the synthesis of a series of styrl substituted semicar-bazides by reaction of trans-styryl isocyanate, which was prepared by Curtius rear-rangement of cinnamoyl azide, with acid hydrazides under microwave irradiation using a one pot procedure. Comparing to the conventional method, this method was advantageous, because of mild conditions, easy handling, and high yield.

11.9   FRIES REARRANGEMENT

Trehan et al. (1997) reported a simple procedure for carrying out the Fries rearrange-ment of aryl esters to ortho- and para-hydroxy acetophenones in dry open media in ordinary glassware using a commercial microwave oven. The Fries rearrangement normally requires a Lewis acid and long reflux times or photochemical conditions. It has been observed that protected N-allylanilines with different substitution on the aryl ring can be transformed by microwave irradiation in the presence of $BF_3 \cdot Et_2O$ within 1–2 min at 170°C.

An ecofriendly solvent-free Fries rearrangement with acyl/alkyl migration has been observed by Selvakumar et al. (2007). Rearrangement of N-acylaniline in pres-ence of $AlCl_3$ adsorbed on neutral alumina under microwave irradiation afforded exclusively the p-rearranged product. Reactions were completed within 3–5 min and the yield varies between 86–96%.

where $R_1 = CH_3$; $R_2 = H$, CH_3, Br; $R_3 = H$, CH_3, Cl

Paul and Gupta (2004) observed selective Fries rearrangement of acetylated phenols catalyzed by zinc powder in presence of N,N-dimethylformamide under the microwave heating. This reaction was completed in about 0.5–7 h under conventional conditions while it is over in 2.25–23 min in presence of microwaves.

Phloroglucinol and β-ketoesters also undergo Fries rearrangement in the presence of microwave irradiation (Seijas et al., 2005). Flavones were obtained as the product in higher yields and that too in a cleaner and faster pathway as compared to classical thermal conditions. This reaction passes though cycloaddition of an α-oxo ketene intermediate.

66-96%

where R_1, R_2, R_3 R_4 = -H, -OCH$_3$, -Cl, -NO$_2$

Eshghi et al. (2003) reported Fries rearrangement of acyloxybenzene and naphthalene derivatives in presence of P_2O_5/SiO_2 reagent under microwave heating in solvent-free media within 5 min. o-Hydroxyarylketones were obtained as products with moderate to excellent yields (47–98%).

Resorcinol diacetate in presence of 1,1,2,2,-tetrachloroethane with $AlCl_3$ produced 2,'4'-dihydroxyacetophenone, 4,6-diacetylresorcinol and 2,4-diacetylresorcinol with 14, 13 and 48%, respectively, in a short time of 15 min and microwave irradiation (700 W) (Kim et al., 2004).

Biswanath et al. (2000) carried out Fries rearrangement of arylsulfonates and sulfonanilides under microwave irradiation. They obtained hydroxy and aminoaryl sulfones, respectively in a very short time with excellent yields. 2-Hydroxyarylsolfones predominates over 4-hydroxyl sulfonates. It was quite interesting to observe that 2-aminoarylsulfones were the only product obtained from arylsulfonanilides.

An $AlCl_3$–$ZnCl_2$ mixture supported on silica gel was found to be a newer and efficient medium for the thia-Fries rearrangement of aryl sulfonates in solvent-free conditions under microwave dielectric heating (Moghaddam and Dakamin, 2000).

11.10   HOFMANN REARRANGEMENT

The microwave-assisted N-bromosuccinimide (NBS) or tribromoisocyanuric acid (TBCA) mediated Hofmann rearrangement in presence of DBU/KOH/MeOH was carried out by Miranda et al. (2011). TBCA is a stable solid that can be easily synthesized from isocyanuric acid and NaBr in the presence of oxone. It was used as an efficient source of electrophilic bromine (Br^+) in basic media (KOH/MeOH) for this rearrangement. Here, high yields and short reaction times were observed for the formation of aromatic benzamides.

11.11 MEYER-SCHUSTER REARRANGEMENT

The Meyer-Schuster rearrangement of tertiary propargylic alcohols gives α,β-unsaturated carbonyl compounds under microwave irradiation (300 W), This rearrangement is promoted by oxovanadium (V) complex $[V(O)Cl(OEt)_2]$. The reaction was completed within 30 min in toluene with excellent yield (97%) (Antinolo et al., 2012).

where R_1, R_2 = Phenyl.

Cadiemo et al. (2009) also reported a simple and ecofriendly Meyer–Schuster rearrangement of propargylic aryl carbinols into α,β-unsaturated carbonyl compounds under microwave irradiation. They catalyzed the rearrangement by $InCl_3$ in aqueous medium.

This rearrangement of 1,1-diphenyl-2-propyn-1-ol (propargylic alcohols) into 3,3-diphenylpropenal under microwave irradiation (300 W) was reported by Garcia-Alvarez et al. (2011). The rearrangement was catalyzed by a Re (I) complex. It is highly efficient catalyst for isomerization where, 99% excellent yield was obtained after only 5 min of reaction time.

Meyer–Schuster rearrangement of both; terminal and internal alkynols was observed using microwave irradiation to afford the α,β-unsaturated carbonyl compounds. The reaction was catalyzed by $AgSbF_6$ without the addition of any cocatalyst. This catalytic system can be recycled up to 10 consecutive runs (1st cycle, 45 min, 99%; 10th cycle, 6 h, 97%) (Garcia-Álvarez et al., 2013).

11.12 NEWMAN-KWART REARRANGEMENT

In the Newman-Kwart Rearrangement (NKR), intramolecular aryl migration of O-thiocarbamates at high temperatures leads to S-thiocarbamates. The NKR allows access to thiophenols from phenols, as O-thiocarbamates are readily prepared and hydrolysis of S-thiocarbamates can readily be achieved (Moseley and Lenden, 2007). The yield ranges between 72–91%.

$$\text{Ar} \overset{O}{\underset{S}{\overset{|}{\text{C}}}} NMe_2 \quad \xrightarrow[\text{MW (300 W) , 15-30 min}]{\text{DMA or NMP, 300°C}} \quad \text{Ar} \overset{S}{\underset{O}{\overset{|}{\text{C}}}} NMe_2$$

Some substrates require so drastic conditions that destruction of the substrate on prolonged contact with the hot reaction vessel walls leads to side products. For example, electron-rich aromatic compounds with higher activation barriers and substrates that bear either thermally sensitive groups or stereocenters (racemization) require specific conditions such as flash vacuum pyrolysis, where the substrate in toluene is passed in a stream of nitrogen through a heated quartz tube (~400 °C). Another solution is provided by the use of thermally stable polar solvents in a microwave-assisted procedure.

High temperature NKR reactions can be conducted more conveniently using pressure-proof microwave equipment, because the temperatures and slightly elevated pressures (10 bar) can be more easily obtained in a very short time, and better controlled. However, there is no specific microwave effect and superheating alone is responsible for the acceleration of the reaction. Reactions leading to only 10% of product in xylene can produce up to 80% in formic acid. DMA, NMP and diphenyl ether are other solvents of choice that are polar and allow high reaction temperatures. The formation of thiols from S-aryl thiocarbamates is readily achieved using 10% aqueous NaOH or methanolic potassium hydroxide. Alternatively, reduction with lithium aluminum hydride under nonhydrolytic conditions affords aryl thiols in good yields.

11.13 PINACOL-PINACOLONE REARRANGEMENT

Gutierrez et al. (1989) have reported a solventless pinacol–pinacolone rearrangement using microwave irradiation. The process involves the irradiation of the gem-diols with Al^{3+}montmorillonite K10 clay for 15 min to afford the rearrangement product in excellent yields (98–99%). These results are compared to conventional heating in an oil bath, where the reaction times are too long (15 hr).

An efficient ring expansion transformation using alumina supported AgBF$_4$ has been described by Villemin and Labiad (1992) under solventless conditions. This solvent-free microwave protocol is superior to the reactions conducted in conventional methanolic solution.

Henderson and Byrne (2011) observed that several protic ionic liquids were tested as potential mediators for pinacol rearrangements employing microwave irradiation. Using hydrobenzoin as a model substrate, the optimal conditions were found to be heating at 80°C for 5 min using H$_2$SO$_4$ and using triethylamine as the ionic liquid. A key feature of this reaction was to keep the microwave power low (20 W) to avoid ionic liquid degradation. Application of these conditions to triphenylethylene glycol gave rearrangement products in high yield and purity, while phenylethylene glycol and styrene oxide gave pinacol products that underwent a cascade aldol condensation. These conditions represent an efficient means by which pinacol rearrangements can be carried out while avoiding the use of strong Bronsted acids, high temperatures and extended reaction times.

where R = H, Phenyl.

11.14   WOLFF REARRANGEMENT

The microwave-assisted Wolff rearrangement of cyclic 2-diazo-1, 3-diketones in the presence of aldehydes and primary amines provides a straightforward access to functionalized bi- and pentacyclic oxazinones following an unprecedented three-component domino reaction (Presset et al., 2009a). Alternatively, in the presence of acyl azides, an efficient Curtius/Wolff/hetero-Diels–Alder sequence allows the direct synthesis of oxazindiones.

The Wolf rearrangement of N^{α}-Boc-/Z-protect aminodiazoketones in the presence of silver benzoate under microwave irradiation has been described. The reaction is found to be rapid, efficient and complete. Verma (2002) observed that the microwave irradiation technique has been used to decrease the time necessary to carry out cleaner reaction with higher selectivity and easier work up. A sample, rapid, and efficient route for the preparation of several amino acid benzyl ester p-toluenesulfonate and hydrochloride salts under microwave irradiation was also demonstrated by Patil et al. (2002).

where X = Boc; R = CH_3, C_6H_5, $C_6H_5CH_2$.

The microwave-assisted Wolff rearrangement of cyclic 2-diazo-1,3-diketones was performed in the presence of a stoichiometric amount of alcohol, amine, or thiol (Presset et al., 2009b). It is an efficient, and environmentally friendly synthetic protocol for the synthesis of α-carbonylated cycloalkanones. This approach proves superior to existing protocols in scope and ecocompatibility.

Sudrik et al. (2002) have studied the specific Wolf rearrangement of 3-Diazoc-amphor in the presence of benzylamine under microwave irradiation.

A variety of α-spirolactones and lactams from 2-diazo-1,3-dicarbonyl compounds, (homo)allylic alcohols or amines and acrylic derivatives, in a single synthetic operation by a Wolff rearrangement/α-oxo ketene trapping/cross metathesis/intramolecular Michael addition sequence has been obtained by Boddaert et al. (2011).

11.15   MISCELLANEOUS

Unsaturated eight membered lactones undergo decarboxylative and nondecarboxylative transannular Ireland–Claisen rearrangement reactions, to give substituted vinylcyclobutanes. A formal synthesis of (±)-grandisol is described by Craig et al. (2011).

where R_1 = H, Me.

The Overman rearrangement converts allylic trichloroacetimidates into trichloroacetamides in presence of o-xylene under microwave irradiation (Gajdosikova et al., 2008). The reaction time was shorter as compared to the conventional thermal rearrangement,

Rearrangement reactions are quite atom economic, as all the atoms of reactants are being used in the product and nothing is lost. These reactions can be carried out under microwave exposure rapidly and with good to excellent yields following a green technology.

KEYWORDS

- **Baker-Venkataraman Rearrangement**
- **Beckmann Rearrangement**
- **Claisen Rearrangement**
- **Curtius Rearrangement**
- **Fries Rearrangement**
- **Wolff Rearrangement**

REFERENCES

Abdel Ghani S. B., Weaver L., Zidan Z. H., Ali H. M., Keevil C. W., & Brown R. C. (2008). Bio-organic & Medicinal Chemistry Letters, 18, 518–522.

Agrawal, Y. P., Agrawal, M. Y., & Gupta, A. K. (2013). Pharmaceutical Sciences, 1, 65–68.

Antinolo, A., Carrillo-Hermosilla, F., Cadierno, V., Garcia-Alvarez, J., & Otero, A. (2012). Chem. Cat. Chem., 4, 123–128.

Bagheri, M., & Karimkoshteh, M. (2013). Iranian Journal of Catalysis, 3, 27–32.

Banerjee, K., & Mitra, A. (2005). Indian Journal of Chemistry, 44B, 1876–1879.

Bennett, C. J., Caldwell, S. T., McPhail, D. B., Morrice, P. C., Duthie, G. G., & Hartley, R. C. (2004). Bioorganic & Medicinal Chemistry, 12, 2079–2098.

Biswanath, D., Purushotham, P., & Bollu, V. (2000). Journal of Chemical Research, 2000, 200–201.

Boddaert, T., Coquerel, Y., & Rodriguez, J. (2011). European Journal of Organic Chemistry, 26, 5061–5070.

Bosch, A. I., De la Cruz, P., Dı́ez-Barra, E., Loupy, A., & Langa, F. (1995). Synlett, 12, 1259–1260.

Bremner, W. S., & Organ, M. G. (2008). Journal of Combinational Chemistry, 10, 142–147.

Cadierno, V., Francos, J., & Gimeno, J. (2009). Tetrahedron Letters, 50, 4773–4776.

Camp, J. S., Craig, D., Funai, K., & Andrew J. P. (2011). Organic & Biomolecular Chemistry, 9, 7904–7912.

Castro, A. M. M. (2004). Chemical Reviews, 104, 2939–3002.

Chai, Zhang, L., Liu, H., Huang, G., Cheng, J., & Qiao-Qiao. (2013). Journal of Chemical Research, 37, 356–358.

Craig, D., Funai, K., Gore, S. J., Kang, A., & Alexander V. W. M. (2011). Organic & Biomolecular Chemistry, 9, 8000–8002.

Craig, D., Paina, F., & Smith, S. C. (2008). Chemical Communications, 3408–3410.

Das, B., Ravindranath, N., Venkataiah, B. & Madhusudhan, P. (2000). Journal of Chemical Research, (S), 10, 482–483.

Daskiewicz, J. B., Bayet, C., & Barron, D. (2001). Tetrahedron Letters, 42, 7241–7244.

Durand-Reville, T., Gobbi, L. B., Gray, B. L., Ley, S. V., & Scott, J. S. (2002). Organic Letters, 4, 3847–3850.

Eshghi, H., Rafie, M., Gordi, Z., & Bohloli, M. (2003). Journal of Chemical Research (S), 763–764.

Gajdosikova, E., Martinkova, M., Gonda, J., & Conka, P. (2008). Molecules, 13, 2837–2847.

Garcia-Alvarez, J., Diez, J., Gimeno J., & Seifried, C. M. (2011). Chemical Communications, 47, 6470–6472.

Garcia-Alvarez, J., Diez, J., Vidal, C., & Vicent, C. (2013). Inorganic Chemistry, 52, 6533–6542.

Goel, V. (2012). Oriental Journal of Chemistry, 28, 1725–1728.

Gonda, J., Helland, A. C., Ernst, B., & Belluš, D. (1993). Synthesis, 7, 729–733.

Gonda, J., Martinkova, M., Zadrosova, A., Sotekova, M., Raschmanova, J., Conka, P., Gajdosikova, E., & Kappe, C. O. (2007). Tetrahedron Letters, 48, 6912–6915.

Gutierrez, E., Loupy, A., Bram, G., & Ruiz-Hitzky, E. (1989). Tetrahedron Letters, 30, 945–948.

Han, X., & Armstrong, D. W. (2005). Organic Letters, 7, 4205–4208.

Hayashi, K., Choshi, T., Chikaraishi, K., Oda, A., Yoshinaga, R., Hatae, N., Ishikura, M., & Hibino, S. (2012). Tetrahedron, 68, 4274–4279.

Henderson, L. C., & Byrne, N. (2011). Green Chemistry, 13, 813–816.

Herrero, M. A., Kremsner, J. M., & Kappe, C. O. (2008). Journal of Organic Chemistry, 73, 36–47.

Hill, R. K., & Gilman, W. (1967). Tetrahedron Letters, 8, 1421–1423.

Kaboudin, B., & Balakrishna, M. S. (2001). Synthetic Communications, 31, 2773–2776.

Kad, G. L., Trehan, I. R., Kaur, J., Nayyar, J. S., Arora, A., & Brar, J. S. (1999). Indian Journal of Chemistry, 35B, 734–736.

Kantharaju, Basanagoud, S. P., & Vommina, V., Babu, S. (2005). Indian Journal of Chemistry, 44B, 2611–2613.

Kim, J. H., Yoon, H. J., & Chae, W. K. (2004). Bulletin of the Korean Chemical Society, 25, 1447–1448.

Kotha, S., Sreenivasachary, N., & Brahmachary, E. (2001). Tetrahedron, 57, 6261–6265.

Kumar, A., & Makrandi, J. K. (2012). Heterocyclic Letters, 2, 271–276.

Li, Z., & Lu, Z. (2008). Letters in Organic Chemistry, 5, 495–501.

Mehl, F., Bombarda, I., Franklin, C., & Gaydou, E. M. (2010). Synthetic Communications, 40, 462–468.

Minetto, G., Raveglia, L. F., Sega, A., & Taddei, M. (2005). Europian Journal of Organic Chemistry, 24, 5277–5288.

Miranda, L. S. M., Silva, T. R., Crespo, L.T., Esteves, P. M., De Matos, L. F., Diederichs, C. C., & De Souza, R. O. M. A. (2011). Tetrahedron Letters, 52, 1639–1640.

Moghaddam, F. M., & Dakamin, M. G. (2000). Tetrahedron Letters, 41, 3479–3481.

Moseley, J. D., & Lenden, P. (2007). Tetrahedron, 63, 4120–4125.

Moseley, J. D., Sankey, R. F., Tang, O. N., & Gilday, J. P. (2006). Tetrahedron, 62, 4685–4689.

Niralwad, K. S., Ghorade, I. B. & Kharat, P. S. (2013). Indian Journal of Applied Research, 3, 47–48.

Niralwad, K. S., Shingate, B. B., & Shingare, M. S. (2011). Letters in Organic Chemistry, 8, 274–277.

Nubbemeyer, U. (2005). Topics in Current Chemistry, 244, 149–213.

Nushiro, K., Kikuchi, S., & Yamada, T. (2013). Chemical Communications, 49, 8371–8373.

Patil, B. S., Vasathakumar, G. R., & Babu, S. V. V. (2002). Letters in Peptide Science, 9, 207.

Patil, B. S., Vasathakumar, G. R., & Babu, S. V. V. (2003). Journal of Organic Chemistry, 68, 7274–7280.

Paul, S., & Gupta, M. (2004). Synthesis, 1789–1792.

Pinto, D. C. G. A., Silva, A. M. S.; Cavaleiro, & J. A. S. (2007). Synlett, 1897–1900.

Presset, M., Coquerel, Y., & Rodriguez, J. (2009a). Organic Letters, 11, 5706–5709.

Presset, M., Coquerel, Y., & Rodriguez, J. (2009b). Journal of Organic Chemistry, 74, 415–418.

Quesada, E., & Taylor, R. J. K. (2005). Synthesis, 3193–3195.

Sampath Kumar, H. M., Anjaneyulu, S., Subba Reddy, B. V., & Yadav, J. S. (2000). Synlett, 1129–1130.

Seijas, J. A., Vazquez-Tato, M. P., & Carballido-Reboredo, R. (2005). Journal of Organic Chemistry, 70, 2855–2858.

Selvakumar, S., Easwaramurthy, M., & Raju, G. J. (2007). Indian Journal of Chemistry, 46B, 713–715.

Sim, J., Yun, H., Jung, J.-W., Lee, S., Kim, N.-J., & Suh, Y.-G. (2012). Tetrahedron Letters, 53, 4813–4815.

Srikrishna, A., & Nagaraju, S. (1992). Journal of the Chemical Society Perkin Transactions, 1, 311–312.

Sudrik, S. G., Chavan, S. P., Chandrakumar, K. R. S., Pal, S., Date, S. K., Chavan, S. P., & Sonawane, H. R. (2002). Journal of Organic Chemistry, 67, 1574–1579.

Sugamoto, K., Matsushita Y., & Matsui, T. (2011). Synthetic Communications, 41, 879–884.

Takano, S., Akiyama, M., & Ogasawara, K. (1984). Chemical Communications, 770–771.

Tanaka, K. (2002). Solvent-free organic synthesis. Weinheim: Wiley-VCH.

Trehan, I. R., Arora, A. K. & Kad, G. L. (1997). Journal of Chemical Education, 74, 324.

Toda, F., Tanaka, K., Kagawa, Y., & Sakaino, Y. (1990). Chemical Letters, 19, 373–376.

Valizadeh, H., & Shockravi, A. (2006). Journal of Heterocyclic Chemistry, 43, 763–765.

Varma, R. S. (2002). Advanced in green chemistry: Chemical synthesis using microwave irradiation. M. K. Sumitra & A. Zeneca (Eds.), Bangalore, Research Foundation India.

Varma, R. S., Dahiya, R., & Saini, R. K. (1997). Tetrahedron Letters, 38, 8819–8820.

Varma, R. S., Naicker, K. P., Kumar, D., Dahiya, R., & Liesen, P. J. (1999). Journal of Microwave Power Electromagnetic Energy, 34, 113–123.

Villemin, D., & Labiad, B. (1992). Synthetic Communications, 22, 2043–2052.

Wang, X., Chai, L., Wang, M., Quan, Z., & Li, Z. (2006). Journal for Rapid Communication of Synthetic Organic Chemistry, 36, 645–652.

Yadav, J. S., Subba Reddy, B. V., Abdul Rasheed, M., & Sampath Kumar, H. M. (2000). Synlett, 4, 487–488.

Yamamoto, T., Wada, Y., Enokida, H., Fujimoto, M., Nakamura, K., & Yanagida, S. (2003). Green Chemistry, 5, 690–692.

Yu, H. M., Chen, S. T., Tseng, M. J., & Wang, K. T. (1999). Journal of Chemical Research (S), 1, 62–63.

CHAPTER 12

COUPLING

SURBHI BENJAMIN, NEELAM KUNWAR, KUMUDINI BHANAT, and
SURESH C. AMETA

CONTENTS

Coupling reaction in organic chemistry is a term for a variety of reactions, where two hydrocarbon fragments are coupled with the aid of a metal, it salts or complexes as a catalyst.

Two types of coupling reactions have been recognized:

- Homo couplings couple two identical partners, for example, the conversion of iodobenzene (PhI) to biphenyl (Ph-Ph) and
- Cross couplings involving reactions between two different partners, for example, bromobenzene (PhBr) and vinyl chloride (CH_2=CH-Cl) to give styrene ($PhCH=CH_2$).

One of the important reaction type of coupling is where a main group organometallic compound of the type RM (R = Organic fragment, M = Main group center) reacts with an organic halide of the type R'X resulting in the formation of a new carbon-carbon bond in the product (R-R') (Bates, 2000). The Nobel prize in Chemistry (2010) was awarded jointly to Richard F. Heck, Ei-ichi Negishi and Akira Suzuki for palladium-catalyzed cross couplings in organic synthesis.

The reaction mechanism usually begins with oxidative addition of an organic halide to the catalyst. Subsequently, the second partner undergoes transmetallation, which places both coupling partners on the same metal center. The final step is reductive elimination of the two coupling fragments to regenerate the catalyst and give the final organic product. Unsaturated organic groups couple more easily because they readily add. The intermediates are also less prone to β-hydride elimination (Hartwig, 2010).

Oxidative addition

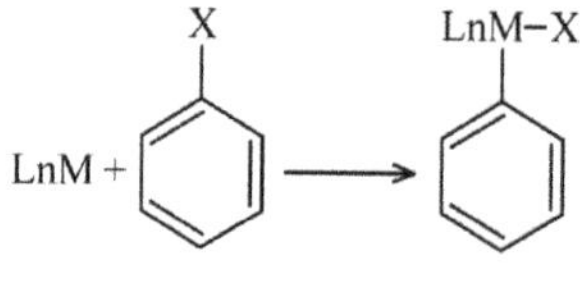

Transmetallation

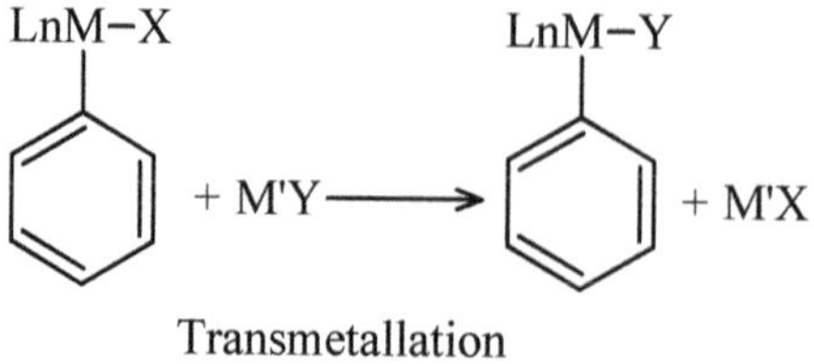

Transmetallation

Reductive elimination

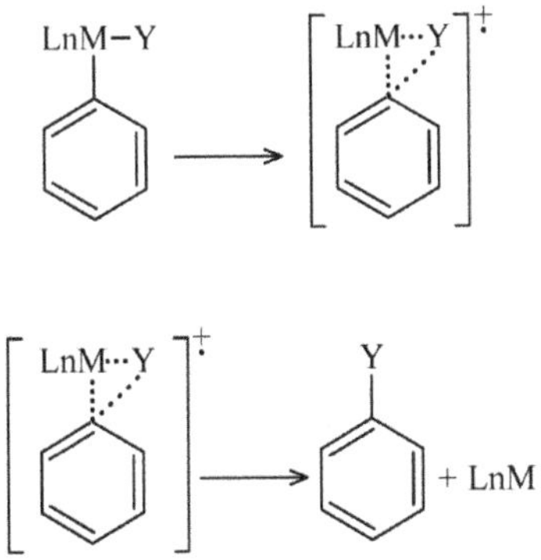

The order of ease of C-Y reductive elimination is:

C-C > C-N > C-O > C-F

The leaving group X in the organic partner is usually a bromide, iodide or triflate, but the ideal leaving group is chloride, since organic chlorides are low cost than related compounds. The main group metal in the organometallic partner is usually tin, zinc, or boron.

While many coupling reactions involve reagents that are extremely susceptible to presence of water or oxygen, it is quite unreasonable to assume that all coupling reactions should be performed in absence of water. It is possible to perform palladium-based coupling reactions in aqueous solutions using the water-soluble sulfonated phosphines (made by the reaction of triphenyl phosphine with sulfuric acid). In general, the oxygen in the air is more able to disrupt coupling reactions, because many of these reactions occur via unsaturated metal complexes that do not have 18 valence electrons.

Some catalysts might be easily poisoned by heterocycles under prolonged reaction at elevated temperatures. To avoid this, chemists often use pressure reactors to accelerate reactions at high temperature and pressure. Q-Tube and microwave synthesizer are available as safe pressure reactors.

Many coupling reactions have found their way into conjugated organic materials (Hartwig, 2010) and pharmaceutical industry (Crabtree, 2005).

12.1   HECK REACTION

High temperature and microwave heating appear to be beneficial for Heck-type coupling of simple alkenes in water. Bulky phosphine ligands were found to increase the rate of this reaction. Heck reaction between aryl halides and alkenes are well established under microwave irradiation. Regioselectivity is the most important aspect in the Heck reaction. Larhed et al. (1997) described a regioselective Heck coupling employing different arylboronic acids with both electron rich and electron poor olefins. Controlled microwave processing was used to reduce reaction time from hours to minutes both; in small scale and in 50-mmol-scale batch process (Lindh et al., 2007).

Another significant point of interest in the microwave enhanced Heck reaction is the asymmetric induction. A highly modular library of readily available phosphate oxazoline ligands was applied in the Pd-catalyzed asymmetric Heck reactions of several substrates and triflates under thermal and microwave conditions. Both enantiomers of the Heck coupling products showed excellent regioselective and enantioselective activities (Mazuela et al., 2010).

Dighe and Degani (2011) demonstrated that 1-(2-cyanoethyl)-3-(2-hydroxyethyl)-1H-imidazol-3-ium tetrafluoroborate (IL-2) in presence of PdCl$_2$ was found to be an efficient and reusable, ligand-free and base-free catalytic system for Heck-coupling of activated and deactivated iodo- and bromoarenes with different olefins. Under microwave irradiation, it was reported to exhibit good efficiency in terms of activity, selectivity and recyclability for six consecutive runs without any significant loss of activity.

(E)-ethyl cinnamate

Dawood and El-Deftar (2010) studied the catalytic activity of benzimidazole-oxime Pd (II) complex towards Suzuki and Heck C-C cross coupling reaction of activated and deactivated aryl and heteroaryl bromides under microwave irradiation as well as thermal heating using water as a green solvent.

Koopmans et al. (2006) reported the first MW assisted synthesis of poly(2,5-dibutoxy-1,4-phenylenevinylene) via Heck-polycondensation as an example for efficient heating.

12.2   HIYAMA REACTION

Shah and Kaur (2012) reported a nonfunctional macroporous commercial resin, Amberlite XAD-4, impregnated with palladium nanoparticles (PdNPs) of size 5–10 nm. These supported Pd nanoparticles were used to catalyze the sodium hydroxide activated Hiyama cross-coupling reaction of phenyltrimethoxysilane with a variety of bromo and chloroarenes under microwave heating. These were found to have very high efficiency (TOF ≈ 3×10^4) and excellent recyclability. The procedure, carried out in the absence of any additional ligands, surfactants or toxic organic solvents, can be used as a sustainable and green procedure for the production of biaryls.

12.3   LIEBESKIND-SROGL COUPLING

The Liebeskind-Srogl coupling is a palladium catalyzed coupling between a thioester and a boronic acid. It was first reported by Liebeskind and Srogl (2000). In this reaction, Pd_2 (dba)$_3$ was used as a palladium precursor and copper (I) thiophene-2-carboxylate (CuTC) as a mediator.

Nutlin-like structures can be achieved by performing this coupling between arylboronic acids and imidazolidine-2-thiones, prepared by a novel MCR. Microwave-assisted Liebeskind-Srogl couplings gave desired Nutlin derivatives (the so-called Nutloids) in fair yields.

Bon et al. (2008) optimized this procedure of Liebeskind-Srogl couplings of imidazolidine-2-thiones with arylboronic acids, affording Nutloids in 41–65% yields.

Prokopcova et al. (2005) performed thioether-boronic acid cross-coupling (Liebeskind-Srogl reaction) using microwave heating involving C-C cross-coupling reaction between thioamides and boronic acids. Microwave-assisted two step synthesis of Bay 41–4109 analogs applying Biginelli multicomponent and Liebeskind-Srogl chemistry was reported. A new possibility of the modification of dihydropyrimidine at the C5 position using Liebeskind-Srogl coupling under microwave conditions was also explored.

12.4 NEGISHI COUPLING

Based on the pioneering studies of Kumada–Corriu about cross coupling reactions using nickel- or palladium-catalysts in combination with Grignard reagents, Negishi developed a procedure employing zinc reagents. The Negishi cross-coupling reaction using organozinc reagents represents a powerful method for this synthesis (Negishi et al., 1998). One major disadvantage of Pd or Ni-catalyzed Negishi couplings is the long reaction times involving hours or even days.

The Negishi coupling is a cross coupling reaction in organic chemistry involving an organozinc compound, organic halide and nickel or palladium catalyst creating a new carbon–carbon covalent bond. The preparation of unsymmetrical biaryls in good yields was reported (King et al., 1977; Negishi et al., 2010; Phapale and Cardenas, 2009).

Lipshutz et al. (2006) have given several examples of Negishi reactions applying heterogeneous catalyst conditions. Cross coupling of substituted arylzinc halides with substituted aryl chlorides employing Ni-on-charcoal (as a cheap and highly efficient alternative catalyst) under microwave irradiation for 15–30 min was reported, where as under conventional heating, the same reactions took approx. 24 h. The required biaryl compounds were obtained in good to excellent yields.

A large number of zinc reagents synthesized at an elevated temperature under conventional heating or microwave irradiation have been shown to be highly thermally stable and functional group tolerant. An interesting approach for the use of $(tmp)_2Zn_2 MgCl_2.LiCl$ (tmp = 2,2,6,6-tetramethylpiperidine) as a zinc cation reagent for the direct ortho-metalation of (hetero)aromatic compounds has been reported (Wunderlich and Knochel, 2008). These newly generated zinc reagents were further used in cross-coupling reactions to generate biaryl compounds.

The palladium-catalyzed alkynylation of aromatic rings has emerged as one of the most reliable methods for the synthesis of arylalkynes (Brandsma et al., 1998). Negishi also developed an effective palladium-catalyzed alkynylation reaction using alkynylzinc reagents (King et al., 1977; Negishi and Xu 2002) and 1-halo-1-alkynes (Negishi et al., 1978).

Molander et al. (2002) reported the successful coupling of aryl bromides with alkynyltrifluoroborates under thermal condition. This synthesis generally requires long reaction times (12 h) and relatively high catalyst loadings (9 mol %). Using microwave irradiation, it was found that the reactions can be carried out only in 20 min in the presence of relatively much less amount of catalyst. Typical reaction conditions involve a catalyst loading of 2 mol% $PdCl_2(dppf)-CH_2Cl_2$ while providing yields comparable to the thermal reactions (Coltuclu et al., 2013).

An application of high-speed Negishi coupling was reported by Kroscsenicsova et al. (2004) for the preparation of enantiopure 2,2'-diarylated 1,1'-binaphthyls.

Walla and Kappe (2004) carried out microwave-assisted Negishi and Kumada cross-coupling reactions of aryl chlorides.

$$Ar-Cl + R-M-X \xrightarrow[\substack{MW, THF/NMP \\ 175°C, 10\ min}]{Pd_2(dpa)_3/Bu_3P.HBF_4} Ar-R$$

where Ar = Alkyl, Aryl; M = Zn, Mg; X = Cl, Br, I.

The organozinc reagents for the Negishi couplings could be prepared by insertion of activated Rieke zinc dust into aryl bromides (or iodides). This transformation normally requires several hours under reflux conditions in THF (Zhu et al., 1991). This process was accelerated by the use of controlled microwave irradiation under sealed vessel conditions to 5–30 min.

where R = 2-CH$_3$, 4-CN, 3-COOC$_2$H$_5$, 3-OCH$_3$; Zn is in the form of Rieke zinc dust.

Mutule and Suna (2004) observed that the aryl zinc reagents could be readily prepared from aryl iodides using a Zn–Cu couple in a microwave environment and used for coupling with p-bromobenzaldehyde.

Moore and Vicic (2008) developed a new method to prepare 2,2'-bipyridines in short reaction times using microwave radiations and heterogeneous catalysts. It was found that Ni/Al$_2$O$_3$-SiO$_2$ afforded 2,2'-bipyridine products in upto 86% yields in 1 h. Palladium supported on alumina also provided comparable yields of 2,2'-bipyridines (72%) to that for homogeneous PEPPSI™ and tetrakis(triphenylphosphine) palladium complexes.

84% yield

12.5   SONOGASHIRA COUPLING

Raut et al. (2009) carried out a rapid and efficient synthesis of novel 2-[6-(arylethynyl)pyridin-3-yl]-1H-benzimidazole derivatives under microwave-assisted Sonogashira coupling conditions. Here, the use of protecting groups for benzimidazole NH during Sonogashira coupling was avoided. This microwave-assisted method offered shorter reaction times, higher yields and it was also applicable to a large set of substrates.

Chen et al. (2009) reported one pot, three-component coupling reaction for the synthesis of indoles. The reaction was carried out in two steps under standard Sonogashira coupling conditions from an N-substituted/N, N-disubstituted 2-iodoaniline and a terminal alkyne, followed by the addition of acetonitrile and aryl iodide.

Huang et al. (2008) developed an efficient method of microwave-assisted cross-coupling of terminal alkynes with various aryl chloride including sterically hindered, electron-rich, electron-neutral, and electron-deficient aryl chlorides. It proceeds with faster rate and generally gave good to excellent yields of the products. This simple catalytic system was also effective for Suzuki coupling, Buchwald Hartwig amination, and Heck coupling reactions with unactivated aryl chlorides.

12.6   STILLE CROSS-COUPLING

The Stille cross-coupling of organic halides with vinyl stannanes has become quite an attractive method in modern organic synthesis (Farina et al., 1997; Stille, 1986). Under classical Stille conditions, most of the substrates cross-couple at reaction temperatures of 45–100°C with reaction times ranging from hours to days. It has been shown that the cross-coupling time for fluorous (Larhed et al., 1997; Olofsson et al., 1999) and organic-phase Stille couplings (Fugami et al., 1999; Han et al., 1999) can be reduced to only minutes by using microwave flash heating. It is crucial to minimize Pd (0)-catalyzed conversion of Bu_3SnH into $Bu_3SnSnBu_3$ to achieve efficient one-pot palladium-mediated hydrostannylation/Stille couplings (Mitchell et al., 1986). A variety of alkynes underwent microwave-accelerated hydrostannylation/Stille couplings with aryl, benzyl, and vinyl electrophiles via this modified procedure (Littke and Fu, 1999). The expected dienes or styrenes were obtained in good yields after microwave irradiation of 8–13 min while, Stille coupling of 2-methyl-3-butyn-2-ol with bromobenzene took 16 h in THF at reflux and afforded the diene in 60% yield.

Stille coupling of 4-acetylphenyl triflate proceeds rapidly with microwave irradiation. In the Stille reaction, the formation of minor amounts of 4-butylacetophenone was encountered, which is formed frequently also on using standard heating procedures (Farina et al., 1993).

The combination of ethanol/water/DME and N-methyl-2-pyrrolidone (NMP) was found to interact sufficiently strongly with microwave and generates the heat required to promote the Stille cross-couplings. Methyl acrylate was converted smoothly to the corresponding cinnamic acid esters in 3.8 min at 60 W in the presence of DMF (Patel et al., 1977).

Fluorinated aryltin, heteroaryltin, and allyltin reactants were treated with organic halides in the presence of lithium chloride and a catalytic amount of bis(triphenylphosphine)palladium (II) chloride in DMF under microwave irradiation. The reaction was completed in less than 2 min with microwave dielectric heating and a microwave power of 50–70 W (Larhed et al., 1997).

Alternating copolymers form a very important class of semiconductors, as they offer the possibility of manipulating their optoelectronic properties in order to maximize their application potential in areas such as organic photovoltaics, field effect transistors and light emitting diode.

where $R_1 = C_2H_4(C_2H_5)C_4H_9$ or C_8H_{17}.

A microwave-assisted Stille coupling was reported as a convenient tool in the synthesis of hypoxia selective 3-alkyl-1,2,4-benzotriazine 1,4-dioxide as anticancer agents. The introduction of an ethyl substituent was a key step in this synthesis. The microwave-assisted Stille reaction was carried out using $Pd(PPh_3)_4$ in MeCN. It furnished the target compounds in 20–60 min with good yields ranging between 54 and 88% (Pchalek and Hay, 2006). Microwave-assisted Stille coupling of chlorides gave dramatically improved yields, superior to those from the corresponding iodides. The application of MW assisted synthesis extended the range of substituted

3-chloro-1,2,4-benzotriazine-1-oxides provided an efficient scalable synthesis of anticancer agent.

Coffin et al. (2009) have shown the importance of the Stille cross-coupling reaction for the generation of conjugated polymers having fused aromatic heterocycles, especially thiophenes in their backbone.

where Z = -C $(C_{12}H_{25})_2$, -Si$(C_{12}H_{25})_2$, -C(2-Ethylhexyl)$_2$.

An efficient Stille cross-coupling reaction catalyzed by ortho-palladated complex of tribenzylamine was carried out by Hajipour et al. (2012) under microwave irradiation. The catalytic activity of [Pd$\{C_6H_2N(CH_2Ph)_2)\}$(m-Br)]$_2$ complex as an efficient and stable catalyst (nonsensitive to air and moisture) in Stille cross-coupling reaction of various aryl halides with phenyltributyltin under microwave irradiation was investigated. The substituted biaryl halides were produced in excellent yields in short reaction times using a catalytic amount of this complex at 100°C.

12.7 SUZUKI COUPLING

The formation of carbon-carbon bonds via palladium catalyzed coupling of organoboranes with organic halides (the Suzuki reaction) has become an integral part of modern organic synthesis (Suzuki, 1999). In 1979, the seminal paper of Miyaura,

Yamada, and Suzuki laid the base for one of the most important and useful transformations for the construction of C–C bonds in the modern day organic chemistry. The Suzuki reaction becomes popular because of the ready availability of a wide range of functionally substituted boron derivatives and the mildness of the coupling reaction itself. Suzuki reactions generally employ organic solvents such as tetrahydrofuran and ethers as well as complex palladium catalysts, which are soluble in these solvents (Miyaura and Suzuki, 1995). The use of microwave heating is a convenient way to facilitate the Suzuki-type reactions in water (Leadbeater and Marco, 2002).

Carbon-carbon bond formation via Suzuki coupling of organoborane compounds with organic halides provides a mild method for synthesis of various functionalized compounds, especially biaryls (Bedford et al., 2003). An other example of Suzuki coupling reaction using microwave irradiation with Pd catalyst is to give biaryl product in water (Leadbeater and Marco, 2003).

The palladium-catalyzed cross-coupling of aryl halides with aryl boronic acids is a powerful reaction for the construction of biaryls (Suzuki, 1999). Its scope has been extended using aryl triflates ($ArOSO_2CF_3$) or aryl nonaflates ($ArOSO_2(CF_2)_3CF_3$) as halide equivalents (Baraznenok et al., 2000). Aryl perfluoroalkyl sulfonates prepared from a wide variety of commercially available phenols have shown high reactivity, good stability for room temperature storage, chromatography, and resistance towards hydrolysis (Zhang and Sui, 2003).

Microwave-assisted synthesis of following difuran and furan-thiophene derivatives via Suzuki coupling was investigated by Lin and Lin (2005).

An efficient microwave-assisted Suzuki reaction using a new pyridine-pyrazole/ Pd (II) species as catalyst in aqueous media was reported by Shen et al. (2013).

Application of Suzuki coupling reactions for parallel and combinatorial synthesis has been explored by conducting the reaction under microwave irradiation or on solid support with a linker such as perfluoroalkylsulfonyl (Pan et al., 2001). The Suzuki reaction is one of the most versatile and used reactions for the selective construction of carbon-carbon bonds, in particular for the formation of biaryls (Hassan et al., 2002).

In Suzuki reaction, cross-coupling of aryl- or vinyl-boronic acid with an aryl or vinyl halide catalyzed by a palladium complex, is one of the most versatile reactions for the construction of carbon-carbon bonds, in particular for the formation of biaryls. Recent developments have expanded the possible applications of this reaction enormously. Microwave-assisted Suzuki reactions can now be performed in many different ways and have been incorporated into a variety of challenging synthesis. Under microwave-heated condition, the Suzuki coupling of aryl chlorides with boronic acids was performed in an aqueous media using the air and moisture-stable palladium catalyst. A drastic reduction of the reaction time to 15 min and the formation of products in good yields were achieved (Miao et al., 2005).

Yields : 64-99%

Microwave activation enables a Suzuki coupling of boronic acids with aryl halides under solvent-free conditions using palladium catalyst system. A large variety of boronic acids and bromo, chloro and iodoaryls could rapidly give biaryls in 10 min (Nun et al., 2009).

$$Ar-B(OH)_2 + X-Ar' \xrightarrow[\substack{Neat \\ MW, 100°C, 10 min}]{\substack{1 \ mol \ \% \ Pd \ catalyst \\ 3 \ eq. \ K_2CO_3}} Ar-Ar'$$

where X = I, Cl, Br; Catalyst = PEPPSI-iPr.

Replacement of the traditionally used Pd with less expensive Ni-based catalyst systems can significantly reduce costs for cross-couplings of this type, especially when it is performed on large scale. A rapid and highly efficient microwave-assisted Ni-catalyzed Suzuki cross-coupling of aryl carbamates and sulfamates with boronic acid was also reported. This protocol takes coupling time of only 10 min. Also, the microwave irradiation worked excellently well with aryl chlorides as electrophilic cross-coupling partners and the scalability of the coupling process has been demonstrated upto 700 mL scale with a multimode microwave reactor (Baghbanzadeh et al., 2011).

The bromoenone of the Ni (II) complexes of β-alanine Schiff's base was successfully synthesized and a practical and highly efficient route to α-aryl and α-heteroaryl substituted β-amino acids using the Suzuki coupling reaction was developed. The heterogeneous solution was stirred at 70°C for 40 min under microwave irradiation. A broad range of aryl and heteroaryl substituents could be used under the operationally simple and effective conditions (Ding et al., 2009).

12.8   SUZUKI-MIYAURA CROSS-COUPLING

Microwave-assisted Suzuki-Miyaura cross-coupling reaction has been used for the generation of aza-analogs of the natural product Steganacin (Chapman and Thompson, 2007). It was reported that the application of microwave irradiation was highly beneficial for the biaryl coupling of electronically rich aryl bromides with some substituted o-formylphenylboronic acids (Beryozkina et al., 2006).

The use of N-vinyl pyridinium and ammonium tetrafluoroborate salts as new and excellent electrophilic coupling partners for the Suzuki–Miyaura cross-coupling reaction has also been described (Buszek and Brown, 2007).

where R = Ar, Vinyl; R_1 = Me, Ar.

The Suzuki-Miyaura cross-coupling reaction between oxazole-4-yl boronate and 2-iodo-oxazole derivatives was carried out under microwave irradiation at 150 W maximum power for 20 min to afford bis-oxazoles in good to excellent yields. A two fold increase in the yield and a six fold decrease in the reaction time were observed as compared to conventional heating (Flegeau et al., 2008).

Pyridazinones are recognized as privileged scaffolds as these are present in a wide range of commercially important drugs and agrochemicals. Cao et al. (2008) have reported an efficient methodology for the functionalization of 6-chloropyridazinone derivatives using the Suzuki–Miyaura cross-coupling reaction with palladium-bis-(di-t-Bu-phosphino-di-hydroxy)- chloride (POPd) or $Pd_2(dba)_3$ as a catalyst.

where R = Methyl, Pyrrolidine or Piperidine and Ar = Sub(hetero)aryl.

Browne et al. (2009) have described the application of microwave irradiation to generate a range of C-4 arylatedsydnones (4-bromo-N-phenyl-NH-1,2,3-oxadiazole derivatives) from the corresponding 4-bromosydnone via the Suzuki–Miyaura reaction.

A microwave-assisted tandem Ir-catalyzed C–H boronylation/Pd-catalyzed Suzuki–Miyaura cross-coupling reaction has been reported (Harrisson et al., 2009). This includes the synthesis of arylboronic acids, direct boronylation of arenes and alkanes, which provides access to synthetically useful compounds without relying on the accessibility of aryl or alkyl halides. They developed an elegant methodology for the direct C–H boronylation using a pinacoldiborane dimer under microwave irradiation at 80°C for 5–60 min.

A facile synthesis of some new substituted aryl and heteroarylflavones by thermal and microwave-assisted Suzuki-Miyaura coupling reaction was also carried out (Joshi and Hatim, 2012).

Synthesis, characterization and microwave-promoted catalytic activity of novel benzimidazole salt bearing silicon-containing substituents in Suzuki-Miyaura cross-coupling reactions under aerobic conditions was studied by Hasan et al. (2012).

Villemin et al. (2001) studied the possibility of the Suzuki–Miyaura cross-coupling reaction of heteroaryl halides in monomethylformamide (MMF), which has a higher polarity than water and other amides under microwave irradiation. The commercially available nontoxic sodium tetraphenylborate has been used in water (or MMF) for the phenylation of heteroarylhalides under mono-mode microwave irradiation. The rate of reaction was enhanced and this afforded products in a more rapid and clean manner.

Blettner et al. (1998) carried out poly (ethyleneglycol) (PEG)-supported liquid-phase synthesis of biaryls by the Suzuki-Miyaura cross-coupling reaction. A higher temperature was necessary to assure the quantitative conversion of the polymer-bound aryl halide into the coupling products as compared to the nonpolymer supported liquid phase reaction. The polymer bound products were isolated in reasonably good yields by either simple precipitation of the soluble support or column filtration. They also studied the effect of microwave irradiation and found that this reduces the reaction time from 2 h to 2–4 min. The recovery of the product and catalyst, conservation of energy using microwave irradiation, and low waste protocols due to the absence of solvents make this method very attractive. The microwave-assisted reaction reduces the reaction time dramatically and has proved to be an efficient method in different areas of chemistry (Larhed and Hallberg, 2001).

Symmetric and nonsymmetric HIV-1 protease inhibitors were also synthesized under microwave irradiation (Alterman et al., 1999; Schaal et al., 2001).

Non-symmetric cyclic sulfonamide (HIV-1 protease inhibitor)

C_2-symmetric HIV-1 protease inhibitors R = Aryl or Heteroaryl.

Coupling reactions of both the types; homo-coupling and cross-coupling have been carried out in the presence of some metal catalysts to yield the coupled product in presence of microwave irradiation in a short period of time and with reasonably good yields.

KEYWORDS

- **Liebeskind-Srogl Coupling**
- **Negishi Coupling**
- **Sonogashira Coupling**
- **Stille Cross-Coupling**
- **Suzuki Coupling**
- **Suzuki-Miyaura Cross-Coupling**

REFERENCES

Alterman, M., Andersson, H. D., Garg, N., Ahlsen, G., Lovgren, S., Classon, B., Danielson, U. H., Kvarnstrom, I., Vrang, L., Unge, T., Samuelsson, B., & Hallserg, A. (1999). Journal of Medicinal Chemistry, 42, 3835–3844.

Baghbanzadeh, M., Pilger, C., & Kappe, C. O. (2011). The Journal of Organic Chemistry, 76, 1507–1510.

Baraznenok, I. L., Nenajdenko, V. G., & Balenkova, E. S. (2000). Tetrahedron, 56, 3077–3119.

Bates, R. (2000). Organic synthesis using transition metals, New Jersey: Wiley.

Bedford, R. B., Blake, M. E., Butts, C. P., & Holder, D. (2003). Chemical Communications, 466–467.

Beryozkina, T., Appukkuttan, P., Mont, N., & Van der Eycken, E. (2006). Organic Letters, 8, 487–490.

Blettner, C. G., Konig, W. A., Stenzel, W., & Schotten, T. (1998). Synlett, 3, 295–297.

Bon, R. S., Sprenkels, N. E., Koningstein, M. M., Schmitz, R. F., de Kanter, F. J. J., Doemling, A., Groen, M. B., & Orru, R. V. A. (2008). Organic and Biomolecular Chemistry, 6, 130–137.

Brandsma, L., Vasilevsky, S. F., & Vekruijsse, H. D. (1998). Application of transition metal catalysis, New York: Springer-Verlag.

Browne, D. L., Taylor, J. B., Plant, A., & Harrity, J. P. A. (2009). Journal of Organic Chemistry, 74, 396–400.

Buszek, K. R., & Brown, N. (2007). Organic Letters, 9, 707–710.

Cao, P., Qu, J., Burton, G., & Rivero, R. A. (2008). Journal of Organic Chemistry, 73, 7204–7208.

Chapman, E. E., & Thompson, A. (2007). Chemtracts, 20, 32–35.

Chen, Y., Markina, N. A., & Larock, R. C. (2009). Tetrahedron, 65, 8908–8915.

Coffin, R. C., Peet, J., Rogers, J., & Bazan, G. C. (2009). Nature Chemistry, 1, 657–661.

Coltuclu, V., Dadush, E., Naravane, A., & Kabalka, G. W. (2013). Molecules, 18, 1755–1761.

Crabtree, R. H. (2005). The Organometallic Chemistry of the Transition Metals, 4th Ed. John Wiley & Sons.

Dawood, K. M., & El-Deftar, M. M. (2010). Arkivoc, (ix), 319–330.

Dighe, M. G., & Degani, M. S. (2011). Arkivoc, (xi), 189–197.

Ding, X., Ye, D., Liu, F., Deng, G., Liu, G., Luo, X., Jiang, H., & Liu, H. (2009). The Journal of Organic Chemistry, 74, 5656–5659.

Farina, V., Krishnan, B., Marshall, D. R., & Roth, G. P. J. (1993). Journal of Organic Chemistry, 58, 5434–5444.

Farina, V., Krishnamurthy, V., & Scott, W. J. (1997). Organic Reactions, 50, 1–652.

Flegeau, E. F., Popkin, M. E., & Greaney, M. F. (2008). Journal of Organic Chemistry, 73, 3303–3306.

Fugami, K., Ohnuma, S. Y., Kameyama, M., Saotome, T., & Kosugi, M. (1999). Synlett, 1, 63–64.

Hajipour, A. R., Karamia, K., & Rafieea, F. (2012). Applied Organometallic Chemistry, 26, 27–31.

Han, X. J., Stoltz, B. M., & Corey, E. J. (1999). Journal of the American Chemical Society, 121, 7600–7605.

Harrisson, M. J., Marder, T. B., & Steel, P. G. (2009). Organic Letters, 11, 3586–3589.

Hartwig, J. F. (2010). Organotransition metal chemistry, from bonding to catalysis, New York: University Science Books.

Hasan, K. U. C., Nihat, S., Akkurt, M., & Baktir, Z. (2012). Turkish Journal of Chemistry, 36, 201–217.

Hassan, J., Sevignon, M., Gozzi, C., Schulz, E., & Lemaire, M. (2002). Chemical Reviews, 102, 1359–1470.

Huang, H., Liu, H. Jiang, H., & Chen, K. (2008). The Journal of Organic Chemistry, 73, 6037–6040.

Joshi. V., & Hatim, J. G. (2012). Indian Journal of Chemistry, 51B, 1002–1010.

King, A. O., Okukado, N. & Negishi, E. (1977). Journal of the Chemical Society, Chemical Communications, 683–684.

Koopmans, C., Iannelli, M., Kerep, P., Klink, M., Schmitz, S., Sinnwell, S., & Ritter, H. (2006). Tetrahedron, 62, 4709–4714.

Krascsenicsova, K., Walla, P., Kasak, P., Vray, G., Kappe, C. O., & Putala, M. (2004). Chemical Communications, 2606–2607.

Larhed, M., & Hallberg, A. (2001). Drug Discovery Today, 6, 406–416.

Larhed, M., Hoshino, M., Hadida, S., Curran, D. P., & Hallberg, A. J. (1997). Journal of Organic Chemistry, 62, 5583–5587.

Leadbeater, N. E., & Marco, M. (2002). Organic Letters, 4, 2973–2976.

Leadbeater, N. E., & Marco, M. (2003). Journal of Organic Chemistry, 68, 888–892.

Liebeskind, L. S., & Srogl, J. (2000). Journal of the American Chemical Society, 122, 11260–11261.

Lin, Y. S., & Lin, C. Y. (2005). Journal of the Chinese Chemical Society, 52, 849–852.

Lindh, J., Enquist, P., Pilotti, Å., Nilsson, P., & Larhed, M. (2007). The Journal of Organic Chemistry, 72, 7957–7962.

Lipshutz, B. H., Frieman, B. A., Lee, C. T., Lower, A., Nihan, D. M., & Taft, B. R. (2006). Chemistry-An Asian Journal, 1, 417–429.

Littke, A. F., & Fu, G. C. (1999). Angewandte Chemie International Edition, 38, 2411–2413.

Mazuela, J., Pàmies, O., & Diéguez, M. (2010). Chemistry - A European Journal, 16, 3434–3440.

Miao, G., Ye, P., Yu, L., & Baldino, C. M. (2005). The Journal of Organic Chemistry, 70, 2332–2334.

Mitchell, T. N., Amamria, A., Killing, H., & Rutschow, D. J. (1986). Organometallic Chemistry, 304, 257–260.

Miyaura, N., & Suzuki, A. (1995). Chemical Reviews, 95, 2457–2483.

Molander, G.A., Katona, B. W., & Machrouhi, F. (2002). Journal of Organic Chemistry, 67, 8416–8423.

Moore, L. R., & Vicic, D. A. (2008). Chemistry-An Asian Journal, 3, 1046–1049.

Mutule, I., & Suna, E. (2004). Tetrahadron Letters, 45, 3909–3912.

Negishi, E. I., Wang, G., Rao, H., & Xu, Z. (2010). Journal of Organic Chemistry, 75, 3151–3182.

Negishi, E., & Xu, C. (2002). Handbook of organopalladium chemistry for organic synthesis, New York: Wiley, 531–549.

Negishi, E., Okukado, N., King, A. O., Van Horn, D. E., & Spiegel, B. I. (1978). Journal of the American Chemical Society, 100, 2254–2256.

Negishi, F., Zeng, X., Tan, X., Qian, M., Hu, Q., & Huang, Z. (1998). in Metal-catalyzed cross-coupling reactions, de Meijere, A., & Diederich, F. (Eds.), New York: Wiley-VCH.

Nun, P., Martinez, J., & Lamaty, F. (2009). Synlett, 11, 1761–1764.

Olofsson, K., Kim, S. Y., Larhed, M., Curran, D. P., & Hallberg, A. J. (1999). Journal of Organic Chemistry, 64, 4539–4541.

Pan, Y., Ruhland, B., & Holmes, C. P. (2001). Angewandte Chemie International Edition, 40, 4488–4491.

Patel, B. A., Ziegler, C. B., Cortese, N. A., Plevyak, J. E., Zebovitz, T. C., Terpko, M., & Heck, R. F. (1977). Journal of Organic Chemistry, 42, 3903–3907.

Pchalek, K., & Hay, M. P. (2006). Journal of Organic Chemistry, 71, 6530–6535.

Phapale, V. B., & Cardenas, D. J. (2009). Chemical Society Reviews, 38, 1598–1607.

Prokopcova, H., Lengar, A., & Kappe, C. O. (2005). 11th Blue danube symposium on heterocyclic chemistry, 28th August-1st September, Brno, Czech Republic.

Raut, C. N., Mane, R. B., Bagul, S. M., Janrao, R. A., & Mahulikar, P. P. (2009). Arkivoc, (xi), 105–114.

Schaal, W., Karlsson, A., Ahlsen, G., Lindberg, J., Andersson, H. O., Danielson, U. H., Classon, B., Unge, T., Samuelsson, B., Hulten, J., Hallberg, A., & Karlen, A. J. (2001). Medicinal Chemistry, 44, 155–169.

Shah, D., & Kaur, H. (2012). Journal of Molecular Catalysis A: Chemical, 359, 69–73.

Shen, L., Huang, S., Nie, Y., & Lei, F. (2013). Molecules, 18, 1602–1612.

Stille, J. K. (1986). Angewandte Chemie International Edition, 25, 508–523.

Suzuki, A. (1999). Journal of Organometallic Chemistry, 576, 147–168.

Villemin, D., Gomez-Escalonilla, M. J., & Saint-Clair, J. F. (2001). Tetrahedron Letters, 42, 635–637.

Walla, P., & Kappe, C. O. (2004). Chemical Communications, 564–565.

Wunderlich, S., & Knochel, P. (2008). Organic Letters, 10, 4705–4707.

Zhang, X., & Sui, Z. (2003). Tetrahedron Letters, 44, 3071–3073.

Zhu, L., Wehmeyer, R. M., & Rieke, R. D. (1991). Journal of Organic Chemistry, 56, 1445–1453.

CHAPTER 13

SYNTHESIS OF HETEROCYCLES

CHETNA AMETA, RAJAT AMETA, and SEEMA KOTHARI

CONTENTS

Heterocyclic compounds have a wide range of applications. They are predominant among the type of compounds used as pharmaceuticals, agrochemicals, veterinary products, etc. The recent introduction of single-mode technology assures safe and reproducible experimental procedures and microwave synthesis has gained acceptance and popularity among the synthetic chemists (Hayes, 2002). Microwaves are highly important in green synthesis of heterocyclics. Heterocyclic ring formation can be classified in many types depending on the size of ring and number nature of hetero atoms (N, O, S., etc.). The major classes are –

Five Membered Heterocyclic rings
- Pyrroles
- Pyrazoles
- Imidazoles
- Oxadiazoles
- Triazoles and tetrazoles
- Isoxazolines and pyrazolines

Benzoderivatives of Five Membered ring
- Benzimidazoles, Benzoxazoles and Benzthiazoles
- Indoles
- γ-Carbolines

Six Membered rings
- Dihydropyridines
- Dihydropyridopyrimidinones
- Dihydropyrimidines
- Tetrazines

Polycyclic Six Membered rings
- Quinolines and substituted Pyrimidines

Medium Sized Heterocycles

13.1 FIVE MEMBERED HETEROCYCLIC RINGS

13.1.1 PYRROLES

The classical Paal–Knorr cyclization of 1.4-diketones to give pyrroles was dramatically speeded up under microwave irradiation with high yields (70–90%).

where R = $CH_2C_6H_5$, 4-$MeOC_6H_4$, 2-ClC_6H_4

The reaction required 12 h in presence of Lewis acid under conventional heating, but it is completed within 0.5 to 2.0 min by solvent-free microwave irradiation technique (Danks, 1999).

The Clauson-Kaas pyrrole synthesis involving the reaction of primary amines with 2, 5-dialkoxytetrahydrofurans was traditionally carried out in refluxing acetic acid (AcOH). Extension to less activated nitrogen nucleophiles often necessitates the use of acidic promoters. Miles et al. (2009) reported that the synthesis of N-substituted pyrroles can be carried out under microwave conditions (10–30 min) using acetic acid or water without additional catalysts. The reaction is successful for all common nitrogen inputs in the case of acetic acid, where as benzamide and benzylamine are resistant to cyclocondensation under aqueous conditions.

where R = Aryl, Sulfonyl, Benzyl, Alkyl, COPh.

13.1.2 PYRAZOLES

Some pyrazole derivatives play a vital role in biological fields because of their analgesic, antiinflammatory, antibacterial and hypoglycaemic activities. Microwave irradiations have been used in the preparation of pyrazoles (45–80%) from hydrazones using the Vilsmeier cyclization method by treatment with $POCl_3$ and DMF (Selvi and Perumal, 2002). The reaction is speeded up by factors of several hundred fold.

The traditional method required 4–5 h and yield was relatively low (40–75%), while it is completed within 35–50 sec with high yield under microwave exposure.

Mistry and Desai (2005) also synthesized the 1-(4'-methylphenyl)-3-methyl-5-(2"-iminosubstituted benzothiazole)pyrazoles by microwave irradiation of 2-amino-6– nitrobenzothazole (0.01 mole) and 1-(4'methylphenyl)-3 methyl-5-pyrazolone for 2–3 min.

A series of pyrazolo[3,4-b]quinolines has been synthesized using one pot water mediated synthetic route under microwave irradiation involving the condensation of 2-chloroquinoline-3-carbaldehydes with semicarbazide or 2,4-dinitrophenyl hydrazine. The same reaction has been carried out in ethanol by conventional method but in microwave-assisted method, water is used as a solvent and it gives 94% yield in 2–5 min (Alam et al., 2013).

A simple protocol for synthesis of a series of 1-(4-substituted phenyl)-3-phenyl-1H-pyrazole-4-carbaldehyde has been developed under microwave irradiation (Selvam et al., 2011).

where R = H, Br, Cl, F, NHCOCH$_3$.

An efficient and extremely fast method for the synthesis of some thiazolidinone derivatives from benzotriazole has been reported under microwave irradiation (Shrimali et al., 2009).

13.1.3   IMIDAZOLES

An important classical preparation of imidazoles is from α-diketones, an aldehyde and ammonia. Usyatinsky and Khmelinilsky (2000) obtained excellent yield (75–85%) in solvent-free reaction within 8–10 min of microwave irradiation where as classical method takes about 4 h reflux with acetic acid.

$$R_1{-}\underset{\underset{H}{|}}{C}{=}O \quad + \quad \underset{R_2}{\overset{O}{\diagup}}C{-}C\underset{R_3}{\overset{O}{\diagdown}}$$

$$\Big\downarrow \begin{array}{l} Al_2O_2 \\ CH_3COONH_4 \end{array}$$

$$\underset{R_1}{\underset{N\diagdown\diagup NH}{R_2\diagup\diagdown R_3}}$$

where R_1 = C_6H_5, 4-ClC_6H_4, 2-Thiophenyl, etc.; R_2 = R_3 = C_6H_5, 4-$CH_3C_6H_4$.

An efficient and a quick microwave-assisted synthesis of benzimidazoles and trisubstituted imidazoles has been developed by Zhao et al. (2005). Three benzimidazoles were obtained as a result of the condensation of 1,2-phenylenediamine with carboxylic acids and acetoacetic ester without catalyst. A series of trisubstituted imidazoles were synthesized by condensation of benzil, aromatic aldehyde and ammonium acetate in the presence of glacial acetic acid.

$$\overset{NH_2}{\underset{NH_2}{\bigcirc}} \quad + \quad \underset{R_1}{\overset{O}{\diagup}}\diagdown_{R_2} \quad \xrightarrow[\text{MW}]{\text{Solvent-free}} \quad \overset{N}{\underset{\underset{H}{N}}{\bigcirc}}{-}R_1$$

The yield was 92 and 90%, where R_2 = OH and R_1 = $C_6H_5OCH_2$- and 2,4-Di-chloro-$C_6H_3OCH_2$-, respectively.

A facile and environmentally benign method for synthesis of trisubstituted imidazoles by a simple component condensation under microwave irradiation was developed. When there is an electron-donating group on the aromatic aldehydes, one can still obtain good yields by prolonging the reaction time. This method has obvious superiority for its short reaction time and easy work-up process.

13.1.4   OXAZOLINES

The biologically active oxazolines can be prepared by the reaction of substituted aldehyde and polyhydroxyamine in solvent-free condition using microwaves within

5–7 min. (Marrero-Trerreo and Loupy, 1996) where as it requires 15–16 h under conventional heating.

$$R-\underset{\underset{O}{\|}}{C}-O-H + H_2N-\underset{\underset{CH_2OH}{|}}{\overset{\overset{CH_2OH}{|}}{C}}-CH_2OH$$

$$\downarrow -H_2O$$

$$R-\underset{\underset{O}{\|}}{C}-NH-\underset{\underset{CH_2OH}{|}}{\overset{\overset{CH_2OH}{|}}{C}}-CH_2OH$$

$$\downarrow -H_2O$$

where R = 2- Furyl, Phenyl, Heptadecanyl.

Microwave-assisted synthesis of 2-oxazolines from carboxylic acids using the open vessel technique has been reported (Sharma et al., 2009). This method involves direct condensation of carboxylic acids with excess 2-amino-2-methyl-1-propanol at 170 °C to give the corresponding 2-oxazolines in moderate to excellent yields. This also proved to be simple and efficient for the conversion of carboxylic acids into 2-oxazolines.

where $R_1 = R_2 = H$, CH_3, CH_2OH.

4,4-Disubstituted 2-oxazolines have been synthesized by a microwave promoted solvent-free direct condensation of carboxylic acids and disubstituted β-amino alcohols in good to excellent yields (Garcia-Tellado et al., 2003). Zinc oxide is a very good solid support, particularly in cases, where a Lewis acid is required. This method is a good, safe, clean, economical, and environmentally friendly alternative to the classical procedures. The zinc oxide seems to play a double role as it creates a polar environment for the microwave catalysis (polar solid support) and activates the carbonyl group for the condensation (Lewis acid catalyst).

Okuma et al. (2013) described a simple, rapid and efficient green synthesis of aryl methoxylated benzamides and 2-oxazolines from renewable *Eucalyptus* biomass-tar derivatives. One pot synthesis was realized using microwave irradiation as an alternative energy source, through a direct condensation reaction of amino alcohol with carboxylic acids as starting materials.

13.1.5 TRIAZOLES AND TETRAZOLES

Five membered triazoles were prepared by microwave exposure within 10 min (Bentiss et al., 2000). Ethylene glycol was used as a solvent. The yield of product was upto 96%. However, in conventional method, the yield is only 45–60% and that too in 60 min.

where Ar = C_6H_5, 4-$CH_3C_6H_4$, 4-$NH_2C_6H_4$, 4-OHC_6H_4, 4-$CH_3OC_6H_4$.

Similarly, tetrazoles were synthesized by Alterman and Hallbarg (2000) on cyclization of substituted aryl nitriles in solvent N,N'-dimethylformamide. The conventional method required 3–9 h while microwave method gives the product tetrazoles in 5–25 min.

where R = OCH_3, NO_2, CH_3.

Cycloaddition of substituted chalcones with hydrazides of benzotriazole in presence of glacial acetic acid and microwave exposure gives pyrazole containing triazole heterocycle. The pyrazole ring is a prominent structure motif found in many pharmaceutically active compounds. When this active moiety was added with different substituted hydrazides of benzotriazole, it gives a greater biological active compound in 10 min with 66 – 82% yield under microwaves (Tiwari et al., 2013b).

A fast and facile procedure for the synthesis of pyridomercaptotriazole and pyridothiazolidinone was reported starting from dihydropyridine (Mehta et al., 2008). Subsequent oxidation with nitrating mixture (HNO_3/H_2SO_4) produced the anticipated 2,6-dimethylpyridine derivatives, which were subsequently condensed with thiosemicarbazide in ethanol to produce the key intermediate 2,2-[4-(4-substituted phenyl)-2,6-dimethylpyridine-3,5-diyl] dicarbonyldihydrazine carbothioamides. In the final step, pyridomercaptotriazole derivatives were synthesized by treating substituted carbothioamides in alkaline media. In parallel, pyridothiazolidinone derivatives were obtained by the reaction of substituted carbothioamides with ClCH$_2$COOH/CH$_3$COONa. All the reactions were completed within 4–10 min of microwave irradiation and give the desired products in high yields and excellent purities.

Various 5-substituted 1H-tetrazole derivatives were synthesized in a simple and environmentally benign method from the reaction of aryl and benzyl nitriles with sodium azide in solvent-free media using montmorillonite K10 clay as solid recyclable heterogeneous acidic catalyst and microwave irradiation in good yields and short reaction times.

The catalyst is inexpensive, nontoxic and reusable, which makes the process convenient, more economic and benign. The recyclability of the K10 clay is another advantage.

Synthesis of triazole derivatives from 2-mercaptobenzothiazole has been achieved in four steps under microwave irradiation. LiBr potentially replace solvents and corrosive acids in this method resulting in a facile, efficient and enviro-economic synthesis of 5-[(1,3-benzothiazol-2-ylsulfanyl)methyl]-4-[(phenyl substituted)amino]-4H-1, 2, 4-triazole-3-thiol with catalytic amount of LiBr using basic alumina as inorganic solid support (Ameta et al., 2010b).

where R_1 = H, 4-NO_2, 2-Cl, 3-Cl, 4-Cl, 2-OH, 3-OH, 4-OH, 2-OCH_3, 4-OCH_3.

13.1.6 OXADIAZOLE

Hayes (2002) reported the synthesis of 1,3,4-oxadiazoles by the dehydration of un-symmetrical diarylhydrazine using Burgess's reagent under microwave irradiation.

where R_1 = Alkyl, Aryl; R_2 = Cl, OCH_3.

The conventional synthesis of oxadiazoles takes about 90 min while this time is reduced to 5–10 min under microwave irradiation and sidewise, the yield is also improved.

2-Aryl-5-(4-pyridyl)-1,3,4-oxadiazole

13.1.7 ISOXAZOLINES AND PYRAZOLINES

Derivatives of isoxazolines and pyrazolines have played a crucial role in the history of heterocyclic chemistry. These have been used extensively as important pharmacophores and synthons in the field of organic chemistry. 1,3-Dipolar cycloaddition reactions gave isoxazolines and pyrazolines by the addition of activated olefins or nitrile imides, respectively. The resulting compounds were obtained in high yield (55–85%) under microwave irradiation than under conventional conditions (22–40% yield) (Kaddar et al., 1999).

where X = O, NPh; Y = -COOMe, -COOEt; Z = H, CN; R = H, Ph, -COOMe.

5-(4-Hydroxy-3-methoxyphenyl)-3-[4-(coumarin-3-yl)-thiazole-2-yl]-2-isoxa-zoline was synthesized using 2-N-(4-hydroxy-3-methoxyphenylchalconyl)-amino-4-(coumarin-3-yl)-thiazole and hydroxylamine hydrochloride by both; the conventional and microwave-assisted method. The conventional method requires about 6–9 h, while microwave irradiation method requires only 2.5–3.5 min and yield is also high (Desai and Desai, 2008).

2-Aryl-1H-indol-3-carbaldehyde derivatives underwent Claisen-Schmidt condensation with different acetophenones in ethylene glycol and few drops of piperidine under microwave irradiation. The reaction proceeded within 3–10 min at 750 W to afford substituted 3-(2-aryl-1H-indol-3-yl)-prop-2-en-1-one derivatives, which react with phenylhydrazine in absolute ethanol and few drops of glacial acetic acid under microwave irradiation at 350 W in 1–2 min with successive periods of 30 sec to afford the corresponding indolylpyrazoline analogs (Zahran et al., 2010). On the other hand, the conventional method requires 9–12 h.

A rapid and efficient method for the preparation of some new pyrimidine, pyrazoline and isoxazoline derivatives by the reaction of chalcones with hydroxylamine, urea and hydrazine hydrate has been reported under microwave exposure. This technique gives "yes or no answer" for a particular chemical transformation within 5 to 10 min as compared to several hours in conventional protocol in industry and academia. Chalcones are convenient starting material for the synthesis of pyrozolines, pyrimidines and isoxazolines due to their α, β-unsaturated moiety. When the chalcone reacts with different moieties like urea, thiourea, hydroxylamine hydrochloride, phenylhydrazine hydrate and other substituted hydrazides, it affords corresponding heterocycles in 6 to 8 min. with 62 to 80% yield, which is much more than classical method (Tiwari et al., 2013a).

13.2   BENZODERIVATIVES OF FIVE MEMBERED RING

13.2.1   BENZIMIDAZOLES, BENZOXAZOLES AND BENZTHIAZOLES

Benzoderivatives of five membered rings are privileged bicyclic ring system. They showed important pharmaceutical utilities as well as find several other industrial uses. Benzimidazoles, benzoxazoles and benzothiazoles were prepared by ring closure reaction of appropriate o–substituted aniline under irradiation of microwaves. The yields were significantly higher (85–96%) under microwave irradiation and that too within 9–10 min (Reddy et al., 1997) than conventional conditions (40–80%). The time taken for completion of the reaction was 3–4 h (Narsaiah et al., 1994).

where Z = NH$_2$, SH, OH; Y = NH, O, S.

2-(7-Bromo-9,9-diethyl–2-fluorenyl)benzothiazoles/benzooxazoles were prepared by condensation of 2-bromo-7-formyl-9,9-diethylfluorene and 2-aminothiophenol/2-aminophenol. 2-(7-Bromo-9,9-diethyl-2-fluorenyl)benzothiazole/benzoxazole with 3-hydroxydiephenylamine, in presence of Pd(dba)$_2$ and diphenylphosphinosphinoferrocene to from 2-(7-(3-benzyloxydiphenylamino)-9,9-diethyl-2-fluorenyl) benzothiazole/benzoxazole. Then these compounds were debenzylated to give 2-(7-(3-hydroxyldiphylamino)-9,9-diethyl-2-fluorenyl) benzothiazole/benzoxazole (Saroja et al., 2012). The time taken under microwave exposure was 2–3 min while the conventional method required a longer time.

where Pd(dba)$_2$ = Dibenzylideneacetone palladium (0); X = O, S.

Yadav et al. (2011) synthesized benzothiazoles by condensation of 2-aminothio-phenol with various saturated and olefinic fatty acids under microwaves in solvent-free condition with the use of catalyst P$_4$S$_{10}$. This reaction was completed within 3–4 min.

where R = $\sim\!\!\diagup\!\!(\diagdown)_n\!\!\diagdown$ with n = 11 and 13.

Organic reactions in water (without using any harmful organic solvent), are of great interest, because water is nontoxic, nonflammable, abundantly available, and inexpensive. Thus, water is generally considered as the reaction medium, which is low cost, safe, and environmentally benign alternative to other synthetic solvents. Furthermore, because of the low solubility of common organic compounds in water, the use of water as solvent often makes the purification of product very easy by simple filtration or extraction. A convenient and clean water-mediated synthesis of a series of 4-amino-2-aryl-1,2-dihydro pyrimido[1,2-a]benzimidazoles has been reported using alternative nonconventional energy sources (Dandia et al., 2008).

The products were obtained in shorter times with excellent yields (78–89%) from the multicomponent reaction of 2-aminobenzimidazole, malononitrile/ethyl-cyanoacetate, and carbonyl compounds. This procedure does not involve the use of any additional reagent/catalyst, produces no waste, and represents a green synthetic protocol with high atom economy. The combination of microwave irradiation, ultra-sonic irradiation, and aqueous-mediated conditions using multicomponent reactions leads to enhanced reaction rates, higher yields of pure products, easier workup, and sometimes selective conversions.

where X = CN, COOEt; Carbonyl compounds =

The reaction occurs in neat condition even in the absence of any solvent and catalyst under microwave irradiation but the product required further purification and recrystallization with suitable solvents, giving relatively lower yield. Further, this reaction occurs smoothly in ethanol also, but requires due precautions as well as some modifications in microwave oven for operational safety.

The reaction was also studied using cetyl trimethyl ammonium bromide as phase-transfer catalyst, but no further improvement in yield was observed. However, the reaction time was reduced slightly and hence, it was concluded that the aqueous medium seems to be a good method for the synthesis of pyrimido[1,2-a] benzimidazoles. There are some advantages, like (i) no additional requirement of reagent/catalyst; (ii) nonflammable and nontoxic reaction medium; (iii) high yields; (iv) virtually no waste generation; and (vi) ease of product isolation/purification, fulfill some of the criteria of green chemistry and thus, makes this methodology environmentally benign in nature.

13.2.2 INDOLES

Microwave-assisted solvent-free synthesis of indoles using Madelung's reaction was carried out by Seijas et al. (2008). They used potassium tert-butoxide as the base.

40% yield

Sridharan et al. (2006) observed that the solid-state reaction between anilines and phenacyl bromides at room temperature for 3 h in the presence of an equimolecular amount of sodium bicarbonate or a second equivalent of the aniline gives an intermediate compound. This compound provides 2-substituted indole under microwave irradiation within 1 min or so in good yields (50–59%). It is a mild, general, and environmentally friendly method for the synthesis of 2-arylindoles.

The Fischer indole one-pot synthesis was also carried out using microwaves (Creencia et al., 2011). 1,2,3,4-tetrahydrocarbazole was obtained with good yield (76%), when a mixture of phenylhydrazine, cyclohexanone and zinc chloride was irradiated with microwaves (600 W) for 3 min. However, when zinc chloride was replaced with p-toluenesulfonic acid (p-TSA), 1,2,3,4-tetrahydrocarbazole was produced with excellent yield (91%). Thus, a series of indoles were prepared using microwaves in the presence of p-TSA catalyst.

13.2.3 γ-CARBOLINES

The Graebe-Ullmann synthesis, converts 1-arylbenzotriazoles into carbazoles or their heterocyclic analogs. It has been accelerated under microwave irradiation, where the 1-(4-pyridyl)-benzotriazole was converted into a γ-carboline (Molina et al., 1993). This synthesis was completed within 2.5 h by conventional method while it takes only 15 min under microwave exposure and the yield was also increased.

where $R_1 = H, CH_3$; $R_2 = R_3 = H, CH_3$.

13.3 SIX MEMBERED RINGS

13.3.1 DIHYDROPYRIDINES

The Hantzsch dihydropyridine synthesis is still one of the most important routes to pyridine ring synthesis. The use of microwaves reduces the heating times (10–15 min) and also increases the yield (51–92%) where as under conventional conditions, it requires a long period of heating (about 12 hr) and poor to moderate yield were obtained (15–61%) (Ohberg and Westmen, 2001).

$Ar = C_6H_5$, 2-$NO_2C_6H_5$, 2-ClC_6H_5.

    The synthesis of various substituted 1,4-dihydropyridines has been achieved by the reaction of aldehydes, ethyl/methyl acetoacetates, and ammonium acetate in water using phase-transfer catalyst under microwave irradiation. This newer method has the advantage of good yields and short reaction times as compared to the classical Hantzsch's reaction conditions (Salehi and Guo, 2004).

    The use of microwave ovens in organic synthesis has been limited because of the selection of solvent for the reaction, possible loss of the solvent and volatile reactants, apart from risks like fire hazards. Water could be a safe alternative to organic solvents in such reactions. Water is a low cost and readily available nontoxic solvent. However, there are some problems with the use of water, such as solubility of substrates in water. These problems can be overcome by using of phase-transfer catalysts like tetrabutyl ammonium bromide (TBAB) and the design of novel heterogeneous catalysts.

The present synthesis of 1,4-dihydropyridines in water using phase-transfer catalyst under microwave irradiation provides an efficient and improved modification of Hantzsch's reaction with high yields and in short reaction times.

13.3.3 DIHYDROPYRIMIDINES

Dihydropyrimidines are associated with broad spectrum of biological activities. Excellent results were obtained for the preparation of dihydropyrimidine derivatives by Biginelle reaction with microwave enhancement (Danks, 1999). An increase in yield was observed to 60–90%, while there was a drastic decrease in reaction time for 12–24 h to 5 min.

where R_1 = H, Me; R_2 = H, Cl, OCH_3.

4-Aryl/alkyl-3, 4-dihydro-2(1H)-pyrimidone esters and 5-acetyl-4-aryl/alkyl-6-methyl-3,4-dihydropyrimidin-2(1)-ones were prepared under microwave radiations efficiently using montmorillonite clay-$Cu(NO_3)_2$.2.5H_2O system (Sitha et al., 2010). The reaction rate was enhanced in presence of microwave irradiation even under solventless conditions. The effect of the molar ratios of reagents and the influence of microwave irradiation on a neat mixture of aldehyde, β-dicarbonyl compound, urea (or thiourea) and montmorillonite clay/Cu $(NO_3)_2$ in the modified Biginelli reaction has been observed. This protocol was able to tolerate the structural variety. Both aromatic and aliphatic aldehydes have been subjected to this condensation very efficiently. Besides the β-ketone ester, β-diketone can also be employed without any reasonable decrease in yield. Thiourea has also been used successfully to provide the corresponding dihydropyrimidines.

$Cu(NO_3)_2.2.5H_2O$ was used as Lewis acid. No special precaution was needed in handling. The catalyst can be reused several times. The advantages of this environmentally benign reaction include the simple reaction set-up, high product yields, short reaction time and solventless conditions. In addition, the catalyst can be recovered and reused and hence, it is valuable from the economic point of view.

where R_1 = OEt, R or Ar; X = O and S.

13.4   POLYCYCLIC SIX MEMBERED RINGS

13.4.1   QUINOLINES

Ranu et al. (2000) reported that the reaction time for the Skraup synthesis of quinolines by microwave irradiation was reduced to a few min (5–12 min) with high yield of the product (80–87%).

where R = H, 2-CH_3, 3-CH_3, 4-CH_3, 2-OCH_3, 2-OH, 4-OCH_3; R_1 = H, CH_3, n-C_3H_7; R_2 = H, Et; R_3 = CH_3, 4-$CH_3OC_6H_4$.

13.4.2 SUBSTITUTED PYRIMIDINES

Pyimido(1,2,a)pyrimidines were prepared from dihydroaminopyrimidines and chromone-3-aldehydes (Eynde et al., 2001). The conventional reaction proceeds with refluxing in ethanol for 4 h with 60–70% yield where as much faster reactions (20 min) and better yields (95%) have been obtained with microwaves and it is solvent-free also.

where Ar = C_6H_5, 4-$CH_3C_6H_4$, 4-ClC_6H_4, CH_3, 4-$CH_3OC_6H_4$, 2-Thienyl.

1-{4'-(4,"6"-Diaryl-(1,"3,"5"-triazine-2"-yl)amino)phenyl}-3-methyl-4-(substituted phenyl)-4,5-dihydropyrazola-[5,4-d]pyrimidine-6-ol and 1-{4'-(4,"6"-diaryl-(1,"3,"5"-triazin-2"-yl)-amino)phenyl}-3-methyl-4-(substituted phenyl)-4,5-di hydropyrazolo [5,4-d] pyrimidin-6-thiol have been synthesized by the reaction of 1-{4'-(4,"6,"-diaryl-(1,"3,"5"-triazene-2"-yl)amino] phenyl}-3-methyl-2-pyrazolin-5-one with urea/thiourea and various substituted aldehydes. The conventional method required 6 h whereas under microwave irradiation technique, it was completed within 3–4 min (Rana et al., 2009).

where Y = O, S; R = -NC_5H_{10}, -NC_4H_8O; R_1 = 3-OCH_3, 4-OCH_3, H, 2-Cl, 3-Cl, 4-Cl.

Pyrimido(4, 5-d)pyrimidine derivatives were synthesized by an efficient, facile and solvent-free procedure. These are synthesized by the reaction between aromatic aldehydes, barbituric acid and urea/ thiourea under microwave irradiation. In this process, solid alumina was used as an energy transfer medium (Kategaonkar et al., 2009).

where X = O, S.

Kidwai et al. (2003) synthesized 4,6-diaryl-2-(4-morpholinyl/1-piperidinyl/1-pyrrolidinyl)pyrimidines by nucleophilic substitution reaction of substituted chalcones, S-benzylthiuronium chloride (SBT) and morpholine/piperidine/pyrrolidine in ethanol. The rate enhancement and high yield was attributed to the coupling of solvent-free conditions and microwaves.

13.5 MEDIUM SIZED RINGS

A diverse range of biologically active natural compounds and pharmaceuticals consists of medium-sized (seven to nine membered) heterocycles as an important structural component, that is, staurosporine analogs (Yang et al., 2007). There are several diverse methods for medium ring synthesis (Yet, 2000; Molander, 1998). In recent times, microwave-assisted chemistry has become a modern concept in organic synthesis as it offers several unique advantages like drastic acceleration of slow transformations, enhanced yields, cleaner reactions, etc. (Eycken and Kappe, 2006). Microwave chemistry already had significant impact on the everyday synthesis of organic molecules. The adoption and integration of microwave technology was considered important for many synthetic transformations, which were very difficult to carry out due to harsh conditions of temperatures and pressures. A series of chemical transformations have also been devised in presence of microwave irradiation with complicated chemical architectures via more expedient routes. Moreover, continuous technological improvements like continuous flow, focused microwave and parallel synthesis, have provided rapid synthesis of some biologically important chemical libraries (Baxendale et al., 2007; Kappe and Matloobi, 2007). Different synthetic methods have been carried out using microwaves for a wide range of N-, O- and S-containing seven to nine membered heterocyclic ring systems.

Seven membered heterocyclic ring systems are of interest to synthetic chemists as these are also predominant in pharmaceutically important compounds. The application of microwave irradiation has been found useful in designing some interesting molecules (Sharma et al., 2012).

Lautens et al. (2005) used allyl acetates and carbonates for a microwave-assisted intramolecular coupling with aryl iodides for the synthesis of some seven membered N- and O-containing heterocycles. Conventional reaction conditions involves reflux heating of allyl carbonates in the presence of $Pd_2(dba)_3$, tri-o-tolylphosphine and N,N-dimethylbutylamine in CH_3CN–H_2O, but with low yield (49%) of the cyclized products. However, increased yield (72%) has been reported under microwave irradiation. This 1,4-benzodiazepine scaffold is an important component of pharmaceuticals used for treatment of neurological disorders.

Ohta et al. (2008) synthesized indole-fused benzo-1,4-diazepines by copper catalyzed domino three component coupling, indole formation and N-arylation under microwave irradiation from a simple N-mesyl-2-ethynylaniline at 170°C for 20–40 min.

The yield increases with the change in group R. It was 56, 71 and 83% for butyl, allyl and benzyl group, respectively.

Application of microwave induced Delepine reaction to the facile one pot synthesis of 7-(fluoro, chloro, bromo, iodo, methyl, methoxy and nitro) and 5, 7-dimethyl substituted, 1, 3-dihydro-2H-[1,4]-benzodiazepin-2-one-5-methyl carboxylate derivatives from the corresponding 1-chloroacetyl isatin and chloroacetyl chloride has been reported by Sharma et al. (2010). They observed that the congenital method required 8–10 h with average yield of 36–42% whereas it is competed in 8–10 min and that too with good yield (70–82%) under microwaves.

A facile synthesis of 4-phenyl-1*H*-1,5-benzodiazepines from *o*-phenylenedi-amines and 1,3-diketones in the presence of a catalytic amount of acetic acid has been achieved, in excellent yields (62–97%) in 2–10 min, under microwave irradiation (Tsoleridis et al., 2008). This method is very simple and the reaction conditions are mild, environmentally friendly and quick.

Kaur et al. (2012) carried out microwave-assisted synthesis of 2-hetryl amino substituted novel analogs of 1,4-benzodiazepine-5-piperidinyl carboxamides. The 2-iminothioethyl ether derivatives undergo smooth nucleophilic displacement reaction with variety of hetryl amines and afforded products in excellent yields (90%) in 2–6 min. under microwave irradiation. They have highlighted the importance of the incorporation of some pharmacophores such as 2-aminopyridinyl, pyrimidinyl, and benzothiazolyl in 1,4-benzodiazepine nucleus. The amide functionality (the CO-NH group) in the seven membered heterocyclic ring of 1,4-benzodiazepine nucleus is the only active functional group present in this molecule, which will provide an important site for the incorporation of a wide variety of heterocycles (and fused heterocycles) and heterocycle appended structures in this molecule.

Thakrar et al. (2013) synthesized a new series of highly functionalized 1,5-benzodiazepine derivatives from 3-[(1E)-N-(2-aminophenyl)ethanimidoyl]-4-hydroxyl-2H-chromen-2-one and pyrazole aldehyde using catalytic amount of trifluoroacetic acid under microwave irradiation. The major advantages of this process is its shorter reaction time (reduced from 8–10 h to 15–20 min), easy work up procedure, and excellent yield with high purity.

A new series of 2,4-disubstituted-2,3-dihydro substituted-1,5-benzodiazepine derivatives was synthesized by the condensation of o-phenylenediamine and various 1-(4'-substituted phenyl)-3-(3-mono-, or 4-mono-, or 3,4-di-, or 3,4,5-tri-substituted phenyl)-2-propene-1-one under microwave irradiation. The modification of coupling microwaves with the solvent-free technique is an easy, rapid, efficient and convenient protocol for the synthesis of 1,5-benzodiazepines. The microwave irradiation is better over the classical reflux heating because the solid-supported reactions gave better yield (58–86%) and these reactions were completed within 4–6 min in comparison to hours under conventional heating. However, it was observed that long exposure time to the microwave irradiation resulted in decomposition of compounds to dark tarry mixtures (Sharma and Joshi, 2012).

Salve and Mali (2013) reported an ecofriendly synthesis of a new series of 2,4-disubstituted-1,5-benzodiazepine derivatives, by the condensation of o-phenylenediamine and various substituted chalcones under microwave irradiation. This method is facile, efficient and environmentally benign. It afforded all the synthesized compounds in good to excellent yields (78–96%) and the reaction time (10–20 min) was also considerably short.

13.5.1 THIAZEPINES, THIADIAZEPINES AND THEIR ANALOGUES

1,5-Benzothiazepine framework has emerged as an important pharmacophore since its derivatives exhibit a wide range of medicinal applications such as anti-HIV, anticancer, angiotensin converting enzyme inhibitors, antimicrobial compounds, etc. Willy and Muller (2010) reported a one pot three component synthesis of 2,4-disubstituted benzo[b][1,5]thiazepines. This method consists of an initial Sonogashira coupling of acid chlorides with terminal alkynes to give the corresponding alkynones, which subsequently undergo a Michael addition-cyclocondensation reaction with o-aminothiophenol in the presence of acetic acid. The cyclocondensation step under conventional as well as microwave conditions was carried out at identical temperature (60 °C), but the yield of reaction was found to be more (45–77%) than double in the case of the latter.

One of the benefits of microwave-assisted synthesis is the remarkable acceleration of reactions conducted in unconventional reaction media. The use of water is an economical and nonflammable option to traditional organic solvents. It is important as the high dielectric constant of the water allows efficient absorption of microwave irradiation. Tu et al. (2009) developed a microwave-assisted reaction under aqueous conditions for an efficient synthesis of benzo[e][1,4]thiazepin-2-ones via a multicomponent reaction between benzaldehyde, an aromatic amine and mercaptoacetic acid at 110 °C in 7–10 min with 94% yield using water as the solvent.

This methodology afforded a suitable route to heterocycles, which were mainly prepared through multiple steps requiring long reaction times or harsh conditions. In addition to the use of unconventional solvents, microwave irradiation also allows the possibility of conducting reactions in the solid phase, that is, under solventless conditions. Dandia et al. (2006) reported a microwave promoted synthesis of thiadiazepines possessing fused benzopyranotriazole rings via a tandem Michael addition-condensation reaction between 3-arylidene flavanones and 4-amino-5-alkyl 3-mercaptotriazole using basic alumina as solid support in 3–5 min.

13.5.2 OXACYCLES

The oxacycles have remained a more challenging synthetic task as compared to the N- and S-containing medium sized rings. The medium sized lactones have also attracted significant attention as these are considered to be a rare class of organic compounds, which are even difficult to produce in natural biosynthetic pathways (Shiina, 2007). Bagley et al. (2009) reported microwave-assisted synthesis of seven membered lactones through a tandem oxidation– cyclocondensation of 1,6-hexanediol.

Amongst the various oxidants screened, $BaMnO_4$ was found to give the best reaction performance in acetonitrile at a ceiling temperature of 150 °C for 60 min.

13.6   EIGHT MEMBERED HETEROCYCLES

Eight membered heterocycles have also attracted extensive synthetic interest due to their presence in a large number of naturally occurring molecules like Balasubramide, Serotobenine, Lundurine A and Buflavine. However, little information are there on generation of the constrained eight membered ring.

Appukkuttan et al. (2005) observed that the synthesis of structurally intriguing natural product analogs bearing highly constrained medium sized rings leads to explore a novel protocol for the synthesis of N-shifted Buflavine analogs. The combination of microwave-assisted Suzuki–Miyaura reaction and ring-closing metathesis

(RCM) reaction was explored to successfully generate the key eight-membered ring system of these molecules.

Microwave irradiation was found to play a key role in promoting the RCM reaction to generate the highly strained ring system of such molecules. The reaction time was also shortened from 3 h to 5 min as well as yields were increased (68–69%) compared with yields under conventional heating (55–58%). An intramolecular carbocyclization mediated by Hg(OTf)$_2$ under microwave irradiation was reported for the highly interesting indoloazocine framework (Donets et al., 2009). It is a novel protocol for the synthesis of N-containing medium sized rings.

The reaction time of 24–40 h under conventional heating were dramatically reduced to only 15–50 min under focused microwave irradiations, while the reaction yields were either similar or even higher.

Cheung et al. (2010) have reported a microwave-assisted ring expansion of N-vinyl β-lactams in a simple way for accessing eight membered ring enamides, making use of the [3,3]-sigmatropic rearrangement between two strategically placed alkenes on a β-lactam. The microwave-assisted rearrangement was carried out in DMF at 160–200 °C for 20–30 min, and the desired enamides were isolated in good yields (48–86%).

The microwave-assisted solvent-free intramolecular cyclization of α-iminoester derivatives leads to the synthesis of a number of seven, eight and ten membered lactams (Zradni et al., 2007). The yield was 70–85% in 15–35 min under microwave irradiation while it takes about 108 h at 140–170°C under conventional condition.

where R_1 = Me, Ph, i-Pr; n = 4, 5, 7; m = 0–2, 4.

An interesting strategy for the generation of 6,7-dihydro-5H-dibenzo[c,e]azepines and 5,6,7,8-tetrahydrodibenzo[c,e]azocines by a microwave-assisted copper catalyzed intramolecular A_3-coupling reaction was given by Bariwal et al. (2010).

13.7   NINE MEMBERED HETEROCYCLES AND BEYOND

The literature about the synthesis of heterocyclic compounds bearing nine- and ten-membered ring systems are hardly available, in spite of their potential applications as valuable pharmacophores (Banfi et al., 2003; Rodrıguez et al., 2004; Qadir et al., 2005). The main reason being the extremely high conformational strain of these rings. Therefore, very high activation energy is required for such a ring closure. The microwave irradiations can be used as tool for such difficult reaction pathways, leading to the synthesis of these target molecules. Appukkuttan et al. (2007) synthesized ring expanded Buflavine analogs, bearing a nine membered N-heterocyclic ring fused with a rigid biaryl system by microwave-assisted Suzuki-Miyaura reaction. The key biaryl intermediate was formed by microwave-assisted Suzuki reaction at 150 °C for 15 min in a mixture of DMF and H_2O (1:1), using $NaHCO_3$ as the base and $Pd(PPh_3)_4$ as the catalyst. The RCM reaction for the generation of the rigid, biaryl bearing nine membered ring system of the dibenzo[c,e]azonine skeleton was carried out under focused microwave irradiation in toluene at 150 °C for 15 min

(54–55% yield). It is quite significant because the RCM has resulted in relatively yield of 15% under conventional heating.

An enantioselective synthesis of the halogenated medium ring ether, natural product (+)-Obtusenyne, was reported involving a ring expansion of a seven membered ketene acetal by means of a Claisen rearrangement to give the core nine membered oxygen heterocycle. The synthesis of ent-obtusenyne from 2-deoxy-D-ribose has been reported (Mak et al., 2008).

The ten membered N-heterocycles have also attracted a lot of interest due to their structural appearance with various biologically interesting molecules like the alkaloid Dysazecine (Aladesanmi et al., 1983), Protopine (Xiao et al., 2008) (a potent antimalarial lead compound) as well as the nanomolar dopamine receptor antagonist LE300 (Mohr et al., 2006). Dunkel et al. (2010) reported a straight forward, microwave-assisted synthesis of tribenzo[b,d,f]- and pyridazino[d]dibenzo[b,f] azecines, employing the 'tert-amino effect.' The desired compounds were furnished via an open-vessel microwave-assisted cyclization of corresponding triphenyl intermediates at 100–200 °C for 2–150 min.

13.8 MISCELLANEOUS

Ameta et al. (2011) reported an environmentally benign, efficient and facile route for the preparation of phthalimide derivatives using LiBr as a catalyst under solvent-free condition and microwave exposure. In comparison to conventional synthesis involving tedious workup, excessive use of solvent and extra labor for separation and purification of compounds, the present method is associated with the advantages like operational simplicity, shorter reaction time and higher yields, which can prove this procedure as a useful alternative for the synthesis of heterocycles.

A rapid and efficient method for the preparation of fused thiazolo quinazoline derivatives by the reaction of chalcones with hydroxylamine, urea and hydrazine hydrate under solvent-free condition and microwave exposure has been reported (Ameta et al., 2010a). Here, toxic, and highly reactive chemicals were replaced with less reactive and less harmful but more selective building blocks, which activate selective chemical reactions by proper catalysis apart from the reduction of by products and easy separation of products. This synthesis was compared with conventional method.

A simple and efficient method has been developed for the synthesis of 2,6-di-substituted-imidazo[2,1-b][1,3,4] thiadiazoles under microwave activation using 2-amino-5-substituted-1,3,4-thiadiazoles and appropriate bromoketones as starting materials. These reactions had the benefits like convenient operation, short reaction time, and good yields (Dhepe et al., 2012).

where R_1 = Benzyl, 4-Chloro benzyl, 4-Nitro benzyl; R_2 = Cl, CH_3, OCH_3, NO_2, Br, F.

Under the framework of green chemistry, an efficient and extremely fast procedure for the synthesis of quinazoline derivatives from 2-arylidenetetralin-1-one through a four-step procedure under microwave irradiation has been reported (Mehta et al., 2007). A considerable increase in the reaction rate has been observed with better yield. An efficient synthesis of quinazoline derivatives carrying potential pharmacophores like thiazolidinone and isoxazole has also been carried out in an environmentally benign microwave protocol. The yields of the products formed under microwave irradiation were high (85%) in comparison to classical method (62%) and time required for completion of these reactions was also less.

A multi component condensation of substituted phenylthiourea/urea, aqueous formaldehyde and substituted aromatic/heterocyclic amines leads to 2-thioxohexahydro-1,3,5-triazines in aqueous medium under microwave irradiation in 30–60 sec in quantitative yield with reasonable purity. Further, triazolo[4,3-a]triazines were also prepared by a one pot reaction of in situ synthesized triazinyl hydrazine with CS_2.

Rapid and highly efficient one pot green chemical synthesis of substituted 6-(2-aminophenyl)-4-(4-substituted   phenyl)-3-thioxo-3,4-dihydro-1,2,4-triazin-5(2*H*)-one and 8-substituted-3,5-dihydro-2*H*-[1,2,4]triazino[5,6-*b*]indole was carried out in aqueous medium under microwave irradiation (Dandia et al., 2004).

Water is potentially a very useful solvent for microwave-mediated synthesis due to its high dielectric constant. The rate enhancement benefits are also there, when water is heated well above its boiling point in sealed vessels, organic substrates may become more soluble. The MDRs were conducted by reacting readily available and inexpensive starting materials in aqueous solution under microwave irradiation with a broad substrate scope and high overall yields (76–93%). The mechanism has been proposed to explain the reaction process and the resulting chemo-, regio- and stereoselectivity. The present green synthesis shows attractive characteristics such as the use of aqueous medium, one-pot conditions, short reaction periods (9–13 min), easy work-up/purification and reduced waste production without using any acids or metal promoters.

where X = N-Ph, O.

All the three classical approaches viz. cyclization, annulation and ring expansion have greatly benefited from microwave-assisted techniques. The application of microwave irradiation on the development of new and efficient synthetic approaches for the generation of heterocycles (with varied ring sizes) had a remarkable impact in last few years. The application of focused microwave irradiation is quite helpful in overcoming the difficult entropic/enthalpic barriers during formation of such ring systems and therefore, a diverse array of N, O and S containing five- to nine-membered heterocycles have been synthesized in higher yields and that too in shorter reaction times as compared to conventional conditions. Microwave-assisted continuous flow synthesis has also enabled fast and efficient synthesis of relatively larger quantities of the order of multigram of such heterocycles.

Heterocyclic moieties are commonly encountered in natural products and pharmaceutical drugs and therefore, more stress is given to synthesize heterocyclic compounds with diverse biological activities. Microwave chemistry is a step towards synthesizing heterocyclic compounds in an ecofriendly manner.

KEYWORDS

- Imidazole
- Indole
- Pyrazole
- Pyrazoline
- Pyrimidine
- Pyrrole
- Quinoline
- Triazole

REFERENCES

Aladesanmi, A. J., Kelley, C. J., & Leary, J. D. (1983). Journal of Natural Products, 46, 127–131.

Alam, M. M., Marella, A., Akhtar, M., Husain, A., Yar, M. S., Shaquiquzzaman, M., Tanwar, O. P., Saha, R., Khanna, S., & Shafi, S. (2013). Acta Poloniae Pharmaceutical-Drug Research, 70, 435–441.

Alterman, M., & Hallbarg, A. J. (2000). Journal of Organic Chemistry, 65, 7984–7489.

Ameta, C., Ameta, R., Tiwari, U., Punjabi, P. B., & Ameta, S. C. (2010a). Afinidad, 67, 394–400.

Ameta, C., Ameta, R., Tiwari, U., Punjabi, P. B., & Ameta, S. C. (2011). Journal of Indian Chemical Society, 88, 827–833.

Ameta, C., Sitha, D., Ameta, R., & Ameta, S. C. (2010b). Indonesian Journal of Chemistry, 10, 376–381.

Appukkuttan, P., Dehaen, W., & Eycken, E. V. (2005). Organic Letters, 7, 2723–2726.

Appukkuttan, P., Dehaen, W., & Eycken, E. V. (2007). Chemistry- A European Journal, 13, 6452–6460.

Bagley, M. C., Lin, Z., Phillips, D. J., & Graham, A. E. (2009). Tetrahedron Letters, 50, 6823–6825.

Banfi, L., Basso, A., Guanti, G., & Riva, R. (2003). Tetrahedron Letters, 44, 7655–7658.

Bariwal, J. B., Ermolatev, D. S., Glasnov, T. N., Hecke, K. V., Mehta, V. P., Meervelt, L. V., Kappe, C. O., & Eycken, E. V. (2010). Organic Letters, 12, 2774–2777.

Baxendale, I. R., Hayward, J. J., & Ley, S. V. (2007). Combinatorial Chemistry High Throughput Screening, 10, 802–836.

Bentiss, F., Lagrenee, M., & Barbry, D. (2000). Tetrahedron Letters, 41, 1539–1541.

Cheung, L. L. W., & Yudin, A. K. (2010). Chemistry– A European Journal, 16, 4100–4109.

Creencia, E. C., Tuskamoto, M., & Horaguchi, T. (2011). Journal of Heterocyclic Chemistry, 48, 1095–1102.

Dandia, A., Arya, K., & Sati, M. (2004). Synthetic Communications, 34, 1141–1155.

Dandia, A., Singh, R., & Khaturia, S. (2006). Bioorganic and Medicinal Chemistry, 14, 1303–1308.

Dandia, A., Singh, R., Jain, A. K., & Singh, D. (2008). Synthetic Communication, 38, 3543–3555.

Danks, T. N. (1999). Tetrahedron Letters, 40, 3957–3960.

Desai, J. T., & Desai, K. R. (2008). Journal of Iranian Chemical Society, 15, 67–73.

Dhepe, S., Kumar, S., Vinayakumar, R., Ramareddy, S. A., & Karki, S. S. (2012). Medicinal Chemistry Research, 21, 1550–1556.

Donets, P. A., Van Hecke, K., Van Meervelt, L., & Eycken, E. V. (2009). Organic Letters, 11, 3618–3621.

Dunkel, P., Turos, G., Benyei, A., Ludanyi, K., & Matyus, P. (2010). Tetrahedron, 66, 2331–2339.

Eycken E. V., & Kappe, C. O. (2006). Microwave-assisted synthesis of heterocycles, Vol. 1, Berlin-Heidelberg: Springer.

eEynde, J. J. V., Hecr, N., Kataeva, O., & Kappe, C. O. (2001). Tetrahedron, 57, 1785–1791.

Garcia-Tellado, F., Loupy, A., Petit, A., & Marrero-Terrero, A. L. (2003). European Journal of Organic Chemistry, 4387–4391.

Hayes, B. L. (2002). In Microwave Synthesis: Chemistry at the speed of light, NC: CEM Publishing.

Kaddar, H., Hamelin, J., & Benhaoua, H. (1999). Journal of Chemical Research (S), 718–719.

Kappe C. O., & Matloobi, M. (2007). Combinatorial Chemistry High Throughput Screening, 10, 735–750.

Kategaonkar, A. H., Sadaphal, S. A., Shelka, K. F., Shingate, B. B., & Shingare, M. S. (2009). Ukrainica Bioorganica Acta, 1, 3–7.

Kaur, N., Sharma, P., Sirohi, R., & Kishore, D. (2012). Archives of Applied Sciences Research, 4, 2256–2260.

Kidwai, M., Rastogi, S., & Saxena, S. (2003). Bulletin of Korean Chemical Society, 24, 1575–1578.

Lautens, M., Tayama E., & Herse, C. (2005). Journal of the American Chemical Society, 127, 72–73.

Mak, S. Y. F., Curtis, N. R., Payne, A. N., Congreve, M. S., Wildsmith, A. J., Francis, C. L., Davies, J. E., Pascu, S. I., Burton J. W., & Holmes, A. B. (2008). Chemistry- A European Journal, 14, 2867–2885.

Marrero-Terrero, A. L., & Loupy, A. (1996). Synlett, 3, 245–246.

Mehta, S., Swarnkar, N., Vyas, M., Vardia, J., Punjabi, P. B., & Ameta, S. C. (2007). Bulletin of Korean Chemical Society, 28, 2338–2342.

Mehta, S., Swarnkar, N., Vyas, R., Vardia, J., Punjabi, P. B., & Ameta, S. C. (2008). Phosphorus, Sulfur, Silicon and Related Elements, 183, 105–114.

Miles, K. C., Mays, S. M., Southerland, B. K., Auvil, T. J., & Ketcha, D. M. (2009). Arkivoc, (xiv), 181–190.

Mistry, K., & Desai, K. R. (2005). Indian Journal of Chemistry, 44B, 1452–1455.

Mohr, P., Decker, M., Enzensperger, C., & Lehmann, J. (2006). Journal of Medicinal Chemistry, 49, 2110–2116.

Molander, G. A. (1998). Account of Chemical Research, 31, 603–609.

Molina, A., Vaquero, J. J., Garcia-Navio, J. L., & Alvarez-Builla, J. (1993). Tetrahedron Letters., 34, 2673–2476.

Narsaiah, B., Sivaprasad, A., & Venkataratnam, R. V. (1994). Fluorine Chemistry, 66, 47–50.

Ohberg, L., & Westmen, J. (2001). Synlett, 8, 1296–1298.

Ohta, Y., Chiba, H., Oishi, S., Fujii, N., & Ohno, H. (2008). Organic Letters, 10, 3535–3538.

Okuma, A. A., Antonio, D. C., Carazza, F., & Duarte L. P. (2013). Green Process Synthesis, 2, 51–56.

Qadir, M., Cobb, J., Sheldrake, P. W., Whittall, N., White, A. J. P., Hii, K. K., Horton P. N., & Hursthouse, M. B. (2005). Journal of Organic Chemistry, 70, 1552–1557.

Rana, P. B., Patel, J. A., Mistry, B. D., & Desai, K. R. (2009). Indian Journal of Chemistry, 43B, 1601–1608.

Ranu, B. C., Hajra, A., & Jana, V. (2000). Tetrahedron, 57, 1785–1791.

Reddy, C. S. A., Rao, S. P., & Venkataratham, R. V. (1997). Tetrahedron Letters, 53, 5847–5854.

Rodrıguez, C., Ravelo, J. L., & Martin, V. S. (2004). Organic Letters, 6, 4787–4789.

Salehi, H., & Guo, Q.-X. (2004). Synthetic Communications, 34, 4349–4357.

Salve, P. S., & Mali., D. S. (2013). Journal of Chemical and Pharmaceutical Research, 5,158–161.

Saroja, N., Laxminarayan, E. Prasad, K. R. S., & Shanmukha, J. V. (2012). Der Pharma Chemica, 4, 490–493.

Seijas, J. A., Vazquez-Tato, M.P., Crecente-Campo, J., & Gomez-Doval, M. A. (2008). 12th International Electronic Conference on Synthetic Organic Chemistry, (ECSOC-12).

Selvam, T. P., Saravanan, G., Prakash, C. R., & Kumar, P. D. (2011). Asian Journal of Pharmaceutical Research, 1, 126–129.

Selvi, S., & Perumal, P. T. (2002). Journal of Heterocyclic Chemistry, 39, 1129–1131.

Sharma, A., Appukkuttana, P., & Eycken, E. V. (2012). Chemical Communications, 48, 1623–1637.

Sharma, N., & Joshi, Y. C. (2012). International Journal of Pharmaceutical and Biomedical Sciences, 3, 55–59.

Sharma, P., Vashistha, B., Tyagi, R., Srivastava, V., Shorey, M., Singh B., & Kishore, D. (2010). International Journal of Chemical Sciences, 8, 41–51.

Sharma, R., Vadivel, S. K., Duclos R. I., & Makriyannis, A. (2009). Tetrahedron Letters, 50, 5780–5782.

Shiina, I. (2007). Chemical Reviews, 107, 239–273.

Shrimali, K., Sitha, D., Vardia, J., Punjabi, P. B., & Ameta S. C. (2009). Afinidad, 66, 173–176.

Sitha, D., Ameta, R., Punjabi, P. B., & Ameta S. C. (2010). International Journal of Chemical Sciences, 8, 1973–1982.

Sridharan, V., Perumal, S., Avendaño, C., & Menéndez, J. C. (2006). Synlett, 91–95.

Thakrar, S., Bavishi, A., Radadiya, A., Parekh, S., Bhavsar, D., Vala, H., Pandya, N., & Shah, A. (2013). Journal of Heterocyclic Chemistry, 50, E73–E78.

Tiwari, U., Ameta, C., Rawal, M. K., Ameta, R., & Punjabi, P. B. (2013b). Indian Journal of Chemistry, 52B, 432–439.

Tiwari, U., Ameta, C., Sharma, S., Sharma, M., Pathak, A. K., & Punjabi, P. B. (2013a). European Chemistry Bulletin, 2, 242–246.

Tsoleridis, C. A., Pozarentzi, M., Mitkidou, S., & Stephanidou-Stephanatou, J. (2008). Arkivoc, (xv), 193–209.

Tu, S.-J., Cao, X.-D., Hao, W.-J., Zhang, X.-H., Yan, S., Wu, S.-S., Han Z. G., & Shi, F. (2009). Organic & Biomolecular Chemistry, 7, 557–563.

Usyatinsky, A. Y., & Khmelnitsky, Y. L. (2000). Tetrahedron Letters, 41, 5031–5134.

Willy, B., & Muller, T. J. J. (2010). Molecular Diversity, 14, 443–453.

Xiao, X., Liu, J., Hu, J., Zhu, X., Yang, H., Wang, C., & Zhang, Y. (2008). European Journal of Pharmacology, 591, 21–27.

Yadav, P. S., Devprakash, & Senthil, G. P. K. (2011). International Journal of Pharmaceutical Science and Drug Research, 3, 1–7.

Yang, S.-M., Malaviya, R., Wilson, L. J., Argentieri, R., Chen, X., Yang, C., Wang, B., Cavender D., & Murray, W. V. (2007). Bioorganic Medicinal Chemical Letters, 17, 326–331.

Yet, L. (2000). Chemical Reviews, 100, 2963–3007.

Zahran, M. A. H., Salama, H. F., Abdin, Y. G., & Eldeen, A. M. G. (2010). Journal of Chemical Sciences, 122, 587–595.

Zhao, N., Wang, Y.-L., & Wang, J.-Y. (2005). Journal of the Chinese Chemical Society, 52, 535–538.

Zradni, F. Z., Hamelin, J., & Derdour, A. (2007). Molecules, 12, 439–454.

CHAPTER 14

NANOMATERIALS

SURBHI BENJAMIN, SHWETA SHARMA, and RAKSHIT AMETA

CONTENTS

14.1 INTRODUCTION

Nanomaterials are defined as materials with at least one external dimension in the size ranging from approximately 1 to 100 nanometers. Nanoparticles are objects with all the three external dimensions at the nanoscale (Buzea et al., 2007). Nanoparticles or nanocrystals made up of metals, semiconductors, or oxides are of particular interest because of their mechanical, electrical, magnetic, optical, chemical and other properties. Nanoparticles have been used as quantum dots and as chemical catalysts such as nanomaterial-based catalysts.

Nanoparticles are of great scientific interest as they are a bridge between bulk materials and atomic or molecular structures. A bulk material should have constant physical properties regardless of its size, but at the nanoscale, this is often not the case. Size-dependent properties are observed in namomaterials such as quantum confinement in semiconductor particles and superparamagnetism in magnetic materials. Nanoparticles exhibit a number of special properties relative to bulk materials. The bending of bulk copper (wire, ribbon, etc.) occurs with movement of copper atoms/clusters at about the 50 nm scale. Copper nanoparticles smaller than 50 nm are considered super hard materials and do not exhibit the same malleability and ductility as bulk copper; however the change in properties is not always desirable. Ferroelectric materials smaller than 10 nm can switch their magnetization direction using room temperature thermal energy; thus, making them useless for memory storage. Suspensions of nanoparticles are possible because the interaction of the particle surface with the solvent is strong enough to overcome differences in density, which usually result in either sinking or floating of a material in a liquid. Nanoparticles often have unexpected visual properties because they are small enough to confine their electrons and produce quantum effects, for example, gold nanoparticles appear deep red to black in solution.

Many forms of nanoparticles/nanoaggregates have been synthesized viz. nanopowders, nanotubes, nanowires, nanonails, nanoarrow, nanobowling, nanobricks, nanotripods, nanorings, nanodisks, nanoflakes, nanopencils, nanorices, nanolines, nanodumbbells, nanoclusters, nanosheets, nanorods, nanodots, nanoalloys, nanobelts, nanoribbons, nanoplates, nanoneedles, nanocones, and nanofilms, which have a lot of applications, particularly as catalysts, magnetic materials, nanocomposites, nanodevices, chemical sensors, degradation of toxic chemicals, possible carriers of isotopes for medical applications. Shapes of some of these nanomaterials are shown here:

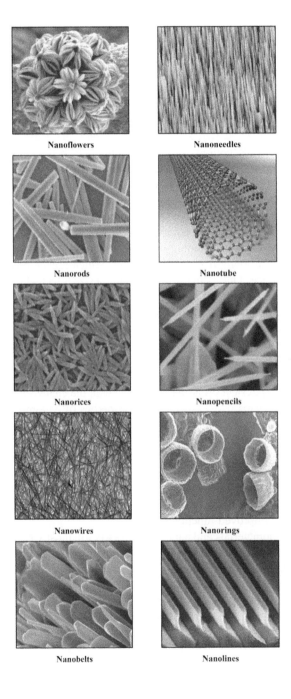

Nanoflowers

Nanoneedles

Nanorods

Nanotube

Nanorices

Nanopencils

Nanowires

Nanorings

Nanobelts

Nanolines

Very high surface area to volume ratio of nanoparticles often provides a tremendous driving force for diffusion, especially at elevated temperatures. Sintering is possible at lower temperatures and over shorter durations for these particles than for larger particles. Theoretically, it does not affect the density of the final product, but flow difficulties and the tendency of nanoparticles to agglomerate do complicate matters. The surface effects of nanoparticles also reduce the melting temperature.

Some nanoparticles are intentionally engineered and produced with very specific properties in mind such as shape, size, surface properties and chemistry. These properties are reflected in aerosols, colloids, or powders. Often, the behavior of nanomaterials may depend more on surface area than particle composition itself. Relative-surface area is one of the principal factors that enhance its reactivity, strength and electrical properties. Some examples of these engineered nanomaterials are carbon buckyballs or fullerenes; carbon nanotubes; metal or metal oxide nanoparticles (e.g., gold, titanium dioxide); quantum dots, etc.

An important aspect of nanotechnology is the vastly increased ratio of surface area to volume present in many nanoscale materials, which makes possible new quantum mechanical effects. One example is the quantum size effect where the electronic properties of solids are altered with great reductions in particle size. Nanoparticles, for example, take advantage of their dramatically increased surface area to volume ratio. Their optical properties, for example, fluorescence, become a function of the particle diameter. This effect does not come into play by going from macro to micro dimensions. However, it becomes pronounced, when the nanometer size range is reached.

A certain number of physical properties also alter with the change from macroscopic to nanometric systems. Novel mechanical properties of nanomaterials may be interesting and these are a subject of nanomechanics research. When brought into a bulk material, nanoparticles can strongly influence the mechanical properties of the material, like stiffness or elasticity, for example, traditional polymers can be reinforced by nanoparticles resulting in novel materials (nanocomposites), which can be used as lightweight replacements for relatively heavy metals. Such nanotechnologically enhanced materials may enable a weight reduction accompanied by an increase in stability and improved functionality

The synthesis of nanomaterials and nanocomposites is still mostly based on conventional techniques, such as wet impregnation followed by chemical reduction of the metal nanoparticle precursors. These techniques based on thermal heating can be time consuming and often lack control of particle size and morphology. Hence, recently interest in microwave technology has been developed, where an alternative way of power input into chemical reactions is through microwaves, that is, dielectric heating.

Nanomaterials can be broadly classified as nanoparticles and nanocomposites. These are:

- **Nanomaterials**

- o Carbon based nanomaterials
- o Dendrimers
- o Metal based nanomaterials
- Nanoncomposites

14.2 CARBON BASED NANOMATERIALS

These nanomaterials are composed mostly of carbon and commonly take the form of a hollow spheres and ellipsoids, or tubes. Spherical and ellipsoidal carbon nanomaterials are referred to as fullerenes, while cylindrical ones are called carbon nanotubes (CNTs). These particles have many potential applications, like improved films and coatings, stronger and lighter materials, and in electronics.

This class of organic/inorganic materials has gained significant attention for potential applications in biomedicine, separations, and magnetic storage. An essential component of this research is the preparation of polymeric surfactants that enable synthesis, passivation, and functionalization of magnetic nanoparticles. Various polymerization techniques, namely, living and controlled polymerizations, such as, ring-opening metathesis polymerization (ROMP) and controlled radical processes have been applied to synthesize core-shell colloids possessing tunable film thickness, composition, and properties.

Metal-free nitrogen and phosphorus dual-doped, electrocatalytically active, functionalized nanocarbon (FNC) and photoluminescent carbon nanodots (PCNDs) were simultaneously synthesized using a facile one pot microwave-assisted process (Prasad et al., 2013). In addition to the electrocatalytic activity, FNC also shows attractive properties as a metal-free oxygen reduction catalyst and it is resistant to methanol crossover effects in alkaline media. The 5–10 nm PCNDs, exhibiting blue fluorescence under UV exposure were successfully used for bioimaging applications.

Carbon nanodots (C-Dots) have also attracted much interest as a new form of carbonaceous nanomaterials and intensive research is going on their inspiring properties in different laboratories all over the globe. These newly emergent nanodots possess a number of advantageous characteristics as compared to traditional semiconductor quantum dots like low-toxicity, which is quite fascinating. Many researchers have focused on various synthesis methods of C-Dots and their biology related applications. Microwave-assisted approaches have attracted much attention, because this treatment can provide intensive and efficient energy, and as a consequence, reduces the reaction time. It also provides rapid, ecofriendly (green) and waste-reused approach to synthesize fluorescent and water-soluble C-Dots from eggshell membrane (ESM) ashes (Wang et al., 2012). ESM was selected as the carbon source as it is a common protein-rich waste in daily life and can be obtained easily and at low cost.

The applications of nanomaterials as photocatalysts for the degradation of pollutants in water or air, removal of Cr (VI) as well as hydrogen evolution was reported by Pan et al. (2013). They demonstrated improved photocatalytic activities as compared with the photocatalysts synthesized by traditional methods.

Glucose derived water-soluble crystalline graphene quantum dots (GQDs) were prepared by a facile microwave-assisted hydrothermal method with an average diameter as small as 1.65 nm. These GQDs exhibit typical excitation wavelength dependent properties as expected in carbon-based quantum dots (Tang et al., 2012).

Carbon nanoparticles have great technological and industrial importance because of their enhanced physicochemical, electrical, thermal and mechanical properties. Hong et al. (2003) successfully synthesized carbon nanotubes by microwave heating of the catalyst loaded on various supports such as carbon black, silica powder, or organic polymer substrates (Teflon and polycarbonate). Acetylene was used as a source of hydrocarbon and 3d transition metals and metal sulfides were used as the catalysts. Different carbon yields were obtained depending on the reaction conditions. A variety of morphologies like fibrous nanocarbons, for example, linear or Y-branched, as well as carbon nanoparticles and amorphous carbon forms were also observed.

Kharissova (2004) and Takagaki (2010) also synthesized carbon nanotubes using a domestic microwave oven. Kharissova developed a highly efficient one-step technique to obtain long and vertically aligned carbon nanotubes with or without iron filling. Ferrocene has also been used to prepare these aligned CNTs. It was observed that CNTs have a metal particle at the tip of each tube. This carbon nanostructure promises to play an important role in fuel cells and in nanoscale engineering.

Graphite has been used in military applications as radar absorbing material and in antielectromagnetic interference coatings for civil purposes since it is a good microwave radiation absorber. In this field, Fan et al. (2009) observed that microwave absorption maxima of milled flake graphite and carbon nanotubes was found in the 10–15 GHz frequency range.

Synthesis of exfoliated graphite as well as the reduction of graphite oxide was carried out by Zhu et al. (2010). A short-time direct exposure to microwave irradiation was used to bring about these reactions.

Nie et al. (2013) synthesized carbon nanotubes within 15–20 sec by a single carbon fiber initiating the pyrolysis of ferrocene under microwave field without any other chemicals. Other carbonaceous materials can be used to trigger the growth of CNTs. Wang et al. (2013) used microwave irradiations to synthesize superhydrophobic multiwall carbon nanotube (MWCNT)–polybenzoxazine (PBZ) nanocomposites and it was observed that the process required only approximately 45 sec to ensure complete polymerization. The MWCNT–PBZ nanocomposites, which do not contain any fluorinated compounds, exhibited superhydrophobicity after pressing and displayed excellent environmental stability. These nanocomposites also had excellent stability in terms of the contact angle with water.

Ikeda et al. (1995) presented the first report of the production of carbon nano-structures under microwave irradiation by synthesizing fullerenes from microwave-induced naphthalene-nitrogen plasma at atmospheric pressure inside a cylindrical coaxial cavity.

Microwave irradiation has been used as an important tool in the functionalization of fullerenes and carbon nanotubes (Langa and De la Cruz, 2007). These compounds have some excellent properties, which can be used for the development of optoelectronic organic devices.

Bhunia et al. (2013) prepared carbon nanoparticle based fluorescent bioimaging probes. The fluorescent nanoparticle based imaging probes have advanced current labeling technology and are expected to generate new medical diagnostic tools based on their superior brightness and photostability compared with conventional molecular probes. The effect of different experimental parameters on the synthesis of carbon nanotube/nanofiber supported platinum by polyol processing techniques was observed by Knupp et al. (2008). Developing corrosion resistant carbon nanotube and carbon nanofiber (CNF) supported Pt catalysts with optimized particle size is important for proton exchange membrane fuel cells.

The fullerenes are a class of allotropes of carbon. Conceptually, these are graphene sheets rolled into tubes or spheres. These include the carbon nanotubes (or silicon nanotubes), which are important because of their mechanical strength and electrical properties. The synthesis of C_{60} has been reported in a conventional microwave oven from the decomposition of camphor resin (Martinez-Reyes et al., 2012). Thermal and microwave-assisted synthesis of Diels−Alder adducts of [60] Fullerene with 2,3-pyrazinoquinodimethanes was carried out by Fernández-Paniagua et al. (1997).

14.3 DENDRIMERS

These nanomaterials are nanosized polymers built from branched units. The surface of a dendrimer has numerous chain ends, which can be tailored to perform some specific chemical functions. This property can also be used for catalysis because three-dimensional dendrimers contain interior cavities, where other molecules could be placed. These may be also useful for drug delivery.

Microwave-assisted reactions for polyamidoamine (PAMAM) dendrimer synthesis was carried out by López-Andarias et al. (2012). The dendrimers were obtained in good yields, with high purity, and near-perfect regioselectivity while multivalent dendrimeric peptides were synthesized via a microwave-assisted Huisgen 1,3-dipolar cycloaddition between azido peptides and dendrimeric alkynes in relatively good yields (46 to 96%) (Rijkers et al., 2005).

Microwave technology was also used for grafting of PAMAM onto the surface of silica by Zhang et al. (2010a). Bovine serum albumin (BSA) was used as a model biological molecule to react with the dendrimer-grafted silica by chemical bonding through glutaraldehyde. It was observed that the immobilization efficiency for BSA increases with increasing PAMAM generation.

Three generations of peptoid-based dendrimers were synthesized by Diaz-Mochon et al. (2008) through solid-phase methods, using N-Fmoc-N-(6-N'-Fmoc-aminohexyl)-glycine as both; the initiator core and the monomer unit. It offers an unusual dendrimeric periphery composed of both secondary and primary amines. The third generation compound (dendrimer) proved to be an efficient mediator of transfection while displaying minimal cytotoxicity

Yoon et al. (2007) worked on the monofunctionalization of dendrimers with use of microwave-assisted 1, 3-dipolar cycloadditions. The monofunctionalized poly-amide-based dendrimers containing either a terminal azide or alkyne moiety were synthesized. This monofunctionalization provides a route for the single attachment of a functional moiety using 1, 3-dipolar cycloaddition, thereby providing a novel pathway for targeted dendrimer functionalization. Dijkgraaf et al. (2007) also synthesized cyclicpeptide dendrimers via 1, 3-dipolar cycloaddition and observed their biological implications for tumor targeting and tumor imaging purposes.

A novel approach for the parallel synthesis of glycopeptides was suggested by Matsushita et al. (2013) by combining solid-phase peptide synthesis and dendrimer-supported enzymatic modifications.

14.4   METAL BASED NANOMATERIALS

Microwave chemistry is a well-established technique in organic synthesis, but its use to synthesize inorganic nanomaterials is still in its early age and still it is far away to reach its full potential. Various new opportunities arising as a result of the unique features of microwave chemistry have been discussed and the expected benefits for nanomaterials have been elaborated (Bilecka and Niederberger, 2010).

These nanomaterials include quantum dots, nanogold, nanosilver and metal oxides, such as titanium dioxide. A quantum dot is a closely packed semiconductor crystal comprised of hundreds or thousands of atoms, and its size is of the order of a few nanometers to a few hundred nanometers. Any change in the size of quantum dots changes their optical properties.

Microwave-assisted nonaqueous synthesis of ZnO nanoparticles was carried out by Ambro and Orel (2011). The microwave heating method provides a better control over experimental parameters compared to classical (conductive) heating. There-fore, it opens some exciting opportunities for a better understanding of the influence of the reaction conditions on the reaction and growth mechanisms of ZnO particles. As the morphology and size of ZnO determine its physicochemical behavior, a variety of polymer-ZnO nanomaterials with tunable optoelectrical and mechanical properties can be prepared by the microwave-assisted synthesis in-situ or ex-situ.

where R = CH$_3$(CH$_2$)$_3$–.

The application of microwave radiations for the preparation of metal oxides (under aqueous and nonaqueous conditions) and polymer/ZnO nanocomposites, has been shown to be a versatile approach to the design novel morphologies of nanoparticles. Particularly the faster reaction rates (shorter reaction times), better product yields and the possibility to automatically combine different experimental parameters makes microwave-assisted synthesis suitable for the studies of the influences of the reaction conditions on the morphology and sizes of ZnO particles.

ZnO in different nanostructures was synthesized via microwave-assisted hydrothermal route by Kondawar et al. (2011). Effect of different experimental conditions such as microwave irradiation power and exposure time on the process of formation of the ZnO nanostructures has been observed. It was found that the microwave exposure time plays a vital role in determining the diameter of the rods. The interaction of microwaves with the growth units of ZnO was investigated to explain formation of different structural geometry of ZnO on nanoscale level. ZnO nanostructures of flower-like, sword-like, needle-like and rod-like structures were prepared by microwave-assisted hydrothermal process at different conditions of microwave power and irradiation time. The obtained ZnO nanostructures were hexagonal in phase. Microwave can interact with growth units of ZnO to generate active centers on the surface of ZnO nuclei so that needle-like ZnO rods are created on those sites, resulting in the formation of the flower-like ZnO nanostructures.

Silver and indium-modified ZnS were synthesized by ambient pressure microwave-assisted technique. Crystalline materials with particle sizes below 10 nm were obtained by this procedure. There were some defects in the crystal, which reduce the band gap leading to indoor light photocatalytic activity in these materials.

Copper oxides have been widely used as catalysts, gas sensors, adsorbents, and electrode materials. CuO nanomaterials were synthesized by a facile hydrothermal process in $Cu(CH_3COO)_2$ (0.1 M)/urea (0.5 M) and $Cu(NO_3)_2$ (0.1 M)/urea (0.5 M) aqueous systems at 150 °C for 30 min under microwave irradiation (Qiu et al., 2011). They observed that copper acetate solution could be hydrolyzed to form urchin-like architecture CuO, and this transformation was accelerated by addition of urea and these exhibited excellent catalytic activities for the epoxidation of alkenes and the oxidation of CO. CO was oxidized to CO_2, when the temperature was kept higher than 115 °C, and 100% conversion was obtained at 130 °C.

A novel microwave-assisted technique was reported for the synthesis of silver and indium-modified ZnS (Synnott et al. 2013). These ZnS samples also show significantly higher photocatalytic activity than the commercially available TiO_2 (Degussa P-25). A simple, fast and efficient route for microwave-assisted synthesis of Pt nanocrystals and its deposition on carbon nanotubes in ionic liquids was reported by Liu et al. (2006). Inorganic salts (such as $H_2PtCl_6·4H_2O$, $RhCl_3·2H_2O$, etc.) dissolved in ionic liquids like, 1,1,3,3-tetramethylguanidinium trifluoroacetate or 1,1,3,3-tetramethylguanidinium lactate, were reduced to metal nanoparticles by glycol under microwave irradiation. These metal nanoparticles could be decorated on CNTs.

Microwave-assisted route for the synthesis of nanomaterials has gained importance in the field of synthetic technology because of its faster, cleaner and cost effectiveness than the other conventional and wet chemical methods for the preparation of metal oxide nanoparticles. Synthesis of metal oxide nanoparticles, for example, γ-Fe_2O_3, NiO, ZnO, CuO, Co-γ-Fe_2O_3 was carried out by microwave-assisted route through the thermal decomposition of their respective metal oxalate precursors employing polyvinyl alcohol as a fuel (Lagashetty et al., 2007).

Iron oxide nanoparticles supported on aluminosilicate catalysts were found to be efficient and easily recoverable materials in the aqueous selective oxidation of alcohols to their corresponding carbonyl compounds using hydrogen peroxide under both conventional as well as microwave heating (Rajabi et al., 2013). A new and versatile approach to synthesize SnO_2 nanocrystals (rutile-type structure) using microwave-assisted hydrothermal method was reported by Mendes et al. (2012). They observed that there are substantial changes in optical absorbance of tin (IV) oxide nanoparticles.

α-MnS hollow spheres and their hybrids with reduced graphene oxide (RGO) had been fabricated by a one pot template-free solvothermal method (Chen et al. 2013). α-MnS spheres have diameters of about 3–5 µm and it was observed that the amount of GO can affect the morphology of α-MnS crystals in hybrids. The study of coating carbon nanotubes with metal/oxides nanoparticles has become a promising and challenging area of research these days. The use of carbon nanotubes can be optimized in various applications, and for that it is necessary to attach some functional groups or other nanostructures to their surface. The combination of the distinctive properties of carbon nanotubes and metal/oxides is expected to be applied in field emission displays, nanoelectronic devices, novel catalysts, and polymer or ceramic reinforcement. Microwave-assisted synthesis of metal oxide nanoparticles supported on carbon nanotubes was carried out and their applications were studied (Motshekga et al., 2012).

TiO$_2$ nanostructures with fascinating morphologies like cubes, spheres, and rods were synthesized by a simple microwave irradiation technique (Suprabha et al., 2008). These different morphologies were achieved by changing the pH and the nature of the medium or the precipitating agent. The synthesized titania nanostructures had shown to possess higher photocatalytic activity than the commercial photocatalyst Degussa P25 TiO$_2$.

Microwave-assisted synthesis of nanocrystalline MgO and its use as a bactericide was demonstrated by Makhluf et al. (2005). The antibacterial activities of the MgO nanoparticles were tested by treating *Escherichia coli* (Gram negative) and *Staphylococcus aureus* (Gram positive) cultures with 1 mg/mL of the nanoparticles. The effect of size, pH, and the form of the active MgO species as a bactericidal agent was also observed. The amount of eradicated bacteria was strongly dependent on the particle size.

Microwave-assisted synthesis of icosahedral nickel nanocrystals was performed by Donegan et al. (2011). Nickel acetylacetonate was used as the metal precursor, while sodium formate and trioctylphosphine oxide were used as the reducing agent and capping ligands, respectively. The nanocrystals, with a mean diameter of 237 ± 43 nm were obtained, which exhibited enhanced ferromagnetic behavior. Microwave-assisted hydrothermal synthesis of nanocrystalline SnO powders was carried out by Pires et al. (2008) using SnCl$_2$·2H$_2$O as a precursor. It was observed that SnO crystals having different sizes and morphologies could be achieved by changing the hydrothermal processing time, temperature, the type of mineralizing agent (NaOH, KOH or NH$_4$OH) and its concentration.

Microwave-assisted fractional precipitation of magnetite nanoparticles was carried out by Edrissi and Norouzbeigi (2010). Highly pure magnetite nanoparticles in its powder form as well as a dispersion in water and an organic solvent (n-hexane) were prepared by the fractional precipitation using ammonium hydroxide and microwave heating in the presence of linoleic acid as a capping agent. This new modified method was able to produce pure magnetite particles in the range of 1–15 nm.

Fast microwave-assisted solvothermal synthesis of metal nanoparticles (Pd, Ni, Sn) supported on sulfonated multiwalled carbon nanotubes (MWCNTs) was carried out by Ramulifho et al. (2012). The results showed that the mixed Pd-based catalysts (obtained by simple ultrasonic-mixing of the individual MWCNT-metal nanocomposites) gave better electrocatalytic activity than their alloy nanoparticles (obtained by coreduction of metal salts) or Pd alone. This Pd-based bimetallic catalysts was used for ethanol oxidation in alkaline medium.

Calcium oxide (CaO) is an important inorganic compound, which can be used across various industries as catalyst, toxic-waste remediation agent, adsorbent, etc. CaO nanoparticles were obtained by the microwave irradiation technique (Roy and Bhattacharya, 2011). It is a simple and efficient method to produce CaO nanoparticles with regular shape, small size, and high purity. The structure of the CaO nano-crystal was found to be cubic.

$$Ca(NO_3)_2 + 2\ NaOH \xrightarrow[\text{10 min}]{\text{MW Irradiation}} CaO + 2\ NaNO_3 + H_2O$$

(i)   Vacuum filtration
(ii) Washed with de-ionized water and absolute ethanol
(iii) Dried in a vacuum at 80°C for 1 hr

CaO nanopowder

Ca(NO$_3$)$_2$ and NaOH were used as starting materials for the formation of CaO nanoparticles in this technique. The particle size of the prepared samples was found to be 24 nm and surface area, 74 m^2/g.

Metallic (Ag, Au, Cu) nanoparticles suspensions are gaining an increasing interest in medical applications. Among the existing synthetic routes available, microwave processing allows to have a better control on particle size and to achieve high purity of the products. The large scale production of nanoparticles requires the development of industrially viable processes, preferably conducted using a green chemical approach, that is, at ambient pressure and relatively low temperature (Veronesi et al., 2011). Synthesis, stability and applications of Pristine nanomaterials has been discussed by Kundu et al. (2013). Li and Komarneni (2006) synthesized Pt nanoparticles and nanorods by microwave-assisted solvothermal technique. The effects of the reaction conditions, such as the molar ratio of the polyvinylpyrrolidone (PVP) repeating unit to the metal sources, the concentration of metal sources, the reaction temperature, and the presence of distilled water were investigated.

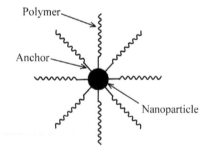

A core-shell nanohybrid

Fatoorehchi et al. (2010) worked on the microwave-assisted synthesis of a special catalyst, namely, Pt/C catalyst comprised of nanoparticles. Owing to its unique characteristics, this catalyst is quite beneficial in the performance of proton exchange membrane fuel cells (PEMFCs).

Metal nanoparticles having interesting shapes can be prepared in aqueous solutions through simple reductions of metal ions with the presence of some additive reagents, such as cetyltrimethylammonium bromide and hexamethylenetetramine. Some successful results for shape-controlled synthesis of metal nanoparticles are summarized, which includes the synthesis of palladium nanocubes, palladium nanobricks and gold nanotripods. In addition, combining with indium tin oxide electrode surfaces, shape-controlled growth is possible to form gold nanoplates and copper oxide nanowires (Oyama et al., 2011).

Benyettou et al. (2012) prepared ultra-small superparamagnetic iron oxide nanoparticles in a fast, efficient and green microwave-assisted synthesis. Ultra-small nanoparticles (2–4 nm) exhibiting robust superparamagnetic behavior were obtained by controlling the temperature and heating mode of a polyol reduction using triethylene glycol as a green and high-boiling solvent, The high coercivity and excellent energy product of $Nd_2Fe_{14}B$ hard magnets have a large number of high value added industrial applications but the chemical synthesis of these nanoparticles had been quite challenging due to the large reduction potential of Nd^{3+} and the high tendency for $Nd_2Fe_{14}B$ oxidation. Swaminathan et al. (2013) carried out a novel microwave-assisted chemical synthesis of $Nd_2Fe_{14}B$ hard magnetic nanoparticles.

Microwave-assisted synthesis of silver nanoparticles using poly N-isopropylacrylamide/acrylic acid microgel particles was reported by Khan et al. (2011). These particles were prepared by emulsion polymerization in the presence of N,N-methylenebisacrylamide. The microwave technique was adopted for the preparation of highly stable and monodisperse silver nanoparticles (Ag NPs) using these microgels as a template via in situ reduction of silver nitrate in the presence of glucose as a reducing agent. Chitosan nanoparticles were prepared based on ionotropic gelation between low molecular weight chitosan and sodium tripolyphosphate (TPP) under microwave irradiation (Kocak et al., 2011). It was determined by the results of the zeta potential measurements that synthesized CHN under microwave irradiation clearly exhibits more homogeneous and stable dispersion.

Cr_2O_3 nanoparticles were prepared by Farzaneh (2011) using $Cr(NO_3)_3.9H_2O$ as starting material, triethanolamine (TEA) as template and water as a green solvent under microwave irradiation. The particle size of Cr_2O_3 was between 25–70 nm. The catalytic activity of the prepared Cr_2O_3 nanoparticles was found to be more than the bulk sample in the epoxidation of norbornene with 70% activity and 95% selectivity while microwave-assisted synthesis and luminescent properties of pure and doped ZnS nanoparticles were studied by Yang et al. (2012).

Microwave-assisted synthesis of nano-sized cadmium oxide as a new and highly efficient catalyst for solvent-free acylation of amines and alcohols was developed (Mazaheriehrani et al., 2010). The biological synthesis of silver nanoparticles is not only a cost-effective and ecofriendly method but also time consuming. A rapid method of silver nanoparticle synthesis using E. coli culture supernatant along with microwave irradiation has been proposed (Mahanty et al., 2013).

ZnO nanoflower morphologies usually include nanowires, nanorods, nanorings, nanoneedles, etc. The synthetic techniques of nanomaterials include oxidation of elemental metals, reduction of metal salts, thermal decomposition of relatively unstable compounds, or electrochemical route. Various other metal nanoflowers have also been synthesized by various workers from time to time.

Well-defined flower-like CdS nanostructures have been synthesized by Tai and Guo (2008) under ultrasound and microwave irradiations, simultaneously. Hexagonal nanopyramids and/or nanoplates were obtained depending on different sulfur sources. The synergistic effect of microwave and sonochemistry has been proposed for the formation of these nanoflowers. They show a large blue-shift upto 100 nm as compared to simple low-dimensional CdS nanostructures. This structure induced shift in optical properties, which may find some potential applications in optoelectronics devices, catalysis, and solar cells.

High density ZnO-nanorod arrays (rod length 1.59 μm) were successfully synthesized by Zhu et al. (2009) via microwave-assisted solution-phase method. Zinc chloride and ammonia solution were used as reactants. The influence of various parameters like concentration of ammonia solution, work power, and microwave irradiation time on the morphology and size of final products were observed.

14.5 NANOCOMPOSITES

Nanoparticles combined with other nanoparticles of other materials or with larger bulk-type materials are termed as composites. Nanoparticles, such as nanosized clays, have already been added to products ranging from auto parts to packaging materials to enhance their mechanical, barrier, thermal, and flame-retardant properties.

Nanocomposites are new materials made with nanosized fillers. These materials have a great potential for their applications in the automotive and aerospace industry as well as in construction, electrical applications and food packing. There is also a tremendous interest for using bionanoparticles like cellulose microfibrils or whiskers to be applied in the new era of biocomposites.

Nanotechnology has been recognized as one of the most promising areas for technological development in the twenty-first century in materials research. The development of polymer nanocomposites is emerging rapidly as a multidisciplinary research field, which can broaden applications of polymers further. Polymer nanocomposites (PNC) are polymers (thermoplastics, thermosets or elastomers) that have been reinforced with small quantities (less than 5% by weight) of nanosized particles having high aspect ratios. Nanotechnology is being applied in the fields of synthesis of single-site catalyst, antimicrobial nanocomposites, fire retardant materials, novel electro-optical devices, sensors, ultra soft magnets and also in the area of drug delivery systems.

William et al. (2005) reviewed various techniques for characterization and trends in the field of nanocomposites. These are new materials made with fillers, which have nanosize and have a big potential for applications in the automotive and aerospace industry as well as in construction, electrical applications and food packing. There is a tremendous interest for using bio-nanoparticles in the new era of biocomposites by using synthetic and natural fillers in polymer nanocomposites. Aranda et al. (1998) studied the microwave-assisted blending-intercalation of ion-conductor polymers into layered silicates. They prepared organo-inorganic hybrid nanocomposites derived from poly(ethylene oxide) and montmorillonite silicate. They observed that ionic conductivity was enhanced as compared to samples prepared by intercalation from solution.

The study of coating carbon nanotubes with metal/oxides nanoparticles is now becoming a promising and challenging area of research. It is necessary to attach different functional groups or other nanostructures to their surface to optimize the use of carbon nanotubes in various applications. The combination of these distinctive properties of carbon nanotubes and metal/oxides may be applied in field emission displays, nanoelectronic devices, novel catalysts, and polymer or ceramic reinforcement.

Wilson et al. (2004) investigated the synthesis and magnetic properties of poly(methylmethacrylate) nanocomposites with embedded iron nanoparticles using microwave plasma technique. Magnetic nanoparticles embedded in polymer matrices have excellent potential for electromagnetic device applications like electromagnetic interference suppression. Hoogenboom et al. (2007) used closed reaction vials, which has opened a completely unexplored area of high-temperature chemistry under microwave irradiation. The closed reaction vessels have also been exploited to replace high-boiling solvents by low-boiling solvents simplifying product isolation.

Microwave-assisted one-step synthesis of polyacrylamide–metal (M = Ag, Pt, Cu) nanocomposites in ethylene glycol was studied by Zhu and Zhu (2006). The advantages of using ethylene glycol are that it acts both; as a reducing reagent as well as solvent and therefore, no additional reductant is needed. Secondly, no initiator for acrylamide metal polymerization and surfactant for stabilization of metal nanoparticles are necessary. These nanocomposites with metal nanoparticles homogeneously dispersed in the polymer matrix had been successfully prepared with the corresponding metal salt and acrylamide monomer in ethylene glycol by microwave heating where as Liao et al. (2007) prepared poly(ε-caprolactone)/clay nanocomposites by microwave-assisted in situ ring-opening polymerization. The preparation of ε-caprolactone was carried out in the presence of either unmodified clay or clay modified by quaternary ammonium cations containing hydroxyl groups. This significantly improved monomer conversion and molecular weight as compared with that produced by conventional heating.

Tian et al. (2006) synthesized polymer electrolyte fuel cells in presence of microwave irradiation. They used carbon black supported polytetrafluoroethylene

(PTFE/C) nanocomposite and PTFE emulsion as the precursor. Incorporation of this nanocomposite into a Pt/C/Nafion catalyst significantly enhances the mass transportation of the catalyst layer without any adverse effect on the electrocatalytic activity of the Pt catalysts. Varma et al. (2008) also investigated microwave-assisted transformations and synthesis of some polymer nanocomposites and nanomaterials.

Liu et al. (2011) worked on the one step microwave-assisted rapid synthesis of Pt/graphene nanosheet composites by reduction and their application for oxidation of methanol. Ethylene glycol was used for the reduction of H_2PtCl_6 in a graphene oxide suspension. This composite exhibited a good catalytic activity toward methanol oxidation.

Microwave-assisted synthesis of Pt/CNT nanocomposite electrocatalysts was carried out by Zhang et al. (2010b) for PEM fuel cells. Microwave-assisted heating of functionalized, single-wall carbon nanotubes (FCNTs) in ethylene glycol solution containing H_2PtCl_6, led to the reductive deposition of Pt nanoparticles (2.5–4 nm) over the FCNTs, yielding an active catalyst for proton-exchange membrane fuel cells (PEMFCs). In single-cell testing, the Pt/FCNT composites displayed a catalytic performance that was superior to Pt nanoparticles supported by raw (unfunctionalized CNTs (RCNTs) or by carbon black (C), prepared under identical conditions. Tsai et al. (2012) observed the efficiency of microwave heating of weakly loaded polymeric nanocomposites. Such a control of thermal and electromagnetic properties could be achieved by using nanocomposites. The heating rate of nanocomposites made of carbon nanofibers was found to be much greater than that of any other nanocomposites. These findings suggested that at 2.45 GHz frequency, the heating rate is mostly controlled by the electrical losses in the fillers.

Ma et al. (2011) synthesized and characterized cellulose/hydroxyapatite nanocomposite under microwave irradiation using LiCl/N,N-dimethylacetamide as a solvent. The effects of heating times, heating temperatures, and the reactant concentrations on the phases, microstructures, and morphologies of the cellulose/HA nanocomposites were also observed. This green cellulose/HA nanocomposites (via green synthetic route) has opened a high value-added applications of biomass. The controlled synthesis of transition metal/conducting polymer nanocomposites was reported by Liu et al. (2012). A novel displacement reaction has been observed between conducting polymers (CP) and metal salts, which could be used to fabricate nanostructured CP–metal composites. It was believed that the CP–metal nanofibril network could be converted to a carbon–metal network by a microwave induced carbonization process and resulted in the sensory enhancement.

Pang et al. (2010) prepared $Mo_2C/CNTs$ nanocomposites, an efficient electrocatalyst, via microwave-assisted method using $Mo(CO)_6$ as single source precursor. Pt electrocatalysts supported on the $Mo_2C/CNTs$ composites were also prepared by using modified ethylene glycol method where as the microwave-assisted preparation of biodegradable water absorbent polyacrylonitrile (PAN)/montmorillonite (MMT) clay nanocomposite was reported by Sahoo et al. (2011). This nanocomposite was prepared in a microwave oven using a transition metal Co (III) complex and ammonium persulfate as an initiator. The water absorption and biodegradation properties were studied for its ecofriendly nature and better commercialization. Polymer layered silicate nanocomposites (PLSNs) are emerging as the most significant and new breed of composite materials due to their extensively enhanced properties. These unique properties of nanocomposites arise from the maximized contact between the organic and inorganic phase, so fillers with high surface-to-volume ratio are commonly used.

Synthesis of PANI/cerium dioxide (CeO_2) nanocomposites was reported by Kumar and Selvarajan (2010). The synthesis of nanocomposite of polymer added cerium dioxide by microwave-assisted solution method was carried out via in situ polymerization of the monomer in the presence of cerium dioxide (ceria) nanoparticles. Among organic conducting polymers, polyaniline (PANI) had attracted intense interest because of its high conductivity, excellent stability and relatively high transparency to visible light. The synthesized PANI/CeO_2 nanocomposites can be used for fabrication of films or coatings, or even in further polymer blending.

Recently, great attention has been drawn towards the synthesis and uses of polymer based nanomagnetic composites as they have become a powerful tool in industries, science, and medicine. Agarwal et al. (2012) studied the fabrication and characterization of iron oxide filled polyvinyl pyrrolidone nanocomposites comprising iron oxide nanoparticles as filler and polyvinyl pyrrolidone as polymer shell. These nanocomposites were synthesized through free radical polymerization under microwave irradiation. The unique magnetic, catalytic and biocompatible properties of nanomagnetic polymer composites (MPCs) make it quite attractive for use in a variety of areas.

Patil and Acharya (2013) prepared ZnS-Graphene (ZnS-GNS) nanocomposite by microwave irradiation method. This composite was explored as photocatalyst for the degradation of methylene blue in aqueous slurry. Under similar conditions, the photocatalytic activity of the pure ZnS was also examined. The ZnS-GNS composite was found to enhance the rate of photodegradation of some toxic dyes as compared to pure ZnS. Graphene based metal sulfide/oxide semiconductor nanocomposites are high potential materials for photocatalytic degradation of toxic dyes. Li et al. (2013) synthesized the Mn_3O_4-encapsulated graphene sheet nanocomposites via a facile, fast microwave hydrothermal method and observed their supercapacitive behavior. Well-crystallized Mn_3O_4-anchored reduced graphene oxide (RGO) nanocomposites had been successfully synthesized via a facile, effective, energy saving, and scalable microwave hydrothermal technique. It has a potential application as supercapacitor material. Its good electrochemical properties indicated that the nanocomposite could be an electrode candidate for supercapacitors.

Microwave synthesis of graphene sheets and graphene-polymer nanocomposites was reported by Hassan et al. (2009). They used hydrazine and copper (II) nitrate for this purpose under microwave irradiation for 30 sec–2 min. These materials are expected to have unique superior properties that combine electrical conductivity, thermal stability, flexibility and possible optical properties.

Aldosari et al. (2013) synthesized and characterized the in situ *bulk* polymerization of PMMA containing graphene sheets using microwave irradiation. The results indicated that this composite had a better morphology and dispersion with enhanced thermal stability as compared with the composites prepared by conventional heating. Kim et al. (2013) carried out the synthesis of $RuO_2.xH_2O$/reduced grapheme oxide (RGO) nanocomposite through the selective heterogeneous nucleation and growth of ruthenium oxide nanoparticles on CNF using microwave-assisted polyol process. The microwave hydrothermal method was used to improve the conductivity of RuO_2/RGO for electrochemical capacitor applications.

Layered silicates are important fillers for improving various mechanical, flame retardant, and barrier properties of polymers, which can be attributed to their sheet-like morphology. Layered silicates can be modified with some organic surfactants to render them compatible with polymer matrices. Organically modified silicates (organoclays) having large surface areas are very cost-efficient and nontoxic nanofillers. These are effective at very low loads and are readily available. Upon amalgamation of organoclays with polymer matrix nanocomposites, polymer chains can penetrate in between the silicate layers. It results in an intercalated structure, where the clay stack remains intact but the interlayer spacing is increased. When penetration becomes more severe, disintegration of clay stacks can occur, resulting in an exfoliated structure. Organoclay particles are mostly intercalated, having a preferred orientation with the clay gallery planes being preferentially parallel to the plane of the pressed film. Preferential orientation of organoclays affects the barrier properties of polymer membranes. Additional fillers like carbon black can induce a change in the orientation of organoclays. The effect of carbon black on the orientation of organoclays was observed and a relationship between orientation and permeability of air through such membranes was established. Nawani (2011) discussed the exfoliation of organoclays using microwave radiation and various characterization techniques were used to evaluate structure, morphology and properties of fillers and polymer nanocomposites.

One-pot microwave-assisted modification of bio-template by Ag-ZnO submicroparticles was investigated by Munster et al. (2012). These submicroscopic particles were synthesized by hydrothermal route in an open vessel system in microwave oven with an external reflux cooling system. Wood dust with size of the particles in range 300–500 μm, bulk density 170–230 g/L was used as a bio-template for microwave-assisted solvothermal modification. Zinc acetate dihydrate $Zn(CH_3COO)_2 \cdot 2H_2O$ and silver nitrate $AgNO_3$ were used as precursors, hexamethylenetetramine (HMTA), $(CH_2)_6N_4$, and aqueous ammonia 25–29 wt.%. were used as precipitating agents. Zinc oxide submicroparticles and silver globular submicro particles were synthesized successfully on the template surface.

The synthesis and comparative study of thermal and microwave cured CNFs /EPON-862 nanocomposites was carried out by Rangri et al. (2011). Other than drastic decrease in processing time, the microwave processing of EPON-862 + 0.3% CNFs showed an increase of 27.51%, 25.98% and 28.63%, respectively for flexure modulus, flexure strength and flexure strain at failure as compared to neat EPON-862/CNFs thermal processing. Eugene and Gupta (2010) synthesized the aluminum and magnesium based nanocomposites using hybrid microwave sintering. The synthesis and characterization of neem chitosan nanocomposites for development of antimicrobial cotton textiles was reported by Rajendran et al. (2012). They used multiple emulsion/solvent evaporation method for this preparation. The size of the spherically nanocomposites was between 50–100 nm.

A new cobalt (II) 4-hydroxobenzoate, $[Co_2(C_7H_5O_3)_2(NO_3)_2(H_2O)_4]\,2H_2O$ was synthesized using microwave technique by Al-Amoudi et al. (2012) as a binuclear complex and characterized. The nanocomposite synthesis and characterization of kesterite Cu_2ZnSnS_4 (CZTS) for photovoltaic applications was studied by Michael (2012). Nanocomposite synthesis methods using binary metal chalcogenide solids have led to the development of kesterite, which is an important emerging material for photovoltaic devices. This synthetic route is more cost effective and environment friendly because it avoids the use of long processing times and harsh solvents. The use of microwave-assisted processing reduces the necessary processing times drastically, which is usually of the order of several hours.

Nanomaterials behave differently than other similarly sized particles. It is, therefore, necessary to develop specialized approaches for testing and monitoring their effects on human health and on the environment. While nanomaterials and nano-technologies are expected to yield numerous health and health care advances, such as more targeted methods of delivering drugs, new cancer therapies, and methods of early detection of diseases, but sidewise, they also may have unwanted effects.

Different nanomaterials, like CNT, metal oxides, and polymers, possess superior mechanical, thermal and electrical properties, which can find a wide range of applications in smart structures, composite materials, energy storage, chemical sensors, nano-electronic devices, etc. The high cost and difficulty in getting high quality nanomaterials on large scale still remains a challenge. It has been shown to be an affordable and scalable microwave approach for the direct growth of CNT, nanostructured metal oxides on a wide range of substrates, like carbon fibers, glass fibers, Kevlar, fly ash, Kaolin, Basalt fibers, etc. The microwave initiated nanomaterial growth will take hardly 20–30 seconds under the microwave irradiation at room temperature in the air. There is no need of any inert gas protection, and additional feed stock gases, so commonly required in chemical vapor deposition technique.

Increased rate of absorption is the main concern associated with manufactured nanoparticles. When nanoparticles of a material are made, then their surface area to volume ratio increases. The greater specific surface area (surface area per unit weight) may lead to increased rate of absorption through the skin, lungs, or digestive tract and may cause unwanted effects to the lungs as well as other human organs. However, the particles must be absorbed only in sufficient quantities in order to pose health risks.

KEYWORDS

- **Dendrimers**
- **Nanocomposites**
- **Nanomagnetism**
- **Nanomaterials**
- **Nanoparticles**
- **Nanotechnology**

REFERENCES

Agarwal, T., Gupta, K. A., Alam, S., & Zaidi, M. G. H. (2012). International Journal of Composite Materials, 2, 17–21.

Al-Amoudi, M. S., Salman, M. S., Megahed, A. S., & Refat, M. S. (2012). European Chemical Bulletin, 1, 293–304.

Aldosari, M. A., Othman, A. A., & Alsharaeh, E. H. (2013). Molecules, 18, 3152–3167.

Ambro, G., & Orel, Z. C. (2011). Materials and Technology, 45, 173–177.

Aranda, P., Galván, J. C., & Ruiz-Hitzky, E. (1998). MRS Proceedings, 519, 375–380.

Benyettou, F., Milosevic, I., Olsen, J. C., Motte, L., & Trabolsi, A. (2012). Journal of Bioanalysis & Biomedicine, doi: 10.4172/1948–593X.S5–006.

Bhunia, S. K., Saha, A., Maity, A. R., Ray, S. C., & Jana, N. R. (2013). Scientific Reports, 3, doi: 10.1038/srep01473.

Bilecka, I., & Niederberger, M. (2010). Nanoscale, 2, 1358–1374.

Buzea, C., Pacheco, I., & Robbie, K. (2007). Biointerphases, 2, 17–71.

Chen, D., Quan, H., Wang, G. S., & Guo, L. (2013). ChemPlusChem, DOI: 10.1002/cplu.201300141.

Diaz-Mochon, J. J., Fara, M. A., Sanchez-Martin, R. M., & Brdley, M. (2008). Tetrahedron Letters, 49, 923–926.

Dijkgraaf, I., Rijnders, A. Y., Soede, A., Dechesne A. C., van Esse, G. W., Brouwer, A. J., Corstens, F. H. M., Boermann, O. C., Rijkers, D. T. S., & Liskamp, R. M. J. (2007). Organic & Biomolecular Chemistry, 5, 935–944.

Donegan, K. P., Morris, M. A., Godsell, J. F., Roy, S., Tobin, J. M., Holmes, J. D., O'Byrne, J. P., & Otway, D. J. (2011). CrystEngComm, 13, 2023–28.

Edrissi, M., & Norouzbeigi, R. (2010). Journal of Nanoparticle Research, 12, 1231–1238.

Eugene, W. W. L., & Gupta, M. (2010). Journal of Microwave Power and Electromagnetic Energy, 44, 14–27.

Fan, Y., Yang, H., Li, M., & Zou, G. (2009). Materials Chemistry and Physics, 115, 696–698.

Farzaneh, F., & Najafi, M. (2011). Journal of Sciences Iran, 22, 329–333.

Fatoorehchi, H., Zamiri, B., & Shams, M. M. (2010). Proceedings of 14th International Electronic Conference on Synthetic Organic Chemistry.

Fernandez-Paniagua, U. M., Illescas, B., Martin, N., Seoane, C., de la Cruz, P., de la Hoz, A., & Langa, F. (1997). Journal of Organic Chemistry, 62, 3705–3710.

Hassan, H. M. A., Abdelsayed, V., Khder, A. E. R. S., Abou Zeid, K. M., Terner, J., El-Shall, M. S., Al-Resayes, S. I., & El-Azhary, A. A. (2009). Journal of Materials Chemistry, 19, 3832–3837.

Hong, E. H., Lee, K. H., Oh, S. H., & Park, C. G. (2003). Advanced Functional Materials, 13, 961–966.

Hoogenboom, R., & Schubert, U. S. (2007). Macromolecular Rapid Communications, 28, 368–386.

Ikeda, T., Kamo, T., & Danno, M. (1995). Applied Physics Letters, 67, 900–902.

Khan, A., El-Toni, M. A., Alrokayan, S., Alsalhi, M., Alhoshan, M., Abdullah S., & Aldwayyan, A. S. (2011). Colloids and Surfaces A: Physicochemical and Engineering Aspects, 377, 356–360.

Kharissova, O. (2004). Reviews on Advanced Materials Science, 7, 50–54.

Kim, J.-Y., Kim, K.-H. Yoon, S.-B., Kim, H.-K., Park, S.-H & Kim, K.-B. (2013). Nanoscale, 5, 6804–6811.

Knupp, S. L., Li, W., Paschos, O., Murray, T. M., Snyder, J., & Haldar, P. (2008). Carbon, 46, 1276–1284.

Kocak, N., Sahin, M., Akin, I., Mahmut, K., & Yilmaz, M. (2011). Journal of Macromolecular Science, 48, 776–779

Kondawar, S. B., Acharya, S. A., & Dhakate, S. R. (2011). Advanced Materials Letters, 2, 362–367.

Kumar, E., & Selvarajan, P. (2010). Journal of Experimental Sciences, 1, 11–14.

Kundu, P., Anumol, E. A., & Ravishankar, N. (2013). Nanoscale, 5, 5215–5224.

Lagashetty, A., Havanoor. V., Basavaraja, S., Balaji, S. D., & Venkataraman, A. (2007). Science and Technology of Advanced Materials, 8, 484–493.

Langa, F., & De la Cruz, P. (2007). Combinatorial Chemistry & High Throughput Screening, 10, 766–82.

Li, D., & Komarneni, S. (2006). Zeitschrift für Naturforschung B, 61b, 1566–1572.

Li, L., Seng, K., Chen, Z., Liu, H., Nevirkovets, I. P., & Guo, Z. (2013). Electrochimica Acta, 87, 801–808.

Liao, L., Zhang, C., & Gong, S. (2007). Macromolecular Rapid Communications, 28, 1148–1154.

Liu, S., Wang, L., Tian, J., Lu, W., Zhang, Y., Wang, X., & Sun, X. (2011). Journal of Nanoparticles Research, 13, 4731–4737.

Liu, Z., Liu, Y., Zhang, L., Poyraz, S., Lu, N., Kim, M., Smith, J., Wang, X., Yu, Y., & Zhang, X. (2012). Nanotechnology, 23, doi: 10.1088/0957–4484/23/33/335603.

Liu, Z., Sun, Z., Han, B., Zhang, J., Huang, J., Du, J., & Miao, S. (2006). Journal of Nanoscience and Nanotechnology, 6, 175–179.

López-Andarias, J., Guerra. J., Castañeda, J., Merino, G. S. V., Ceña, V., & Sánchez-Verdú, P. (2012). European Journal of Organic Chemistry, 12, 2331–2337.

Ma, M.-G., Jia, N., Li, S.-M. & Sun, R.-C. (2011). Iranian Polymer Journal, 20, 413–421.

Mahanty, A., Bosu, R., Panda, P., Neatm, S. P., & Sar, B. (2013). International Journal of Pharma and Bio Sciences, 4, 1030 –1035.

Makhluf, S., Dror, R., Nitzan, Y., Abramovich, Y., Jelinek, R., & Gedanken, A. (2005). Advanced Functional Materials, 15, 1708–1715.

Martinez-Reyes, J., Barriga-Arceo, L. G. D., Rendon-Vazquez, L., Martinez-Gwrrero, R., Romero-Partida, N., Palacios-Gonzalez, E., Garibay-Febles-Febles, V., & Ortiz-Lopez, J. (2012). World Journal of Nano Science and Engineering, 2, 213–218.

Matsushita, T., Handa, S., Naruchi, K., Garcia-Martin, F., Hinou, H., & Nishimura, S.-I. (2013). Polymer Journal, 45, 1101–1106.

Mazaheritehrani, M., Asghari, J., Lotfi, O. R., & Pahlavan, S. (2010). Asian Journal of Chemistry, 22, 2554–2564.

Mendes, P. G., Moreira, M. L., Tebcherani, S. M., Orlandi, M. O., Andre´s, J., Li, M. S., Diaz-Mora, N., Varela, A., & Longo, E. (2012). Journal of Nanoparticle Research, 14, 750–762.

Michael, E. K. (2012). Journal of Advance Renewal Energy Research, 1, 448–451.

Motshekga, S. C., Pillai, S. K., Ray, S. S., Jalama, K. & Kraus, R. W. M. (2012). Journal of Nano-materials, doi: 10.1155/2012/691503.

Munster, L., Klofac, J., Sedlak, J., Bazant, P., Machovsky, M., & Kuritka, I. (2012). NANCON Conference Brno. Czech.

Nawani, P. (2011). Polymer layered silicate nanocomposites: Structure, morphology, and proper-ties, Michigan: ProQuest.

Nie, H., Cui, M., & Russell, T. P. (2013). Chemical Communications, 49, 5159–5161.

Oyama, M., Umar, A. A., Saleh, M. M., & Majlis, B. Y. (2011). Sains Malaysiana, 40, 1345–1353.

Pan, L., Liu, X., Sun, Z., & Sun, C. Q. (2013). Journal of Materials Chemistry A, 1, 8299–8326.

Pang, M., Li, C., Ding, L., Zhang, J., Su, D. S., Li, W., & Liang, C. (2010). Industrial and Engineer-ing Chemistry Research, 49, 4169–4174.

Patil, B. N., & Acharya, S. A. (2013). Advanced Materials Letters, doi: 10.5185/amlett.2013. fdm.16.

Pires, F. I., Joanni, E., Savu, R., Zaghete, M. A., Longo, E., & Varela, J. A. (2008). Materials Let-ters, 62, 239–242.

Sudhakara Prasad, K., Ramjee Pallela, Dong-Min Kim, Yoon-Bo ShimPrasad, K. S., Pallela, R., Kim, D. M., & Shim, Y. B. (2013). Particle & Particle Systems Characterization, 557–564.

Qiu, G., Dharmarathna, S., Zhang, Y., Opembe, N., Huang, H., & Suib, S. L. (2011). *Journal of Physical Chemistry C*, 116, 468–477.

Rajabi, F., Pineda, A., Naserian, S., Balu, A. M., Luque, R., & Romero, A. A. (2013). Green Chem-istry, 15, 1232–1237.

Rajendran, R., Radhai, R., Balakumar, C., Hasabo, A., Mohammed, A., Vigneshwaran, C., & Vaid-eki, K. (2012). Journal of Engineered Fibers and Fabrics, 7, 137–141.

Ramulifho, T., Ozoemena, K. I., Modibedi, R. M., Jafta, C. J., & Mathe, M. K. (2012). Electrochi-mica Acta, 59, 310–320.

Rangri, V. K, Bhuyan, M. S., & Jeelani, S. (2011). Comosite A, 42, 849–858.

Rijkers, D. T., Van Esse, G. W., Merkx, R., Brouwer, A. J., Jacobs, H. J., Pieters, R. J., & Liskamp , R. M. (2005). Chemical Communication, 36, 4581–4583.

Roy, A., & Bhattacharya, J. (2011). International Journal of Nanoscience, 10, 413–418.

Sahoo, P. K., Biswal, T., & Samal, R. (2011). Journal of Nanotechnology, doi: 10.1155/2011/143973.

Suprabha, T., Roy, H. G., Thomas, J., Kumar, K. P., & Mathew, S. (2008). Nanoscale Research Letters, 4, 144–152.

Swaminathan, V., Deheri, P. K., Bhame, S. D., & Ramanujan, R. V. (2013). Nanoscale, 5, 2718–2725.

Synnott, D. W., Seery, M. K., Hinder, S., Colreavy, J., & Pillai, S. C. (2013). Nanotechnology, 24, doi: 10.1088/0957–4484/24/4/045704.

eTai, G., & Guo, W. (2008). Ultrasonic Sonochemistry, 15, 350–356.

Takagaki, Y., Nguyen-Tran, H. D., & Ohta, K. (2010). Bulletin of the Chemical Society of Japan, 83, 1100–1106.

Tang, L., Ji, R., Cao, X., Lin, J., Jiang, H., Li, X., Teng, K. S., Luk, C. M., Zeng, S., Hao, J., & Lau, S. P. (2012). American Chemical Society Nano, 6, 5102–5110.

Tian, Z. Q., Jiang, S. P., Wang, X. L., Zhang, H. M., & Yi, B. L. (2006). Electrochemistry Communications, 8, 1158–1162.

Tsai, C. C., Rubin, B., Tatartschuk, E., Owens, J. R., Luzinov, I., & Kornev, K. G. (2012). Journal of Engineered Fibers and Fabrics, 42–49.

Varma, R. S., Nadagouda, M. N., & Polshettiware, V. (2008). Presented at The 236th American Chemical Society (ACS) National Meeting, Philadelphia, PA.

Veronesi, P. (2011). Microwave Symposium Digest, 1–4.

Wang, C. F., Chen, H. Y., Kuo, S. W., Lai, Y. S., & Yang, P. F. (2013). RSC Advances, 3, 9764–9769

Wang, Q., Liu, X., Zhang, L., & Lv, Y. (2012). Analyst, 137, 5392–5397.

William, G. E., Aldo, B. A. & Jinwen, Z. (2005). Maderas Ciencia y Tecnología, 7, 159–178.

Wilson, J. L., Poddar, P., Frey, N. A., Srikanth, H., Mohomed, K., Harmon, J. P., Kotha, S., & Wachsmuth, J. (2004). Journal of Applied Physics, 95, 1439–1443.

Yang, H., Huang, C., & Tang, X. (2012). Journal of Nanoscience and Nanotechnology, 2, 148–158.

Yoon, K., Goyal, P., & Weck, M. (2007). Organic Letters, 9, 2051–2054.

Zhang, C., Ping, S., Farooq, M. U., Yang, Y., XiangGao, E., & Hongjun. (2010a). Reactive and Functional Polymers, 70, 129–133.

Zhang, W., Chen, J., Swiegers, G. F., Ma, Z. F., & Wallace, G. G. (2010b). Nanoscale, 2, 282–286.

Zhu, J. F., & Zhu, Y. J. (2006). *Journal of Physical Chemistry B*, *110*, 8593–8597.

Zhu, J. Y., Zhang, J. X., Zhou, H. F., Qin, W. Q., Chai, L. Y., & Hu, Y. H. (2009). Transactions of Nonferrous Metals Society of China, 19, 1578–1582.

Zhu, Y., Murali, S., Stoller, M. D., Velamakanni, A., Piner, R. D., & Ruoff, R. S. (2010). Carbon, 48, 2118–2122.

CHAPTER 15

POLYMERIZATION

DIPTI SONI, MONIKA TRIVEDI, and RAKSHIT AMETA

CONTENTS

Microwave irradiation has been based as an alterative heat source in polymer synthesis and day-by-day, it is becoming more and more popular. In this synthesis, reaction time is reduced from days/hours to minute/seconds. Thus, there can be significant time saving in polymerization reaction, in contrast to the conventional heating, which needs much longer reaction times.

The main area, where microwave irradiation is used, are the step growth polymerization, ring opening polymerization, radical polymerization, emulsion polymerization, coordination polymerization and RAFT polymerization. Microwave-assisted polymer synthesis is very interesting because of rapid volumetric heating, suppressed side reaction, energy saving, direct heating, decreased environmental pollution and safe operation.

15.1 STEP-GROWTH POLYMERIZATION

Step-growth polymerization is based on the coupling of two multifunctional, monomers. The resulting product also contains the functional groups and thus, reacts in the same manner as the monomer, eventually leading to polymeric materials. Mostly step-growth polymerization methods are known as polycondensations due to the release of water or any other small molecule during these coupling reactions. These reactions include the synthesis of nonvinyl polymers such as polyamides, polyimides, polyesters, polyureas, polyethers, poly(amideimide)s, poly(amide-ester)s, poly(ester-imide)s, poly(ether-imide)s, poly(ether-ester)s, poly(amideimide-urethane)s and poly(amide-ether-urethane)s via polycondensation reactions.

15.1.1 POLYAMIDES

Imai (1996a) developed a new facile method for the rapid synthesis of aliphatic polyamides and polyimides from polycondensation of ω-amino acids and nylon salts. The polymerization reactions were carried out in domestic microwave oven in the presence of a small amount of a polar organic medium. Suitable organic media for the polyamide synthesis are tetramethylenesulfone (TMS) amide-type solvents such as N-cyclohexyl-2-pyrrolidone (CHP), 1, 3-dimethyl-2-imidazolidone (DMI), phenolic solvents like m-cresol and o-chlorophenol, etc. and for the polyimide synthesis amide-type solvents such as N-methyl-2-pyrrolidone (NMP), CHP and DMI. In polyamide synthesis, the polycondensation was almost complete within 5 min, producing a series of polyamides with inherent viscosities around 0.5 dL/g, and the polyimides having the viscosity values above 0.5 dL/g were obtained within 2 min.

Reactions of chiral 5-(3-methyl-2-phthalimidylpentanoylamino)isophthalic acid and 5-(4-methyl-2-phthalimidyl-pentanoylamino)isophthalic acid with several aromatic and aliphatic diisocyanates like 4,4'-methylenebis(phenyl isocyanate), toluene-2,4-diisocyanate, isophorone diisocyanate and hexamethylene diisocyanate under microwave irradiation were studied by Mallakpour et al. (2008a, 2008b). The resulting novel optically active polyamides have inherent viscosities in the range of 0.25–0.63 dL/g.

where R = -CH(CH$_3$)C$_2$H$_5$, -CH$_2$CH(CH$_3$)$_2$; R$_1$ = -(CH$_2$)$_6$-, -C$_6$H$_4$CH$_2$C$_6$H$_5$, -CH$_2$-C$_6$H$_3$(CH$_3$)$_2$

Pielichowski et al. (2003) synthesized poly(aspartic acid) from maleic anhydride under microwave irradiation. This procedure is multistep in nature and it includes hydrolysis of the maleic anhydride, condensation with ammonium hydroxide followed by polycondensation resulting in poly(anhydroaspartic acid). This on hydrolysis yielded desired poly(aspartic acid) at room temperature The use of microwave irradiation (multimode microwave reactor) accelerated this process by a factor of ten without influencing the yield.

Faghihi and Hagibaygi (2003) used microwave irradiation for the synthesis of polyamides containing azo-benzene moieties. 4,4'-Azobenzoyl chloride was treated with eight different 5,5'-disubstituted hydantoin derivatives in the presence of a small amount of a polar organic medium such as o-cresol. This reaction was carried out in a domestic microwave oven. Polyamides with high yields and inherent viscosities between 0.35 and 0.60 dL/g were obtained within 7–12 min.

Caouthar et al. (2005) synthesized some new aromatic polyamides based on diphenylamino isosorbide. Diphenylamino isosorbide reacts with several diacyl chlorides in the presence of a small amount of N-methylpyrrolidone (NMP) in a monomodal microwave reactor. Microwave-assisted polymerizations in the presence of NMP led to faster polymerizations and higher molecular weight products as compared to standard polymerization methods. The polymers were obtained with inherent viscosities between 0.22 and 0.73 dL/g, corresponding to molecular weight up to 140,000 g/mol.

Najun et al. (2005) used microwave irradiation for the step-growth polymerization of benzoguanamine and pyromellitic anhydride resulting in the formation of the p-π conjugated poly(amic acid).

15.1.2   POLYIMIDES

Aliphatic polypyromellitimides were synthesized from the reaction of the salt monomers composed of aliphatic diamines (the length of the diamines chain was 6–12-CH₂-moieties) and pyromellitic acid or its diethyl ester in domestic microwave oven (Imai et al., 1996b). Polar organic media such as NMP, CHP, DMI and TMS were used. 1,3-Dimethyl-2-imidazolidone was the best polymerization medium for this polycondensation. Polymers were obtained with inherent viscosities of about 0.70 dL/g or above only in 2 min.

Tang et al. (2003) prepared polyimides from the polycondensation of sodium tetrazodiphenylnaphthionate and pyromellitic dianhydride (PMDA) under microwave irradiation. This polymer has third-order nonlinear optical properties.

Poly(amic acid) (PAA) was synthesized with 3,3′,4,4′-benzophenonetetracarboxylic dianhydride (BTDA) as a dianhydride monomer and 4,4′-diaminodiphenylmethane and 4,4′-oxydianiline (ODA) as diamine monomers under microwave irradiation in dimethylformamide. Then this PAA was used to make polyimide by imidization at a low temperature (Quantao et al., 2008).

A copolycondesation-type poly (amic acid) was synthesized using pyromellitic dianhydride and 3,3′,4,4′-benzophenonetetracarboxylic dianhydride as dianhydride monomers, and 4,4′-oxydianiline as a diamine monomer under microwave irradiation in dimethylformamide. PAA was then converted into a polyimide by an imidization (Li et al., 2008).

15.1.3 POLYESTERS

Nagahata et al. (2007) carried out microwave-assisted direct polycondensation of lactic acid by use of a binary catalyst of $SnCl_2$/p-TsOH using a single-mode microwave synthesizer (2.45 GHz frequency and 300 W maximum power). Microwave irradiation accelerated this reaction and produced poly(lactic acid) (PLA) with a molecular weight higher than 10,000 g/mol. The reactions were performed with different catalysts like $SnCl_2$, $Sn(Oct)_2$, Bu_2SnCl_2, DCTB, $SnCl_2$/p-TsOH, p-TsOH, phosphoric acid and sulfuric acid. Under optimized conditions, the $SnCl_2$/p-TsOH binary catalyst gave white polymer with a molecular weight of 16,000 g/mol within 30 min of microwave irradiation at 200°C and the reaction time was reduced significantly compared to conventional polycondensation at the same temperature.

Velmathi et al. (2005) synthesized poly(butylenes succinate) by direct polyesterification of succinic acid and butane-1,4-diol in the presence of 1,3-dichloro-1,1,3,3 tetrabutyldistannoxane as a catalyst. Under optimum conditions, poly(butylene succinate) with a weight-average molecular weight of 2.35×10^4 g/mol was obtained within reaction time of 20 min. On the other hand, this reaction was conducted via conventional heating in 5 h and hence, microwave irradiation showed a ten-fold increase in the rate of polymerization.

15.1.4   POLYUREAS

High-molecular-weight polyureas and polythioureas were prepared from the reaction of aromatic and aliphatic amines with ureas and thioureas in the presence of a catalytic amount of p-toluenesulfonic acid by using a microwave oven in 12 min at 400 W (Banihashemi et al., 2004). The reactions were performed in different solvents such as dimethyl sulphoxide, N,N-dimethylacetamide (DMAc) chlorobenzene and dioxane. DMAc was found to be the most appropriate solvent owing to the greater solubility of the substrates, the higher rate of the reactions, and the excellent energy-transfer properties. When the reaction time was increased up to 15 min, partial degradation of the aliphatic polymers, led to polymers with lower viscosities.

Polycondensation reactions of some urazole derivatives like 4-(4'-N-1,8-naphthalimidophenyl)-1,2,4-triazolidine-3,5-dione,   4-(4'-acetamidophenyl)-1,2,4 triazolidine-3,5-dione and 4-(4'-t-butylphenyl)-1,2,4-triazolidine-3,5-dione with hexamethylene diisocyanate, isophorone diisocyanate and toluene-2,4-diisocyanate in DMAc or NMP solution in the presence of pyridine, triethylamine or dibutyltin dilaurate as a catalyst under microwave activation were carried out. Polyureas with inherent viscosities of 0.06 to 0.30 dL/g were obtained after irradiation for 8–18 min. (Mallakpour and Rafiee, 2003).

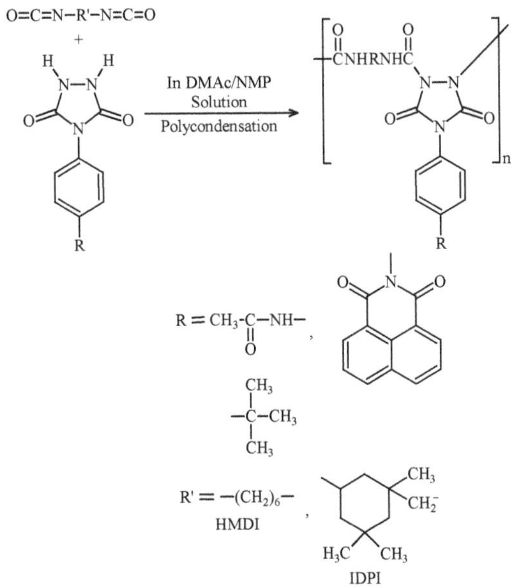

O=C=N–R'–N=C=O

In DMAc/NMP
Solution
Polycondensation

$R = CH_3-\underset{\underset{O}{\|}}{C}-NH-$,

$-\underset{\underset{CH_3}{|}}{\overset{\overset{CH_3}{|}}{C}}-CH_3$

$R' = -(CH_2)_6-$
HMDI

,

IDPI

15.1.5 POLYETHERS

Alimi et al. (2001) prepared microwave-assisted poly(phenylene vinylene)-ether (PPV ether) derived from homopolycondensation of 1-chloro-4-methoxylbenzene in solution in alkaline dimethyl sulphoxide. The mixture was exposed to 600 W microwaves for a few min until a temperature of 470 K was achieved and then continuously stirred at room temperature for 6 h. This raw mixture was composed of three fractions.

(i)   a yellow fraction of PPV-ether insoluble in all common organic solvents with 43% yield.

(ii) a white fraction soluble in $CHCl_3$ with 52% yield.

(iii) a yellow-green fraction soluble in the CH_2Cl_2 with 5% yield.

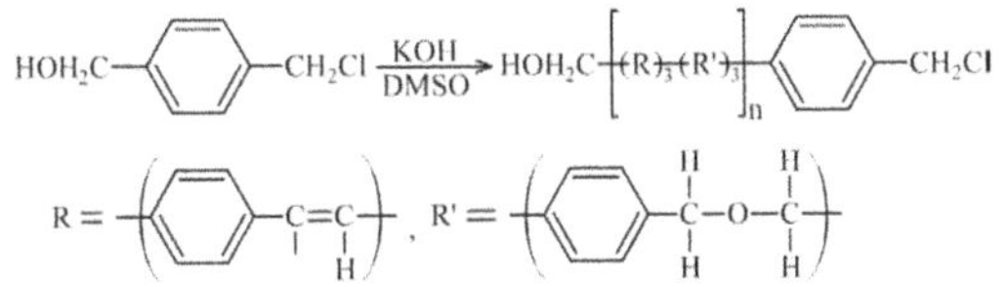

Tetrabutylammonium bromide catalyzed polycondensation of 3,3-bis(chloromethyl) oxetane and various bisphenol derivatives was studied under microwave irradiation (Hurduc et al., 1997). They found that microwave irradiation did not have a considerable influence on the molecular weights of polymers, but it reduced reaction times.

Chatti et al. (2002, 2003) synthesized new polyethers from microwave-assisted polycondensation of isosorbide or isoiodide with 1,8-dibromo- or 1,8-dimesyl-octane in the presence of a small amount of toluene under phase-transfer catalytic conditions within 30 min. It was observed that microwave-assisted polymerization showed an increased yields (total) (68–76% against only 28–30%) as compared with conventional heating, in the case of isosorbide.

15.1.6 POLY(AMIDE-IMIDE)S

Some novel optically active poly(amide-imide)s were synthesized with tetrahydro-pyrimidinone and tetrahydro-2-thioxopyrimidine moieties by microwave-assisted polycondensation, in the presence of a small amount of o-cresol (Mallakpour et al., 2001). Poly(amide-imide)s were produced within 10 min with inherent viscosities of about 0.25–0.45 dL/g in high yields.

Khoee et al. (2007a, 2007b) carried out the synthesis of photoactive poly(amide-imide)s containing anthracenic or naphthalenic pendent groups with fluorescent property in a domestic microwave. The reactions were performed in o-cresol solvent and yielded photoactive polymers in high yields. The microwave-assisted polycondensation reactions proceeded rapidly and were completed within 8–10 min.

Faghihi and Hajibeygi (2004) reported the formation of poly(amide-imide)s under microwave irradiation through the polycondensation reactions of N,N'-(4,4'-diphenylether)bistrimellitimide diacid chloride with six different derivatives of 5,5-disubstituted hydantoin in o-cresol solvent. The reaction mixture was irradiated for 7–10 min to afford polymers in 84–94% yield.

A number of flame-retardant poly(amide-imide)s containing phosphine oxide and hydantoin moieties in main chain were prepared (Faghihi, 2006). Irradiation of the reaction mixtures in a microwave oven led to the desired poly(amide-imide)s within a short time of 7–12 min while classical thermal polycondensation required almost a day.

Mallakpour and Rafiemanzelat (2004) reported step-growth polymerization re-actions of monomer bis(p-aminobenzoic acid)-N-trimellitylimido-L-leucine with different diisocyanates via direct polycondensation under microwave irradiation, solution polymerization under gradual heating and reflux conditions in the pres-ence of pyridine, dibutyltin dilaurate, and triethylamine as a catalyst and without a catalyst also. The optically active poly(amide-imide)s were obtained within a short time of 3 min in good yields (53–95%) and inherent viscosities in the range of 0.17 to 0.61 dL/g.

15.1.7   POLY(AMIDE-IMIDEURETHANE)S

Mallakpour and Rafiemanzelat (2005a, 2005b) synthesized poly(amide-imi-deurethane)s by using central unit L-valine base asymmetric diacid under micro-wave irradiation. The diacid and poly(ethylene glycol)s were copolymerized using 4,4'-methylene-bis(4-phenylisocyanate) as coupling agent in a one-step or two step procedure. For the two-step procedure, the diisocyanate was first reacted with the diol or the diacid and subsequently with the other reagent. The properties of the polymers were found to be influenced by changes in catalyst, microwave power, irradiation time as well as the molecular weight of the used poly(ethylene glycol). The resulting thermoplastic poly(amide-imideurethane)s have good thermal stability and phase mixing. The thermal stability of these polymers was more than conventional poly(urethanes).

15.1.8 POLY(AMIDE-ESTER)S, POLY(ESTER-IMIDE)S, POLY(ETHER-IMIDE)S, AND POLY(ETHER-ESTER)S

Borriello et al. (2007) synthesized poly(amide-ester)s by polycondensation of sebacic acids and ω-amino alcohols in the presence of stannous 2-ethyl hexanoate, $Sn(Oct)_2$, as a catalyst under microwave irradiation.

$$HOOC\,(CH_2)_8\,COOH \;+\; H_2N\,(CH_2)_n\,OH$$

Sebacic acid   Amino alcohol
                          n = 2, 3, 6

$$Sn\,(Oct)_2 \Big|\, \text{-}H_2O$$

$$\text{-----}OC\,(CH_2)_8\,CONH\,(CH_2)n\,O\text{------}$$

Facile and rapid polycondensation reactions of N,N'-(pyromellitoyl)-bis-L-leucine diacid chloride, N,N'-(pyromellitoyl)-bis-L-phenylalanine diacid chloride, 4,4'-carbonyl-bis(phthaloyl-L-alanine) diacid chloride or 4,4'-(hexafluoroisopropylidene)-N,N'-bis(phthaloyl-Lleucine) diacid chloride with several aromatic diols such as phenol phthalein, bisphenol-A, 4,4'-hydroquinone, 1,8-dihydroxyanthraquinone, 1,5-dihydroxy naphthalene, 4,4-dihydroxy biphenyl and 2,4-dihydroxyacetophenone were reported by using a domestic microwave oven in the presence of a small amount of a polar organic solvent such as o-cresol (Mallakpour and Habibi, 2003). The polymerization reactions occurred rapidly and are completed within 10–20 min, producing a series of optically active poly(ester-imide)s with good yields and moderate inherent viscosities.

where R = $\overset{H_3C}{\underset{H_3C}{\diagdown}}CH\text{-}CH_2\text{-}$, $\bigcirc\!\!-\!CH_2\text{-}$;

Ar =

Gao et al. (2004) reported the polycondensation reaction of disodium bisphenol A with bis(chlorophthalimide)s in a domestic microwave oven in o-dichlorobenzene by phase transfer catalysis. Poly(ether-imide)s with inherent viscosities between 0.55 and 0.92 dL/g were obtained within 25 min.

Chatti et al. (2006) described the synthesis of poly(ether-ester)s based on diol-ether of isosorbide and adipoyl chloride or terephthaloyl chloride. The polymerization reactions almost completed within 5 min with weight average molecular weights of 4200 g/mol and high yield (approx. 95%) while under conventional heating, the polymer was obtained only with 19% yield and that too with weight average molecular weight of 4050 g/mol.

15.1.9 OTHER STEP-GROWTH POLYMERIZATIONS

Bezdushna et al. (2005) reported the step-growth polymerization of dimethyl hydrogen phosphonate with poly(ethylene glycol) resulting in biodegradable poly(alkylene hydrogen phosphonate). Thermal and microwave heating both resulted in comparable polymers, but the microwave polymerization was performed in 55 min at 140–180°C where as the thermal polymerization required 9.5 h at 130–140°C. However, the higher temperatures (above 140°C) used for the microwave polymerization could not be applied in the thermal polymerization process due to thermal degradation of the dimethyl hydrogen phosphonate.

Seong et al. (2013) synthesized π-conjugated polymers by polycondensation via direct arylation of 3,4-ethylenedioxythiophene with 9,9-dioctyl-2,7-dibromofluorene under microwave irradiation. Under the optimized conditions, the microwave-assisted direct arylation of 3,4-ethylenedioxythiophene with dibromofluorene for 30 min with catalyst (1 mol %) gave the corresponding polymer with an extremely high molecular weight up to 147,000. The high molecular weight of the polymer enables formation of a large and flexible self-standing film, leading to an advantage in fabricating organic thin-film devices. It was revealed that the polymer has high purity and no bromo- and metallo-terminals. Because only microwave-assisted direct arylation provided these features in the polymer, this method is superior to conventional methods.

Triarylamine polymers can be rapidly assembled by microwave-assisted amination of aryldibromides (Horie et al., 2008). A series of polymers was formed with backbones containing 4,4'-biphenyl, 2.9-fluorene and 3,6-carbazole repeating units. These polymers have been used to fabricate organic field-effect transistors. The devices show very stable operation under ambient conditions and p-type mobilities up to 2.3×10^{-3} cm^2 V^{-1} s^{-1}, close to the highest mobility reported to date for this class of amorphous semiconductors.

15.2   FREE RADICAL POLYMERIZATION

The free radical polymerization of vinylic monomers is one of the major processes for the industrial production of bulk polymers like polystyrene and poly(methyl methacrylate).

Iannelli et al. (2005a, 2005b) reported the microwave-assisted direct synthesis and polymerization of a series of chiral (meth)acrylamides. The direct synthesis of chiral meth(acrylamide) from (meth)acrylic acid and 1-phenylethylamine under microwave irradiation yielded the desired vinyl monomers, whereas the same reaction under thermal heating resulted mainly in the formation of the Michael addition product.

where X = H or CH$_3$.

The microwave acceleration for the direct synthesis of N-phenylmaleimide from maleic anhydride and aniline based on specific microwave absorption of the ionic intermediates has been reported (Bezdushna and Ritter, 2005). Fischer et al. (2005) investigated the free radical polymerization of N-alkylacrylamides with 3-mercaptopropionic acid as chain transfer agent in methanol with thermal heating at ambient pressure and under superheated conditions as well as under microwave irradiation. While the chain transfer polymerization could be accelerated from 5 h to 1 h when going to superheated conditions with thermal heating, it was further accelerated down to several seconds under microwave irradiation.

Free radical copolymerization of methyl methacrylate and styrene as well as butyl methacrylate with styrene or isoprene in toluene under microwave irradiation (monomode microwave reactor) has also been carried out (Fellows, 2005). However, no changes in reactivity ratios were observed although more detailed studies were required for the copolymerization of butyl methacrylate and isoprene. The microwave-assisted polymerization procedure accelerated the polymerizations by a factor of 1.7, may be due to an increase in radical flux. It was proposed that the increased radical flux under microwave irradiation is due to rapid orientation of the radicals that are formed from decomposition of the azoisobutyronitrile. This orientation would reduce the number of direct terminations by recombination of the two radical fragments under microwave irradiation and thus, cause a higher radical flux.

Stange et al. (2006) investigated the free radical copolymerization of methyl methacrylate and styrene with different initiators in different solvents using both microwave (monomode microwave reactor) and thermal heating. Agarwal et al. (2005) studied the copolymerization of 2,3,4,5,6-pentafluorostyrene and phenylmaleimide. They revealed that a higher initial polymerization rate and a lower final monomer conversion for the microwave-assisted procedure were there as compared to thermal heating.

The copolymerization of N,N-dimethylaminoethyl methacrylate with allyl thiourea was performed under microwave irradiation (domestic microwave oven) by Lu et al. (2005). The influence of reaction time and microwave power on the copolymerization was studied. Copper was coordinated to this polymer by microwave irradiation. This polymer-copper system was used as heterogeneous catalyst for the polymerization of methyl methacrylate.

Super-absorbing materials have been prepared under microwave irradiation by the copolymerization of cornstarch, sodium acrylate and poly(ethylene glycol) diacrylate by Zheng et al. (2005). Microwave-assisted polymerization method resulted in the preparation of a super-absorber with a swelling ratio of 520 to 620 g water per gram polymer. Optimal conditions were applied for the preparation of the hydrogel using conventional heating. The material exhibited lower swelling ratios than microwave heating. Xu et al. (2005) prepared super-absorbers by the copolymerization of starch, sodium acrylate, and 2-acrylamido-2-methylpropanosulfonic acid.

The material produced under microwave irradiation consisted of many evenly distributed pores, whereas the material prepared with thermal heating contained irregular pores. This observation was correlated to the different temperatures that were used. A higher temperature with microwave heating causes in situ drying and thus, the pores resulted from water evaporation during the preparation while thermal heating gave a hydrogel that was dried afterwards.

15.3 RING-OPENING POLYMERIZATION

15.3.1 POLYMERIZATION OF ε-CAPROLACTONE

The ring-opening polymerization of ε-caprolactone was reported by Liao et al. (2002). This polymerization was carried out smoothly and effectively with constant microwave powers (170, 340, 510, and 680 W). The range of temperature was maintained from 80 to 210 °C for this process. 90% yield of poly(ε-caprolactone) was obtained at 680 W, when 0.1% (mol/mol) stannous octanoate catalyst was used for 30 min and its weight-average molar mass was 124,000 g/mol. But when the polymerization was carried out at 680 W for 270 min, and catalyzed by 1% (w/w) zinc powder weight-average molar mass was reduced to 92,300 g/mol.

The effect of microwave irradiation by chain propagation of the benzoic acid initiated polymerization of e-caprolactone. They used closed ampoules in a domestic microwave oven. The chain propagation was studied as a function of microwave power, monomer to initiator ratio, and polymerization temperature. The results were quite interesting. It indicated that microwave heating formed chain growth in the initial stage and number of polymer chains was limited. While formation of growing centers in initial stage was favored by thermal heating (Yu and Liu, 2004).

The enzyme-catalyzed (novozym435) polymerization of ε-caprolactone was investigated (Kerep and Ritter, 2006). They used both; thermal and microwave (monomode microwave reactor) heating under reflux conditions in toluene, benzene, and diethyl ether.

Polymerization of ε-caprolactone with lanthanide halide catalysts using different heat profile was studied by Barbier-Baudry et al. (2003). Broader molecular weight distributions were obtained as compared to the use of an initial power boost (300 W) when 200 W microwave power was applied. This was observed that the faster heating with the higher initial microwave power, inhibits secondary transfer reactions.

Sinnwell et al. (2006) demonstrated ring-opening polymerization of ε-caprolactone under microwave irradiation in the present of methacrylic acid. From this process, they obtained radical polymerizable polyester macromonomers. This process required only one step and high functionality. The melting point of the macromonomers was between 46 and 51°C.

A catalyst free microwave-assisted polymerization was reported for the preparation of poly(e-caprolactone)-block-poly(ethylene glycol)- block-poly(e-caprolactone) copolymers (Yu and Liu, 2005). The resulting monomer conversion and the molecular weight of the polymers could be adjusted by changes in the microwave irradiation time, microwave power and length and amount of added poly(ethylene glycol).

15.3.2 POLYMERIZATION OF OTHER ACIDS AND ESTERS

Ring-opening polymerization of poly (D,L-lactide) (PDLLA) was also catalyzed by stannous octoate (Sn(Oct)$_2$) under ambient atmosphere (Shu et al., 2006). In this process, they used carborundum (as heating medium), 99.9% pure lactide and

0.15% catalyst and applied 450 W microwave power. 85% yield of PDLLA with a viscosity-average molecular weight (M_η) over 2.0×10^5 was obtained with 30 min irradiation.

Poly(1-azabicyclo[4.2.0]octane) (polyconidine) was synthesized with bromo-acetic acid using microwave irradiation (Karabulut et al., 2008). 1-Azabicyclo[4.2.0] octane contains unsubstituted four-membered azetidine ring. It was synthesized with microwave-assisted methods and used in cationic ring-opening polymerization as a monomer.

The effect of microwave power on the polymerization process under microwave irradiation (domestic microwave oven) was studied by Zhang et al. (2004). They found that molecular weight of the polymer was increased and reached a maximum, when 90% conversion was reached up to 255 W. When a higher microwave energy (510 W) was applied, the molecular weight first increased and subsequently de-creased in time due to transesterification reactions.

The microwave-assisted ring-opening polymerization of poly(p-dioxanone) (PPDO) from the monomer p-dioxanone (PDO) was observed by Li et al. (2006), with constant microwave power (90, 180, 270 and 360 W) and temperature main-tained at 158 to 198°C. The polymerization at 270 W for 25 min using $Sn(Oct)_2$ as a catalyst gave 63% yield of PPDO with a viscosity-average molecular weight (Mv) of 156,000 g/mol. It was observed that time for polymerization was 14 h in conven-tional polymerization method.

The synthesis of biodegradable β-tricalcium phosphate (β-TCP)/poly(lactide-coglycolide) (PLGA) composites was carried out in situ polymerization with mi-crowave irradiation (Kim et al., 2006). The β-TCP content increased up to 10 wt.% while the molecular weight of composites decreased. It was observed that the bend-ing strength and Young's modulus of the β-TCP/PLGA composites were propor-tional to the molecular weight of PLGA.

Arazi et al. (2010) reported microwave-assisted synthesis and ring-opening po-lymerization of optically active α-hydroxy acids derived from amino acids and their polymerization to some new biodegradable polyesters. The positively charged side was hydrophobic and negatively charged side was hydrophilic. Hydroxy acids were prepared from diazotization of α-amino acids. The range of molecular weight of the prepared polymers from ring-opening polymerization in conventional method is between 2000 to 5000 Da. While under microwave irradiation, this range is 20,000–30,000 Da.

The ring-opening polymerization of poly(lactic acid) was carried out by micro-wave irradiation catalyzed by stannous octoate ($SnOct_2$) under vacuum (Singla et al., 2012). Polymerization of lactide as carried out in two different environments in the reaction vessel, an inert gas cover and partial vacuum. The rate of microwave heating was much faster than in conventional heating.

15.3.3 CATIONIC RING-OPENING POLYMERIZATION

Living cationic ring-opening polymerization of 2-ethyl-2-oxazoline under microwave irradiation was first reported by Wiesbrock et al. (2004). They demonstrated that the polymerization in acetonitrile tremendously accelerated up to 180°C. The cationic ring-opening polymerization under the microwave irradiation of 2-phenyl-2-oxazoline and 2-phenyl-2-oxazine could be accelerated in both open and closed reactors (Sinnwell and Ritter, 2005, 2006). They proved that the acceleration was due to nonthermal effects.

The superheated cationic ring-opening polymerization of 2-ethyl-2-oxazoline was reported by Hoogenboom et al. (2006). They used both; mono-mode and multimode microwave reactors. This polymerization was performed at scales ranging from 4.0 mmol to 1.0 mol. Guerrero (2006, 2007) also investigated the cationic ring-opening polymerization of 2-ethyl-2-oxazoline and 2-phenyl-2-oxazoline using ionic liquids as solvent to exclude the use of organic solvents under microwave irradiation. The use of ionic liquids enhanced the polymerization rate in comparison to the reaction in common organic solvents.

The synthesis of poly(2-alkyl-2-oxazoline)s, quasi-diblock and diblock copoly (2-oxazoline)s with improved microwave-assisted polymerization was reported (Hoogenboom et al., 2005, 2007; Wiesbrock et al., 2005a).

Hoogenboom and Schubert (2006) observed the microwave-assisted cationic ring-opening polymerization of a soy-based 2-oxazoline monomer (SoyOx). In presence of acetonitrile and without additional solvent, poly(2-oxazoline)s with fatty acid side chains in a living manner was formed. In this polymerization procedure, unsaturated side of fatty acids was unaffected. However, unsaturated side chains could be cross-linked by UV-irradiation.

The living cationic ring-opening polymerizations of 2-methyl-, 2-ethyl-, 2-nonyl-, and 2-phenyl-2-oxazoline were performed in acetonitrile (Wiesbrock et al., 2005b) at high temperatures up to 200°C in a single-mode microwave reactor. The polymerization was carried out in highly concentrated solutions or bulk conditions. A maximum number of 300 monomers (100 in the case of 2-methyl- and 2-nonyl-2-oxazoline) could be incorporated in the polymer chain. The average molecular weight of polymers was near about polydispersity index (PDI).

15.3.4 POLYMERIZATIONS VIA C–C COUPLING REACTIONS

C–C coupling reactions are often based on the metal catalysis, which might got some benefits from microwave irradiation by specific absorption of the metal ions. Khan and Hecht (2004) investigated the palladium-catalyzed synthesis of poly(m-phenyleneethynylene)s under both thermal heating as well as microwave heating (multimode microwave reactor). Tierney et al. (2005) investigated the synthesis of polythiophenes via Stille-type cross-coupling in a mono-mode microwave reactor.

Soluble semiconducting polythiophenes were obtained using both thermal and microwave heating. The microwave-assisted polymerization method yielded higher molecular weight polymers and slightly lowers polydispersity indices.

The Suzuki C–C coupling method was also studied under microwave irradiation (monomode microwave reactor) by Nehls et al. (2005). Three differently substituted naphthalene boronic ester monomers were coupled to 4,4'-didecyl-2,'5'-dibromoterephthalophenone using a palladium catalyst. The polymerization of the less sterically hindered 2,6-diboronic ester naphthalene gave similar results under both microwave irradiation and thermal heating while sterically hindered 1,5-diboronic ester naphthalenes could not be polymerized using conventional heating, but the use of microwave irradiation allowed their polymerization. All these resulting conjugated polymers served as precursors for the formation of ladder-type polymers, which have desirable optical and emitting properties. Palladium-catalyzed C–C coupling procedure (namely the Heck reaction) was carried out for the formation of conjugated polymers under both microwave (mono-mode microwave reactor) and thermal heating by Koopman et al. (2006).

The previous examples were all based on C–C coupling reactions that require two different functional groups. On the contrary, Ni (0) catalyzed coupling reactions can be used for homocoupling reactions.

Yamamoto et al. (2003) used Ni (0) catalyzed polymerization procedure for the synthesis of poly(pyrazine-2,5-diyl) starting from 2,5-dibromopyrazine within 10 min. The polymer obtained under microwave irradiation had a higher molecular weight. Beinhoff et al. (2005) reported the use of a Ni (0) catalyzed C–C coupling for the synthesis of poly(biphenylmethylene)s starting from bistriflate monomers under microwave irradiation (monomode microwave reactor). The microwave-assisted polymerizations were performed in only 10 min at 200°C, whereas the conventional polymerizations require for 16–24 h at 80°C. Surprisingly, end capping the polymerization by the presence of 4-bromostyrene could also be performed at 200°C under microwave heating without coupling or degradation of the vinyl groups.

The Ni (0) mediated polymerization was also carried out by Galbrecht et al. (2005) for preparing polyfluorenes with electrophosphorescent platinum-salen chromophores. This polymerization required 3 h under thermal heating in tetrahydrofuran (THF) at 80°C and could be accelerated down to 12 min under microwave irradiation at 115 or 220°C in THF or a mixture of N,N-dimethylformamide and toluene, respectively. The resulting copolymers revealed high electroluminescence efficiencies due to energy transfer from the polyfluorene to the salen complex.

15.4   CONTROLLED RADICAL POLYMERIZATION

Radical termination reactions is suppressed in controlled radical polymerization techniques. The suppression of termination reactions is achieved by lowering the concentration of free radicals by the following processes:
- Reversible addition fragmentation chain-transfer (RAFT) polymerization
- Atom-transfer radical polymerization (ATRP)
- Surface-initiated free radical polymerization (SI-FRP)
- Nitroxide-mediated radical polymerization (NMP)

15.4.1   REVERSIBLE ADDITION FRAGMENTATION CHAIN-TRANSFER (RAFT) POLYMERIZATION

The microwave-assisted RAFT polymerization of both styrene and methyl methacrylate was investigated by Perrier et al. (2005). An increase in polymerization rate was observed under microwave irradiation for both monomers. Zhu et al. (2006) also reported similar observations for the RAFT polymerization of styrene using a domestic microwave oven. The synthesis of well-defined homopolymers and block copolymers of acrylamido and acrylate monomers via reversible addition-fragmentation chain transfer (RAFT) polymerization have been reported (Roy et al., 2009). The rates of polymerization of N,N-dimethylacrylamide and N-isopropylacrylamide were significantly higher than those observed under conventional heating conditions. The resulting homopolymers retained thiocarbonylthio end group functionality.

Homopolymers and block copolymers of vinyl esters were also synthesized under microwave irradiation by RAFT/macromolecular design via interchange of xanthates polymerization without any significant inhibition or retardation, which is often observed under conventional heating conditions. Poly(vinyl acetate) (PVAc) with molecular weights of 1000–10,000 g/mol was prepared in less than 15 min under microwave irradiation at an apparent temperature of 70 °C in the presence of the commercially available chain transfer agent ethylxanthogen acetic acid (Roy and Sumerlin, 2011).

Grigoras and Negru (2012) synthesized star poly(N-vinylcarbazole) by microwave-assisted polymerization. The chain transfer agent 1,3,5-benzyl tri (diethyldithiocarbamate), was used. Chain transfer agent, containing a 1,3,5-trisubstituted benzene ring as core and three dithiocarbamate functionalities attached through an intermediate for fragmenting covalent bonds, led to poly(N-vinylcarbazole) (PVK) with star architecture. Polymerizations were carried out in 1,4-dioxane as solvent, at 70 °C, and studied for different polymerization times and monomer/CTA/initiator ratios.

15.4.2 ATOM-TRANSFER RADICAL POLYMERIZATION (ATRP)

Cheng et al. (2005) investigated the ATR polymerization of styrene under pulsed microwave irradiation. Controlled polymerization of styrene revealed that polymerization rates were three times larger than those for thermal heating.

Delfosse et al. (2005) reported that the iron (II)-catalyzed ATR polymerization of acrylonitrile could be accelerated under microwave irradiation. The Ru (II)-catalyzed ATR polymerization of methyl methacrylate (MMA) was also enhanced by the use of microwave irradiation in mono-mode microwave reactor. Acceleration of the polymerization under microwave irradiation at temperatures in the range of 100–140°C was observed and there was a retardation at temperatures above 150°C.

It was reported that the ATR polymerization of methyl methacrylate with both microwave (mono-mode microwave reactor) and thermal heating revealed comparable polymerization rates (Zhang and Schubert, 2004).

15.4.3 SURFACE-INITIATED FREE RADICAL POLYMERIZATION (SI-FRP)

Surface initiated controlled radical polymerization (CRP) without chemical additives, based on local heating due to microwave absorption by the substrate was observed. A simple model was developed to predict the temperature gradient at the interface between a microwave absorbing substrate and a nonabsorbing medium. Microwave irradiation may increase throughput of SI-FR polymerization, affording either faster brush growth, thicker brushes, or both, compared with conventional heating by an oil bath. Experimental results of microwave SI-FR polymerization was compared against conventional heating on silicon wafers, quartz slides, particles, and in bulk. Reproducibility of heating for silicon wafers was found to depend on orientation relative to the incident radiation (Erich et al., 2013).

15.4.4 NITROXIDE-MEDIATED RADICAL (NM) POLYMERIZATION

Li et al. (2006) observed a clear acceleration of the NM polymerization of styrene in bulk under microwave irradiation in a monomode microwave reactor.

Rigolini et al. (2010) reported microwave-assisted NM radical polymerization of acrylamide in aqueous solution. Reasonable results were obtained for a nitroxide-mediated radical polymerization with a combination of a conventional hydrosoluble radical initiator and a phosphorylated nitroxide. The microwave enhancement of the polymerization was found to depend on the mode of irradiation, whether it is a dynamic (DYN) mode or pulsed power mode (SPS).

15.5 EMULSION POLYMERIZATION

The emulsion polymerization of styrene was observed in the presence (Gao and Wu, 2005) and absence (Ngai and Wu, 2005) of surfactants under microwave heating. The ultrahigh molecular weight polystyrene was synthesized in mini emulsion through fast heating and cooling by modern microwave reactors (Holtze and Antonietti, 2006).

O'Mealey et al. (2006) studied the effect of microwave irradiation (multimode microwave reactor) on the emulsion polymerization of styrene and methyl methacrylate. The emulsion polymerization of methyl methacrylate was faster with microwave irradiation. It was observed that the obtained molecular weights for polystyrene and poly(methyl methacrylate) were higher when the polymerization was performed under microwave irradiation. It was ascribed to higher degree of branching, when microwave irradiation was applied.

Bao and Zhang (2004) synthesized thermosensitive poly(methyl methacrylate) coated poly(N-isopropylacrylamide) particles using a microwave-assisted two-step emulsion polymerization. A comparison was made between the microwave-assisted emulsion polymerization of methyl methacrylate and thermal heating (Sierra et al., 2006).

Yi et al. (2005) and Deng et al. (2006) reported the surfactant-free emulsion copolymerization of styrene and N-isopropylacrylamide using microwave heating where as Cristiane et al. (2013) used ionic liquid as surfactant in microwave-assisted emulsion polymerization. 1-n-Dodecyl-3-methylimidazolium chloride (ionic liquid) was selected to replace the surfactant dodecyltrimethylammonium bromide (DTAB) in methyl methacrylate emulsion polymerizations. The conversion evolutions and the final average diameter of polymeric particles were quite similar for reactions using the surfactant DTAB or the ionic liquid, showing that the ionic liquid also acted as efficiently as a surfactant in emulsion polymerizations.

15.6 GRAFTING

Grafting process is a convenient method to add new properties to a natural polymer with minimum loss of the initial properties of the substrate. Grafting modification of the polysaccharide has been done by the use of microwave irradiation (Kumar et al., 2012). Water was used as a solvent in most of the polysaccharide grafting reactions, as many polysaccharides are soluble in water.

Mostly vinylic polymers were grafted on to natural polymers in domestic microwave ovens. Grafting of acrylonitrile and acrylamide onto guar gum under both thermal heating and microwave irradiation was carried out by Singh et al. (2004a, 2004b). In thermal heating, grafting was performed in the presence of redox initiating systems (potassium persulfate and ascorbic acid) at 35°C. Under microwave irradiation, grafting on to natural polymer was done without initiating system at 97°C.

They grafted acrylonitrile, acrylamide, and methyl methacrylate onto chitosan using microwave heating. (Singh et al., 2005). These grafted copolymers showed increased zinc (II) binding (methyl methacrylate and acrylamide) grafted polymer and/or calcium (II) binding acrylamide grafted polymer. They removed these ions from waste water.

Prasad et al. (2006) used microwave irradiation as heat source for the grafting methyl methacrylate on K-carrageenan using a potassium persulfate initiating system. The most important advantages of the microwave-assisted grafting procedure were the very short polymerization times and the simplicity of the procedure. The microwave-assisted graft copolymerization of acrylamide on *Mimosa pudica* seed mucilage was also reported (Ahuja et al., 2011).

The graft polymerization of ε-caprolactone onto magnetite was carried out under microwave irradiation in the presence of tin (II) 2-ethylhexanoate. (Nan et al., 2009). The molar ratio of ε-caprolactone to tin (II) 2-ethylhexanoate was 300, whereas the molar ratio of ε-caprolactone to magnetite was 5. The chemical structures of the obtained poly(ε-caprolactone) coated magnetic nanoparticles were characterized by FTIR and XPS spectroscopy.

15.7 DEPOLYMERIZATION

Depolymerization is the process of converting a polymer into a monomer or a mixture of monomers. Thioglycolysis, thiolysis and phloroglucinolysis are reactions used to study condensed tannins by means of their depolymerization. Alavinikje and Nazari (2006) carried out microwave-assisted hydroglycolysis of poly(ethylene terephthalate) using an excess of methanol, ethanol, 1-butanol, 1-pentanol, and 1-hexanol in the presence of different simple basic catalysts, namely, potassium hydroxide, sodium hydroxide, sodium acrtate, and zinc acetate while Navarro et al. (2007) desulfated several sulfated polysaccharides by heating their pyridinium salts dissolved in dimethyl sulfoxide in a microwave oven. The procedure was applied to different products like a λ-carrageenan, a partially cyclized μ/ν-carrageenan, an agar-like product like corallinan, a fucoidan and a chondroitin sulfate. Parab et al. (2012) carried out the aminolytic depolymerization of PET bottle waste with ethanolamine by conventional heating and microwave irradiation heating method with heterogeneous, recyclable acid catalysts such as beta zeolite (SiO_2/AlO_2 = 15 Na-form) and montmorillonite KSF.

Bis(2-hydroxyethyl)terelphtalate

Kamimura et al. (2011) used ionic liquids for depolymerization of waste fiber-reinforced plastic (FRP) and unsaturated polyesters under heating conditions. The use of microwaves for heating effectively progressed depolymerization, whereas the conventional heating method was ineffective for this purpose. Pingale and Shukla (2009) investigated depolymerization of polyethylene terephthalate (PET) fiber waste and disposable soft drink bottle via aminolysis using excess of ethanolamine. The reaction was carried out under nonconventional microwave energy in the presence of different simple chemicals as catalysts namely, sodium acetate, sodium bicarbonate and sodium sulfate.

Siddiqui et al. (2012) proposed a recycling method of poly(ethyleneterepthlate) (PET) with methanolic pyrolysis. High degree of depolymerization was found at 180°C with in 5–10 min.

Fan et al. (2013) carried out direct microwave-assisted hydrothermal depolymerization of cellulose. The interaction of MW irradiation with micro crystalline cellulose has been carried out, covering a broad temperature range (150–270°C). The yield of glucose depends on the applied microwave density. The highest glucose yield was 21%.

These days, almost all materials like wood, metal, glass, etc., are being replaced by polymers and therefore, this period is called 'Era of Polymers.' However, the degradation and disposal of polymers is still a challenging problem for the chemists. Microwave irradiation can provide a solution to these problems.

KEYWORDS

- **Controlled Radical Polymerization**
- **Depolymerization**
- **Emulsion Polymerization**
- **Free Radical Polymerization**
- **Grafting**
- **Polymerization**
- **Ring-Opening Polymerization**
- **Step-Growth Polymerization**

REFERENCES

Agarwal, S., Becker, M., & Tewes, F. (2005). Polymer International, 54, 1620–1625.

Ahuja, M., Kumar, S., & Yadav, M. (2011). Polymer Bulletin, 66, 1163–1175.

Alavinikje, M. M., & Nazari, F. (2006). Advances in Polymer Technology, 25, 242–246.

Alimi, K., Molinie, P., Majdoub, M., Bernede, J. C., Fave, J. L., Bouchriha, H., & Ghedira, M. (2001). European Polymer Journal, 37, 781–787.

Arazi, N. C., Hagag, I., Kolitz, M., Domb, A. J., & Katzhendler, J. (2010). Advances in Science and Technology, 76, 30–35.

Banihashemi. A., Hazarkhani, H., & Abdolmaleki, A. (2004). Journal of Polymer Science Part A: Polymer Chemistry, 42, 2106–2111.

Bao, J., & Zhang, A. (2004). Journal of Applied Polymer Science, 93, 2815–2820.

Barbier, B. D., Brachais, L., Cretu, A., Gattin, R., Loupy, A., & Stuerga, D. (2003). Environmenal Chemistry Letters, 1, 19–23.

Beinhoff, M., Bozano, L. D., Scott, J. C., & Carter, K. R. (2005). Macromolecules, 38, 4147–4156.

Bezdushna, E., & Ritter, H. (2005). Macromolecule Rapid Communications, 26, 1087–1092.

Bezdushna, E., Ritter, H., & Troev, K. (2005). Macromolecule Rapid Communications, 26, 471–476.

Borriello, A., Nicolais, L., Fang, X., Huang, S. J., & Scola, D. A. (2007). Journal of Applied Polymer Science, 103, 1952–1958.

Caouthar, A. A., Loupy, A., Bortolussi, M., Blais, J-C., Dubreucq, L., & Meddour, A. (2005). Journal of Polymer Science Part A: Polymer Chemistry, 43, 6480–6491.

Chatti, S., Bortolussi, M., Bogdal, D., Blais, J. C., & Loupy, A. (2006). European Polymer Journal, 42, 410–424.

Chatti, S., Bortolussi, M., Loupy, A., Blais, J. C., Bogdal, D., & Roger, P. (2003). Journal of Applied Polymer Science, 90, 1255–1266.

Chatti, S., Bortolussi, M., Loupy, A., Blais, J. C., Bogdal, D., & Majdoub, M. (2002). European Polymer Journal, 38, 1851–1861.

Cheng, Z., Zhu, X., Zhou, N., Zhu, J., & Zhang, Z. (2005). Radiation Physics and Chemistry, 72, 695–701.

Cristiane, C., Verusca, H. S., Santos, C. D., Alexandre, F. S., Montserrat, F., Pedro, H. H. A., & Claudia, S. (2013). Journal of Applied Polymer Science, 127, 448–455.

Delfosse, S., Wei, H., Demonceau, A., & Noels, A. F. (2005). Polymer Preprints American Chemical Society, Division of Polymer Chemistry, 46, 295–296.

Deng, Z. W., Hu, X. X., Li, L., Xu, Z. S., & Yi, C. F. (2006). Journal of Applied Polymer Science, 99, 3514–3519.

Erich, B., Xinfang, H., Christopher, G., & Jan, G. (2013). American Physical Society, 18–22.

Faghihi, K., & Hagibeygi, M. (2003). European Polymer Journal, 39, 2307–2314.

Faghihi, K. (2006). Journal of Applied Polymer Science, 102, 5062–5071.

Faghihi, K., & Hajibeygi, M. (2004). Journal of Applied Polymer Science, 92, 3447–3453.

Fan, J., Bruyn, D. M., Budarin, L. V., Gronnow, J. M., Shottleworth, S. P., Breeden, S., Macquarrie, J. D., & Clark, H. J. (2013). Journal of American Chemical Society, 135, 11728–11731.

Fellows, C. M. (2005). Central European Journal of Chemistry, 3, 40–52.

Fischer, F., Tabib, R., & Freitag, R. (2005). European Polymer Journal, 41, 403–408.

Galbrecht, F., Yang, X. H., Nehls, B. S., Neher, D., Farrell, T., & Scherf, U. (2005). Chemical Communication, 18, 2378–2380.

Gao, C., Zhang, S., Gao, L., & Ding, M. (2004). Journal of Applied Polymer Science, 92, 2415–2419.

Gao, J., & Wu, C. (2005). Langmuir, 21, 782–785.

Grigoras, M., & Negru, O. I. (2012). Polymer, 4, 1183–1194.

Guerrero-Sanchez, C., Hoogenboom, R., & Schubert, U. S. (2006). Chemical Communications, 3797–3799.

Guerrero-S., C., Lobert, M., Hoogenboom, R., & Schubert, U. S. (2007). Macromolecular Rapid Communications, 28, 456–464.

Holtze, C., & Antonietti, K. T. (2006). Macromolecules, 39, 5720–5728.

Hoogenboom, R., & Schubert, U. S. (2006). Green Chemistry, 8, 895–899.

Hoogenboom, R., Fijten, M. W. M., Thijs, H. M. L., van Lankvelt, B. M., & Schubert, U. S. (2005). Designed Monomers & Polymers, 8, 659–671.

Hoogenboom, R., Fijten, M. W. M., Thijs, H. M. L., van Lankvelt, B. M., & Schubert, U. S. (2007). Journal of Polymer Science Part A: Polymer Chemistry, 45, 416–422.

Hoogenboom, R., Paulus, R. M., & Pilotti, A., Schubert, U. S. (2006). Macromolecular Rapid Communications, 27, 1556–1560.

Horie, M., Luo, Y., Morrison, J. J., Majewski, L. A., Song, A., Saunders, B. R., Michael L., & Turner, M. L. (2008). Journal of Materials Chemistry, 18, 5230–5236.

Hurduc, N., Abdelylah, D., Buisine, J-M., Decock, P., & Surpateanu, G. (1997). European Polymer Journal, 33, 187–190.

Iannelli, M., & Ritter, H. (2005a). Macromolecular Chemistry and Physics, 206, 349–353.

Iannelli, M., Alupei, V., & Ritter, H. (2005b). Tetrahedron, 61, 1509–1515.

Imai, Y. (1996a). ACS Symposium Series, 624, 421–430.

Imai, Y., Nemoto, H., & Kakimoto, M-A. (1996b). Journal of Polymer Science Part A: Polymer Chemistry, 34, 701–704.

Kamimura, A., Yamamoto, S., & Yamada, K. (2011). Chem Su Schem., 4, 644–649.

Karabulut, E., Özdemir1, Z. O., Yolaçan, C., Sarıçay1, Y., & Mustafaeva1, Z. (2008). Hacettepe Journal of Biology & Chemistry, 36, 319–328.

Kerep, P., & Ritter, H. (2006). Macromolecular Rapid Communication, 27, 707–710.

Khan, A., & Hecht, S. (2004). Chemical Communications, 3, 300–301.

Khoee, S., & Zamani, S. (2007b). European Polymer Journal, 43, 2096–2110.

Khoee, S., Sadegh, I. F., & Zaman, I. S. (2007a). Journal of Photochemistry and Photobiology A: Chemistry, 189, 30–38.

Kim, H. S., Li, Y. B., & Lee, S. W. (2006). Materials Science Forum, 510–511, 758–761.

Koopmans, C., Iannelli, M., Kerep, P., Klink, M., Schmitz, S., Sinnwell, S., & Ritter, H. (2006), Tetrahedron, 62, 4709–4714.

Kumar, R., Setia, A., & Mahadevan, N. (2012). International Journal of Recent Advances in Pharmaceutical Research, 12, 45–53.

Li, J., Zhu, X., Zhu, J., & Cheng, Z. (2006). Radiation Physics and Chemistry, 75, 253–258.

Li, Q., Yang, X., Chen, W., Yi, C., & Xu, Z. (2008). Macromolecular Symposia, 261, 148–156.

Li, Y., Wang, X.-L., Yang, K.-K., & Wang, Y.-Z. (2006). Polymer Bulletin, 57, 873–880.

Liao, L.Q., Liu, L. J., Zhang, C., He, F., Zhuo, R. X., & Wan, K. (2002). Journal of Polymer Science Part A: Polymer Chemistry 40, 1749–1755.

Lu, J., Wu, J., Wang, L., & Yao, S. (2005). Journal of Applied Polymer Science, 97, 2072–2075.

Mallakpour, S., & Dinari, M., (2008a). Polymers for Advanced Technologies, 19, 1334–1342, 2008.

Mallakpour, S., & Habibi, S. (2003). European Polymer Journal, 39, 1823–1829.

Mallakpour, S., & Rafiee, Z. (2003). Journal of Applied Polymer Science, 90, 2861–2869.

Mallakpour, S., & Rafiemanzelat, F. (2005a). Iranian Polymer Journal, 14, 909–919.

Mallakpour, S., & Rafiemanzelat, F. (2005b). Journal of Applied Polymer Science, 98, 1781–1792.

Mallakpour, S., & Taghavi, M. (2008b). European Polymer Journal, 44, 87–97.

Mallakpour, S. E., Hajipour, A. R., Faghihi, K., Foroughifar, N., & Bagheri, J. (2001). Journal of Applied Polymer Science, 80, 2416–2421.

Mallakpour, S., & Rafiemanzelat, F. (2004). Journal of Applied Polymer Science, 93, 1647–1659.

Nagahata, R., Sano, D., Suzuki, H., & Takeuchi, K. (2007). Macromolecular Rapid Communications, 28, 437–442.

Nagi, T., & Wu, C. (2005). Langmuir, 21, 8520–8525.

Li, N., Lu, J., & Yao, S. (2005). Macromolecular Chemistry and Physics, 206, 559–565.

Nan, A., Turcu, R., Craciunescu, I., Pana, O., Scharf, H., & Liebscher, J. (2009). Journal of Polymer Science Part A: Polymer Chemistry, 47, 5397–5404.

Navarro., D. A., Flores., M. L., & Stortz., C. A. (2007). Carbohydrate Polymers, 69, 742–747.

Nehls, B. S., Fueldner, S., Preis, E., Farrell, T., & Scherf, U. (2005). Macromolecules, 38, 687–694.

O'Mealey, J., Phillips, J., & Pilcher, S. C. (2006). Journal of Undergraduate Chemical Research, 5, 89–93.

Parab, S. Y., Shah, V. R., & Shukla, R. S. (2012). Current Chemistry Letters, 1, 81–90.

Perrier, S., Vettier, F., & Brown, S. I. (2005). Polymer Preprints American Chemical Society, Division of polymer chemistry, 46, 291–292.

Pielichowski, J., Dziki, E., & Polaczek, J. (2003). Polish Journal of Chemical Technology, 5, 3–4.

Pingale., N. D., & Shukla., S. R. (2009). European Polymer Journal, 45, 2695–2700.

Prasad, K., Meena, R., & Siddhanta, A. K. (2006). Journal of Applied Polymer Science, 101, 161–166.

Quantao, Li., Zushun, X., & Changfeng, Yi. (2008). Journal of Applied Polymer Science, 107, 797–802.

Rigolini, J., Grassl, B., Reynaud, S., & Billon, L. (2010). Journal of Polymer Science Part A: Polymer Chemistry, 48, 5775–5782.

Roy, D., & Summerlin, B. S. (2011). Polymer, 52, 3038–3045.

Roy, D., Ullah, A., & Sumerlin., B. S. (2009). *Macromolecules*, 42, 7701–7708.

Seong, J. C., Kuwabara, J., & Kanbara, J. (2013). ASC Sustainable Chemistry Eng., doi:10.1021/SC 4000576

Shu, J., Wang, P., Zheng, T., & Zhao, B. (2006). Journal of Applied Polymer Science, 100, 2244–2247.

Siddiqui, M. N., Redhwi, H. H., & Achilias, D. S. (2012). Journal of Analytical and Applied Pyrolysis, 98, 214–220.

Sierra, J., Palacios, J., & Vivaldo-Lima, E. (2006). Journal of Macromolecular Science Part A: Pure Applied Chemistry, 43, 589–600.

Singh, V., Tiwari, A., Tripathi, D. N., & Sanghi, R. (2004a). Journal of Applied Polymer Science, 92, 1569–1575.

Singh, V., Tiwari, A., Tripathi, D. N., & Sanghi, R. (2004b). Carbohydrate Polymer, 58, 1–6.

Singh, V., Tripathi, D. N, Tiwari, A., & Sanghi, R. (2005). Journal of Applied PolymerScience, 95, 820–825.

Singla, P., Kaur, P., Mehta, R., Berek, D., & Upadhyay, S. N. (2012). Procedia Chemistry 4, 179–185.

Sinnwell, S., & Ritter, H. (2006). Macromoleculor Rapid Communication, 27, 1335–1340.

Sinnwell, S., & Ritter, H. (2005). Macromoleculor Rapid Communication, 26, 160–163.

Sinnwell, S., Schmidt, A. M., & Ritter, H. (2006). Journal of Macromolecular Science Part A: Pure Applied Chemistry, 43, 469–476.

Stange, H., Ishaque, M., Niessner, N., Pepers, M., & Greiner, A. (2006). Macromolecule Rapid Communications, 27, 156–161.

Tang, X., Lu, J., Zhang, Z., Zhu, X., Wang, L., Li, N., & Sun, Z. (2003). Journal of Applied Polymer Science, 88, 1121–1128.

Tierney, S., Heeney, M., & McCulloch, I. (2005). Synthetic Metals, 148, 195–198.

Velmathi, S., Nagahata, R., Sugiyama, J. I., & Takeuchi, K. (2005). Macromolecular Rapid Communications, 26, 1163–1167.

Wiesbrock, F., Hoogenboom, R., Abeln, C. H., & Schubert, U. S. (2004). Macromolecular Rapid Communication, 25, 1895–1899.

Wiesbrock, F., Hoogenboom, R., Leenen, M., van Nispen, S. F. G. M., van der Loop, M., Abeln, C. H., van den Berg, A. M. J., & Schubert, U. S. (2005a). Macromolecules, 38, 7957–7966.

Wiesbrock, F., Hoogenboom, R., Leenen, M. A. M., Meier, M. A. R., & Schubert, U. S. (2005b). *Macromolecules, 38*, 5025–5034.

Xu, K., Zhang, W. D., Yue, Y. M., & Wang, P. X. (2005). Journal of Applied Polymer Science, 98, 1050–1054.

Yamamoto, T., Fujiwara, Y., Fukumoto, H., Nakamura, Y., Koshihara, S.-Y., & Ishikawa, T. (2003). Polymer, 44, 4487–4490.

Yi, C., Deng, Z., & Xu, Z. (2005). Colloid Polymer Science, 283, 1259–1266.

Yu, Z., & Liu, L. (2005). Journal of Biomaterials Science, Polymer Education, 16, 957–971.

Yu, Z. J., & Liu, L. J. (2004). European Polymer Journal, 40, 2213–2220.

Zhang, C., Liao, L., & Liu, L. (2004). Macromolecular Rapid Communications, 25, 1402–1407.

Zhang, H., & Schubert, U. S. (2004). Macromolecular Rapid Communications, 25, 1225–1230.

Zheng, T., Wang, P., Zhang, Z., & Zhao, B. (2005). Journal of Applied Polymer Science, 95, 264–269.

Zhu, J., Zhu, X., Zhang, Z., & Cheng, Z. (2006). Journal of Polymer Science Part A: Polymer Chemistry, 44, 6810–6816.

CHAPTER 16

OTHER REACTIONS

SURBHI BENJAMIN, PARAS TAK, and RAKSHIT AMETA

CONTENTS

Microwave-assisted organic synthesis has been found equally important to carry out some other chemical reactions like protection, deprotection, hydrolysis, esterification, cyclization, etc.

16.1 PROTECTION

Protection reactions are important when one needs to protect one of the functional group and wish to convert another functional group. Then certain protecting reagents are used for this purpose. In last few years, with the development of microwaves, protecting reactions are now being carried out with the help of microwave with the same or even better efficiency.

Godoi et al. (2013) have reported an efficient, quick, and sustainable method for the protection of amines with a 9-fluorenylmethoxycarbonyl (Fmoc) group. It has been found that under microwave irradiation this solvent-free approach resulted in good to excellent isolated yields of the desired products within only 5 min.

95% yield

Primary amines can also be protected as N-substituted 2,5-dimethylpyrroles. In spite of stable nature of this protecting group towards strong bases and nucleophiles, longer reaction times are required for both; the protection and deprotection steps, which normally results in low deprotection yields. Walia et al. (2013) have attempted to reduce protection and deprotection reaction times by using microwave irradiation. Reaction yields are increased in the presence of dilute hydrochloric acid in ethanol in deprotection process. Diverse deprotection conditions have been developed in conjunction with microwave irradiation, so that protection as an N-substituted 2,5-dimethylpyrrole can be orthogonal to other standard amine protecting groups, such as tert-butyloxycarbonyl (Boc), carbobenzyloxy (Cbz), and 9-fluorenylmethyloxycarbonyl (Fmoc).

Complete acetylation of totally O-unprotected mono- and disaccharides was carried out under microwave heating in 90 sec at 720 W under closed vessel conditions using indium (III) chloride catalyst. Das et al. (2005) studied reactions in acetonitrile with stoichiometric amounts of acetic anhydride, which were quantitative and afforded predominantly the α-peracetylated form. Normally partial acetylation of carbohydrates is achieved with acetyl chloride and pyridine.

A novel procedure for the selective acetylation of hydroxyls groups was reported by Witschi and Gervay-Hague (2010). This procedure was based on a protecting group exchange strategy under microwave irradiation starting from per-O-trimethylsilylated (TMS) pyranosides. It was found that 1,6-O-diacetate monosaccharide was formed under microwave exposure, while 6-O-monoacetate adduct was formed preferentially under classical heating conditions in an oil bath. It was also observed that the reactions were completed, when these were conducted using acetic acid as a catalyst in neat acetic anhydride for 3 ×25 min under microwaves. 52% Yield was obtained in diacetylated adduct using 2 equivalents of acetic acid for per-O-TMS-galactoside. The reaction showed only 50% completion, when the galactoside was stirred for 2 days with two equivalents of acid with 35% yields of 6-O-monoacetate after column chromatography purification.

Microwave methodology suffers from some limitations due to the sensitivity of some carbohydrate derivatives at elevated temperatures, notably for hydroxyls protection processes through acetal formation. Salanski et al. (1998) have reported the monoacetylation of O-unprotected sucrose in the presence of p-toluenesulfonic acid as the catalyst. Transacetylation provides 83% yield of a mixture of geranial and neral sucrose acetals (E and Z isomers of citral acetals) using citral dimethylacetal, after 2 min at 100 °C under classical heating conditions. When microwave was applied as the heating source, yield drops to 42% after an identical runtime in open vessel conditions. Cleavage of the glycosidic linkage is denoted and leads also to the formation of unwanted side products. Only 17–26% yield was obtained after 2 to 10 min of microwave exposure.

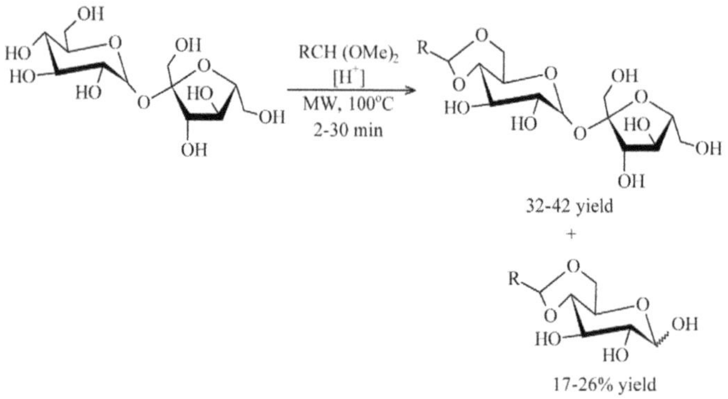

32-42 yield

+

17-26% yield

Soderberg et al. (2001) have selected 1,2,5,6-di-O-isopropylidene-α-D-glucofuranose as carbohydrate model compound for its acetylation with acetyl chloride (AcCl) and pyridine, or N,N-(diisopropyl)aminoethylpolystyrene (PS-DIEA) or N-(methylpolystyrene)-4-(methylamino)-pyridine (PDS-DMAP) as conventional base (Monosaccharide:Acetyl chloride = 1:2) to give the acetyl derivative.

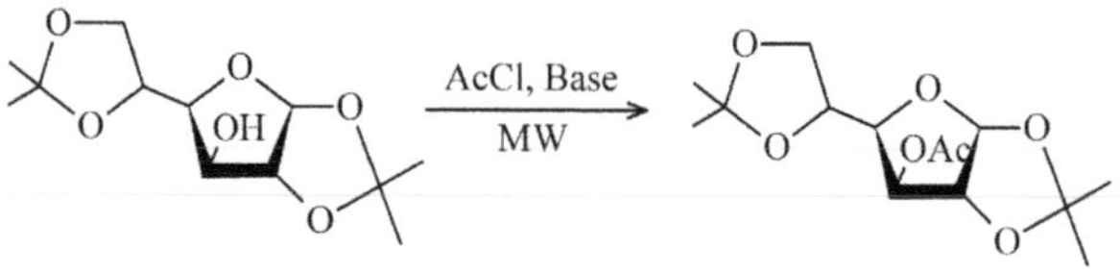

Limousin et al. (1997) have observed peracetylation of D-glucose to give the acetyl derivative with a small excess of acetic anhydride in presence of catalyst anhydrous potassium or sodium acetate, or zinc chloride quantitatively in less than 15 min with microwave heating. These reactions were found highly β-selective with potassium or sodium acetate under classical conditions. In presence of zinc dichloride and microwave irradiation, a 1:1 mixture of α/β pentacetates was obtained, where as a ratio of 7:3 was resulted under conventional oil bath heating conditions.

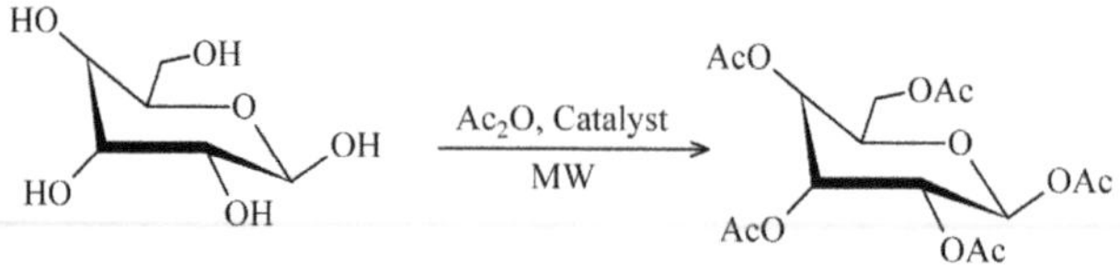

Saxena et al. (2003) have carried out the reactions of alcohols or phenols with tert-butyldimethylsilyl chloride (TBDMSCl) or trimethylsilyl chloride (TMSCl) in presence of catalytic amount (20 mol%) of iodine in a microwave oven for 2 min, which resulted into corresponding silyl ethers in excellent yield. It was also observed that under similar reaction conditions, iodine in methanol deprotects the silyl ether into its parent alcohol or phenol.

Beregszaszi and Molnar (1997) have reported that Envirocat® supported reagents (EPZG, EPZ10, and EPIC) were efficient in catalyzing the acetylation of carbonyl compounds with 1,2-ethanediol under microwave irradiation in solvent-free conditions. It was also observed that the reagents could be used in repeated experiments to perform the reaction with the same activity.

In a novel method, the reaction of hydroxylamine hydrochloride with a number of aldehydes and ketones under microwave irradiation and solventless 'dry' condition gave oximes in excellent yield (Hajipour, 1999).

16.2   DEPROTECTION

As the protection of a particular functional group is necessary along with the conversion of another functional group; similarly, deprotection process is equally important to get the actual functional group back in the product after completion of protecting reaction. Microwaves have been successfully employed for this purpose.

Srinivasan et al. (2005) reported a rapid method for the deprotection of N-Boc-protected amines using microwave irradiation. This procedure used five equivalents of trifluoroacetic acid (TFA) and dichloromethane as solvent. The protected amines were heated at 60 °C for 30 min in the cleavage cocktail. The free base amines were isolated rapidly by scavenging the crude reaction mixture with basic Amberlyst A-221 ion exchange resin.

28-99% yield

It is well known that certain fast methods are required for the removal of permanent amide exo-cyclic protective groups widely used in phosphoramidite-method DNA synthesis, which are used for many genomics and proteomics applications. In this context, Culf et al. (2008) reported a method for the deprotection of a range of N-acyl deoxyribonucleosides (T, dA Bz, dC Bz, dC Ac, dG ibu, dG PAC) and synthetic oligodeoxyribonucleotides, ranging in length from 5-mer to 50-mer. Oligodeoxyribonucleotides were synthesized using standard amide protecting groups (dA Bz, dC Bz, dG ibu) and phosphoramidite chemistry on cis-diol solid phase support. A 29% of aqueous ammonia solution was used to perform deprotection at 170°C for 5 min under mono-mode microwave irradiation at a 20-nmole-reaction scale.

Ramesh et al. (2003) have studied deprotection of several aryl acetates under microwave irradiation, which were rapidly and selectively deprotected to the corresponding phenols in excellent yields using silica gel supported ammonium formate. This process was found environmentally benign in nature.

98% yield

Perumal et al. (2004) have reported a new mild and efficient method for the cleavage of oximes to carbonyl compounds using readily available urea nitrate in acetonitrile–water (95 : 5), under microwave irradiation. This reaction was not only completed within 2 min, but also with good yields of the products.

Kumar and Gupta (1997) have also developed a novel method for the deprotection of oligodeoxyribonucleotides under microwave irradiation. The oligodeoxynucleotides having base labile, phenoxyacetyl (pac), protection for exocyclic amino functions were fully deprotected in 0.2 M sodium hydroxide (Methanol:Water 1:1 v/v) (A) and 1 M sodium hydroxide (Methanol:Water 1:1 v/v) (B) using microwaves in 4 and 2 min, respectively. The deprotection of oligodeoxyribonucleotides carrying conventional protecting groups, dAbz, dCbz and dGpac, for exocyclic amino functions was achieved in 4 min in (B) without any side product formation. On the basis of retention time on HPLC and biological activity, the deprotected oligonucleotides were also compared with the oligomers deprotected using standard deprotection conditions (29% aq. ammonia, 16 h, 55°C).

Heravi et al. (1999) reported that alcohols and phenols are tetrahydropyranylated in the presence of sulfuric acid adsorbed on silica gel in high to excellent yields in solvent-free conditions and is expedited by microwave irradiation. Addition of methanol performs the complete deprotection.

A variety of thioacetals, dithiolanes and dithianes are deprotected into their carbonyl compounds using clay-supported ammonium nitrate (Clayan) under microwave irradiation. The present method avoids the use of toxic oxidants and excess of solvent (Meshram et al., 1999a).

Meshram et al. (1999b) also observed the deprotection of a variety of tetrahydropyranyl ethers (THP), acetonides and acetals into their parent compounds using clayan under microwave irradiation. The ecofriendly nature of the reagent and solvent-free conditions are the important features of the procedure.

Regeneration of carbonyl functions from oximes can be accomplished by use of pyridinium chlorochromate under microwave irradiation within a short time with excellent yields (Chakraborty and Bordoloi, 1999).

Primary and secondary trimethylsilyl ethers were converted to their corresponding carbonyl compounds efficiently and rapidly with supported potassium ferrate under microwave irradiation in solventless system (Tajbakhsh et al., 2003).

Wettergren et al. (2003) have studied in situ deprotection and ω-methoxylation of TMS-protected aryl alkynes under microwave exposure. They found that using microwave technology, rapid ω-methoxylation of aryl alkynes was possible.

1-Benzyl-4-aza-1-azoniabicyclo[2.2.2]octane dichromate (BAABOD) is reported as a useful reagent for the selective cleavage of trimethylsilyl ethers, tetrahydropyranyl ethers, ethylene acetals and ketals to their corresponding alcohols, aldehydes and ketones. This method is very simple and efficient and the reaction has been carried out under microwave irradiation (Hajipour et al., 2002).

16.3 HYDROLYSIS

Zhang et al. (2011) have developed a microwave-enhanced hydrolysis of amides using KF/Al_2O_3 in the absence of solvents. Amines were produced in excellent yields along with the corresponding carboxylic acids.

Orozco et al. (2011) have investigated a microwave reactor system as a potential technique to maximize sugar yield for the hydrolysis of municipal solid waste for ethanol production. They have studied dilute acid hydrolysis of α-cellulose and waste cellulosic biomass (grass clippings) with phosphoric acid within the microwave reactor system. Their observations indicated that the use of a microwave reactor system is useful in facilitating dilute acid hydrolysis of cellulose and waste cellulosic biomass, producing high yields of total sugars in short reaction times.

It was found that maximum yield of reducing sugars was obtained at 7.5% (w/v) phosphoric acid and 160 °C, corresponding to 60% of the theoretical total sugars, with a reaction time of 5 min. When a very low acid concentration (0.4% w/v) for the hydrolysis in the microwave reactor was used, it was observed that 10 g of total sugars/100 g dry mass was produced, which was significant considering the low acid concentration. The optimum conditions were an acid concentration of 2.5% (w/v), 175 °C with a 15 min reaction time, giving 18 g/100 g dry mass of total sugars, with xylose being the sugar with the highest yield. They observed that pentose sugars were not more easily formed but also easily degraded and being significantly affected by increases in acid concentration and temperature. Kinetic studies showed that the use of microwave heating was good as compared to conventional systems especially in terms of rate constant.

Engelhardt et al. (1990) reported that the detection of sensitivity for proteins can be improved significantly by microwave enhanced hydrolysis and subsequent post-column OPA derivatization. It was observed that the detection limit can be lowered by a factor between 60 and 120 depending on the amino acid composition of the protein in comparison to native fluorescence detection. A hydrolysis time of 45 sec with microwave energy of 750 W was sufficient to reduce the detection limit to 0.04 μg/mL for BSA. It was also shown that protein hydrolysis was not only enhanced by the higher temperature achieved but microwave irradiation seems to play a major role in improving the efficiency of the hydrolysis.

A rapid microwave-assisted protein digestion technique based on classic acid hydrolysis reaction with 2% formic acid solution has been investigated by Hua et al. (2006). In this mild chemical environment, proteins were hydrolyzed to peptides. It was observed that dilute formic acid cleaved proteins specifically at the C-terminal of aspartyl (Asp) residues within 10 min of exposure to microwave irradiation. On the basis of the observations of the single fragmentation of myoglobin at the C-terminal of any of the Asp residues, they found that the extent of protein fragmentation could be controlled as compared to native fluorescence detection.

Lukasiewicz et al. (2009) investigated enzymatic hydrolysis of potato starch by γ-amylase to reveal the potential coupling mechanism of MIECC. The MIECC effect on increasing initial reaction rate ~2.5 times was observed in case of low viscous reaction system, that is, low substrate concentration. It is known that amylases are microwave sensitive enzymes, that are strongly deactivated, when placed in microwave field.

Orozco et al. (2007) have carried out dilute acid hydrolysis of grass and cellulose with phosphoric acid in a microwave reactor system. The experimental results indicated that it was a potential process for cellulose and hemi-cellulose hydrolysis, due to a rapid hydrolysis reaction at moderate temperatures. The optimum conditions for grass hydrolysis were found to be 2.5% phosphoric acid at a temperature of 175°C. It was found that sugar degradation occurred at acid concentrations greater than 2.5% (v/v) and temperatures greater than 175°C. They observed that dilute acid hydrolysis of cellulose with high yields of glucose in short reaction times was easily achieved in presence of microwave irradiation. 90% Yield of glucose was obtained at optimum conditions. Hence, the use of microwave heating provided high rate constant at moderate temperatures and prevent 'hot spot' formation within the reactor.

Mathews' reaction is a one-pot preparation of carboxylic acids from their corresponding nitriles or amides by a dry hydrolysis with phthalic acid or anhydride in the absence of water and solvent. Excellent isolated yields and selectivity (upto 99%) were attained within short reaction times (typically 30 min) when the reaction was performed under microwave heating (Chemat, 2002).

Moghaddam and Ghaffarzadeh (2001) have successfully transformed aldehydes and aryl alkyl ketones to thioamides with the same number of carbon atoms via Willgerodt-Kindler reaction under microwave irradiation in solvent-free conditions.

It was observed that the obtained thioamides were hydrolyzed to corresponding carboxylic acids with microwave dielectric heating in one min. It was also observed that both these reactions were very fast and obtained yields were also very good.

16.4 ESTERIFICATION

Microwaves irradiations have been found useful in carrying out esterification reactions also, which has been applied at large-scale syntheses. Amore and Leadbeater (2007) have reported microwave promoted esterification reaction. The water generated during the course of the reaction was allowed to remove and as a consequence, the process could be driven toward completion. The reported reactions have been run on scales up to 3 mol.

It is well known that in esterifications of phosphinic acids, it does not normally react with alcohols under thermal conditions, but in the presence of microwave exposure, the esterifications took place, which shows the importance of microwaves. It was observed that maximum 12–15% conversions were attained on traditional heating. It was found that due to the consequence of the hydrophobic medium established by the long chain alcohol/phosphinic ester, microwave-assisted esterifications were not reversible. The potential of the microwave technique in the synthesis of phosphinates can be understood on the basis of the energetics of the esterification of phosphinic acids under microwave conditions. A series of new cyclic phosphinates with lipophilic alkyl groups was synthesized by Keglevich et al. (2012).

Esters are biologically active against a range of potential targets. In addition, esterification has also been used successfully to facilitate the penetration of polar compounds into cells, where the ester group can then be removed by nonspecific cellular esterases unmasking drug molecules. While direct esterification of carboxylic acids with alcohols using acid catalysts can prove effective in many cases, there are instances when substrates are acid sensitive or where the use of strong acids leads to some side reactions and even product decomposition. Other methods are needed for the preparation of esters to overcome this problem.

Devine et al. (2010) have presented a methodology for titanium catalyzed esterification and transesterification using microwave irradiation. It was observed that the reactions were completed within 1 h of heating at 160 °C. It was found feasible to use aromatic, aliphatic and heteroaromatic acids in esterification. Some acid sensitive alcohols like furfuryl alcohol were also found as suitable substrates. Transesterification reactions are also possible in presence of microwaves.

Mazo and Rios (2010) have attempted to obtain alkyl ester from crude palm oil, using microwaves as heating source, in a process of two stages by means of heterogeneous catalysis. In the first stage of esterification, Dowex 50X2, Amberlyst 15 and Amberlite IR-120 resins catalysts were used to diminish the acid value of the oil, avoiding the soap formation and facilitating the separation of the phases while in the second stage of transesterification, potassium carbonate catalyst was used. It was found that the obtained biofuels were able to fulfill the requirements of the American standards for biodiesel. The proposed methodology for the synthesis has shown some environmental advantages and also an increase in the reactivity, in context to the traditional methods of heating. The reactivity of different alcohols have shown the order as:

Isobutyl alcohol > Isopentyl alcohol > 2-Butyl alcohol > Isopropyl alcohol

The conversion percentage of isoprophyl alcohol, isobutyl alcohol, 2-butyl alcohol and isopentyl alcohol was 14.85, 25.78, 9.55 and 23.47% under conventional heating, which was 49.51, 67.59, 52.00 and 54.59%, respectively, under microwave irradiation.

Cirin-Novta et al. (2006) have carried out the synthesis of esters of natural petroleum acids of the naphthenic type in presence of microwave irradiation under the conditions of acid catalysis with various alcohols like methanol, ethanol, n-butanol and tert-butyl alcohol. The esters of the naphthenic acid were prepared under microwaves irradiation by using naphthenic acids with sulfuric and p-toluenesulfonic acid. The yield of naphthenic esters from 31.25 to 88.90% was achieved, which was dependent on the catalyst and the steric and nucleophilic properties of the alcohols. It was also found that the esterification time was reduced from 6–10 h to 5 min in microwaves.

Where conventional method took about 6–10 h for esterification of naphthenic acids, this reaction was completed with high conversion in the first minute (with methanol : 60.39% in the presence of H_2SO_4 and 50.31% in the presence of p-TsOH), and with very high yields after 5 min (86.14% and 68.10% in the presence of H_2SO_4 and p-TsOH, respectively) under microwave exposure. The best yield of 88.08% was obtained in 5 min for the reaction of naphthenic acids in the presence of H_2SO_4 under microwave irradiation.

Although microwave heating is widely used in organic chemistry for synthesis purposes, as it normally shortens the reaction time and enhances the reaction rate, but there are very few reports available for its effect on enzymatic esterification reactions in ionic liquid media. Major et al. (2008) have made a comparison of the ethyl lactate synthesis in different media, where two organic solvents and 20 ionic liquids were tested. Toluene and seven ionic liquids were found suitable media. The reaction conditions of the enzymatic synthesis were optimized in toluene and in Cyphos 104. The highest yield, that is, 80% was achieved using toluene in a reaction mixture consisting of 1 mmol lactic acid, 5 mmol ethanol and 4.5 w/w% initial water content diluted by organic solvent to 5 cm³. 250 mg amount of the enzyme was required. 0.8 mmol ionic liquid, 2 mmol lactic acid, 7 times ethanol excess, 2 w% initial water content and 25 mg immobilized *Candida Antarctica* lipase B in Cyphos 104 medium were desired to carry out the reaction up to 95% yield in 24 h on 40 °C. The results showed that the smaller enzyme amount was sufficient in ionic liquid than in toluene and the enzyme stability was also much better in it. The same synthesis was studied under microwave conditions as well, and it was observed that the optimal initial water content was shifted from 3.7 w/w% to 3 w/w% as well as the same yield was achieved. It was concluded that microwave heating accelerated the hydrolysis of lactoyllactic acid providing the mixture with fresh lactic acid and enhancing the reaction rate.

Ebringerova and Srokova (2007) studied the hydrophobic modification of xylans, where the synthesis of water-soluble amphiphilic beechwood GX derivatives by various transesterification reactions. They have compared the effect of classical and microwave heating on the surface-active properties of the derivatives and reported the application of microwave radiation in transesterification reactions. They have shown that by applying microwave radiation as heating source in transesterification reactions using vinyl laurate and methyl laurate as acylation agents, water-soluble GX fatty acid ester derivatives of very low DS, and of acceptable emulsifying efficiency for O/W-type emulsions and washing power can be prepared rapidly and efficiently.

Both types of transesterification reactions were studied in the range between 90 and 120°C. The reaction time was appreciably shortened to 2–5 min at a power of 200–300 W; thus, limiting side reactions like decomposition of reactants, depolymerization of the polysaccharide, where as the amount of the solvent DMF was reduced by 50%.

Li et al. (2009) have studied the synthesis of alkyl ferulates under microwave irradiation. They observed the reactions time ranged from 3–5 min, which was much shorter than the traditional synthetic methods, as well as the alkyl ferulates were obtained in higher yields. It was the first time, when they have reported an efficient microwave-assisted esterification of ferulic acid with alcohols.

It is an important method to convert free fatty acids (FFA) into valuable ester and obtain a FFA-free oil that can be further transesterified using alkali bases by esterification of FFA in vegetable oils with alcohol using an acid catalyst. Suppalakpanya et al. (2011) have investigated the direct esterification reaction of FFA in crude palm oil to ethyl ester by continuous microwave along with the effects of the main variables involved in the process like amount of catalyst, reaction time and the molar ratio oil/alcohol, etc. The optimum condition obtained for the continuous esterification process was; a molar ratio of oil to ethanol 1:6, using 1.25%wt of H_2SO_4/oil as a catalyst, microwave power of 78 W and a reaction time of 90 min. The observations showed that the amount of FFA was reduced from 7.5% wt. to values around 1.4%wt. The same results were obtained by using conventional heating at 70 °C, where the reaction was completed in a longer time (4 hr).

Mazzocchia et al. (2004) have prepared fatty acids methyl esters (FAME); both under conventional heating and microwave irradiation. They have studied catalytic tests in two-phase systems in presence of barium hydroxide monohydrate and at different temperatures and pressures. As a result, it was observed that microwaves irradiation were fast for transesterification reactions (alcoholysis of triglycerides with methanol) leading to high activity and yields of fatty acid methyl esters as compared to conventional heating.

It was observed that the esterification of storage proteins from sunflower seeds renders these molecules more hydrophobic and increases internal plasticization, facilitating their use in the fabrication of materials by thermo-mechanical processes (Orliac and Silvestre, 2003). They have carried out the reaction in solvent-free conditions to simplify the process and to reduce costs. They have observed that microwave heating was far better to reduce the reaction time and the hydrolysis of protein chains. 89% Esterification was obtained in 18 min (amount of catalyst = 3.4 meq 5 N HCl/g protein and microwave power of 560 W) where as for the same optimization level of esterification by the classical method, heating in a thermostat-regulated oil bath, was achieved in 84% esterification for a reaction time of 4 h (amount of catalyst = 3.9 meq 5 N HCl/g protein, T = 90 °C).

16.5 CYCLIZATION

Figueroa-Villar and de Oliveira (2011) have developed a thermal cyclization reaction of o-halobenzylidene barbiturates, which was an efficient and simple method for the preparation of oxadeazaflavines. The products were prepared in 5 min with 47 to 98% yield using solid-state reaction conditions and under microwave irradiation. The results showed the molecular modeling mechanism simulation, indicating that this reaction occurred through an intramolecular hetero-Diels-Alder cyclization followed by fast rearomatization.

Airiau et al. (2010) have carried out microwave-assisted domino hydrofor-mylation/ cyclization reactions and they have also reported the scope and limitations of these reactions.

Farran and Bertrand (2012) have reported two methods for the synthesis of sub-stituted [2,3]-dihydro-2-methyl-benzofuran-3-ones from corresponding salicylate esters under microwave irradiation. It was observed that a two step sequence via ether intermediates was convenient for various substituted salicylate derivatives, while the second strategy involving a one pot procedure was efficient for electron-donating substituted salicylates.

Crawford et al. (2003) have studied intramolecular cyclization reactions of 5-halo- and 5-nitro-substituted furanylamides under microwave heating. The rear-ranged dihydroquinone in 36% yield was produced from 2-alkoxy-5-bromofuran derivative, which was a product from the rearrangement of the intermediate oxabi-cycle. The 5-halo substituted furoyl amide was converted into the polyfunctional oxabicycle in 82% yield at a much faster rate than the unsubstituted furanyl system. 1,4-Dihydro-2H-benzo[4,5]furo[2,3-c]pyridin-3-one was produced by 5-nitro-substituted furfuryl amide by isomerization-cyclization reaction under microwave irradiation.

A microwave procedure for the synthesis of a focused library of 3-amino-imidazopyridines has been developed by DiMauro and Kennedy (2007). Imidazopyridine products were obtained via an Ugi-type cyclization, starting from 2-aminopyridine-5-boronic acid pinacol ester, where an intermediate was formed. It was further converted into the final product by a Suzuki coupling in a one-pot procedure. It was interesting to note that it was neither possible to perform the reaction sequence in one step nor the sequence can be reversed.

42-68% yield

Heravi et al. (2006) have studied the synthesis of 2-phenyl-3-hydroxy-quinolin-4(1H) under microwave irradiation in solventless system. It was a two-step synthesis. Anthranilic acid, phenacyl bromide and potassium carbonate were exposed under microwaves in solventless media to produce phenacyl anthranilate (within 2 min), where as the same product was obtained in 1 h under conventional heating. Final product 2-phenyl-3H-benz-[e] [1,4] oxazepin-5-one was prepared in very short time of only 2 min when phenacyl anthranilate was treated further with polyphosphoric acid supported on silica gel.

Cao and Xiao (2005) have carried out rapid palladium catalyzed carbonylative cyclization reactions of 2-iodophenol with various alkynes. The reactions were conducted in the presence of DIEA and DMAP in 1,4-dioxane at 160 °C for 30 min under microwave exposure. As a result, the corresponding chromen-2-one derivatives were obtained in good yields. These microwave-assisted reactions have shown their importance due to short reaction time, simplicity, and the fact that no addition of carbon monoxide was required from outside.

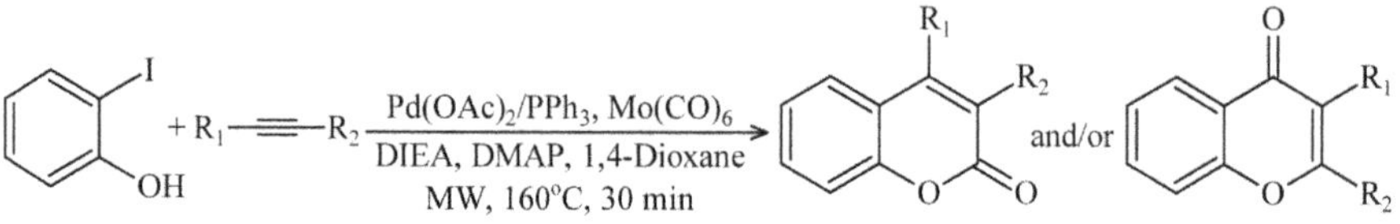

where R_1, R_2 = H, Alkyl or Aryl

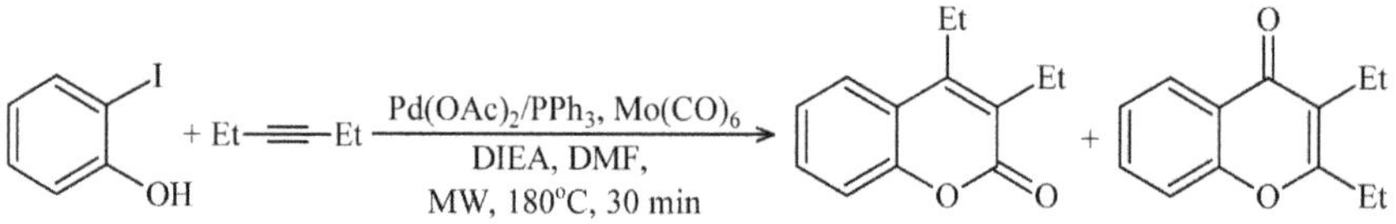

Thus, microwave irradiation can assist us in carrying out various reactions like cyclization, esterification, hydrolysis, protection, deprotection, etc. in preparing organic molecules of interest in limited time and with good yields.

KEYWORDS

- **Cyclization**
- **Deprotection**
- **Esterification**
- **Hydrolysis**
- **Protection**

REFERENCES

Airiau, E., Chemin, C., Girard, N., Lonzi, G., Mann, A., Petricci, E., Salvadori, J., & Taddei, M. (2010). Synthesis, 17, 2901–2914.

Amore, K. M., & Leadbeater, N. E. (2007). Macromolecular Rapid Communications, 28, 473–477.

Beregszaszi, T., & Molnar, A. (1997). Synthetic Communications, 27, 3705–3709.

Cao, H., & Xiao, W.-J. (2005). Canadian Journal of Chemistry, 83, 826–831.

Chakraborty V., & Bordoloi, M. (1999). Journal of Chemical Research (S), 120–121.

Chemat, F. (2002). Tetrahedron Letters, 43, 5555–5557.

Cirin-Novta, V., Kuhajda, K., Kevresan, S., Kandrac, J., Grbovic, L., & Vujic, D. (2006). Journal of Serbian Chemical Society, 71, 1263–1268.

Crawford, K. R., Bur, S. K., Straub, C. S., & Padwa, A. (2003). Organic Letters, 5, 3337–3340.

Culf, A. S., Cuperlović-Culf, M., Laflamme, M., Tardiff, B. J., & Ouellette, R. J. (2008). Oligonucleotide, 18, 81–92.

Das, S. K., Reddy, K. A., Krovvidi, V. L. N. R., & Mukkanti, K. (2005). Carbohydrate Reasearch, 340, 1387–1392.

Devine, W. G., Leadbeater, N. E., & Jacob, L. A. (2010). Future Medicinal Chemistry, 2, 225–230.

DiMauro, E., & Kennedy, J. (2007). Journal of Organic Chemistry, 72, 1013–1016.

Ebringerova, A., & Srokova, I. (2007). Preparation of beechwood xylan surfactants by microwave-assisted transesterification, Workshop: Production and functionalization of hemicelluloses for sustainable advanced products, Hamburg.

Engelhardt, H., Kramer, M., & Waldhoff, H. (1990). Chromatographia, 30, 523–526.

Farran, D., & Bertrand, P. (2012). Synthetic Communications, 42, 989–1001.

Figueroa-Villar, J. D., & de Oliveira, S. C. G. (2011). Journal of Brazilian Chemical Society, 22, 2101–2107.

Godoi, M., Botteselle, G. V., Rafique, J., Rocha, M. S. T., Pena, J. M., & Braga, A. L. (2013). Asian Journal of Organic Chemistry, 2, 746–749.

Hajipour, A. R., Mallkpour, S. E., Mohammadpoor-Baltork, I., & Khoee, S. (2002). Synthetic Communications, 32, 611–620.

Hajipour, A. R., Mallkpour, S. E., & Imanzadeh, G. (1999). Journal of Chemical Research (S), 228–229.

Heravi, M. M., Ajami, D., & Ghassemzadeh, M. (1999). Synthetic Communications, 29, 1013–1016.

Heravi, M. M., Oskooie, H. A., Bahrami, L., & Ghassemzadeh, M. (2006). Indian Journal of Chemistry, 45B, 779–781.

Hua, L., Low, T. Y., & Sze, S. K. (2006). Proteomics, 6, 586–591.

Keglevich, G., Kiss, N. Z., Mucsi, Z., & Kortvelyesi, T. (2012). Organic and Biomolecular Chemistry, 10, 2011–2018.

Kumar, P., & Gupta, K. C. (1997). Nucleic Acids Research, 25, 5127–5129.

Li, N.-G., Shi, Z.-H., Tang, Y.-P., Li, B.-Q., & Duan, J.-A. (2009). Molecules, 14, 2118–2126.

Limousin, C., Cleophax, J., Petit, A., Loupy, A., & Lukacs, G. J. (1997). Journal of Carbohydrate Chemistry, 16, 327–342.

Lukasiewicz, M., Marciniak, M., & Osowiec, A. (2009). Microwave-assisted enzymatic hydrolysis of starch, 13th International Electronic Conference on Synthetic Organic Chemistry (EC-SOC-13).

Major, B., Nemestothy, N., Belafi-Bako, K., & Gubicza, L. (2008). Hungarian Journal of Industrial Chemistry, 36, 77–81.

Mazo, P. C., & Rios, L. A. (2010). Latin American Applied Research, 40, 337–342.

Mazzocchia, C., Modica, G., Kaddouri, A., & Nannicini, R. (2004). Comptes Rendus Chimie, 7, 601–605.

Meshram, H. M., Reddy, G. S., Sumitra G., & Yadav, J. S. (1999a). Synthetic Communications, 29, 1113–1119.

Meshram, H. M., Sumitra G., Reddy, G. S., Ganesh, Y. S. S., & Yadav, J. S. (1999b). Synthetic Communications, 29, 2807–1815.

Moghaddam, F. M., & Ghaffarzadeh, M. (2001). Synthetic Communications, 31, 317–321.

Orliac, O., & Silvestre, F. (2003). Bioresource Technolology, 87, 63–68.

Orozco, A. M., Al-Muhtaseb, A. H., Albadarin, A. B., Rooney, D., Walker, G. M., & Ahmad, M. N. M. (2011). RSC Advances, 1, 839–846.

Orozco, A., Ahmad, M., Rooney, D., & Walker, G. (2007). Process Safety and Environmental Protection: Transactions of the Institution of Chemical Engineers Part B, 85, 446–449.

Perumal, P. T., Anniyappan, M., & Muralidharan, D. (2004). Journal of Chemical Sciences, 116, 261–264.

Ramesh, C., Mahender, G., Ravindranath, N., & Das, B. (2003). Green Chemistry, 5, 68–70.

Salanski, P., Descotes, G., Bouchu, A., & Queneau, Y. (1998). Journal of Carbohydrate Chemistry, 17, 129–142.

Saxena, I., Deka, N., Sarma, J. C., & Tsuboi, S. (2003). Synthetic Communications, 33, 4185–4191.

Soderberg, E., Westman, J., & Oscarson, S. (2001). Journal of Carbohydrate Chemistry, 20, 397–410.

Srinivasan, N., Yurek-George, A., & Ganeshan, A. (2005). Molecular Diversity, 9, 291–293.

Suppalakpanya, K., Ratanawilai, S., Nikhom, R., & Tongurai, C. (2011). Journal of Science and Technology, 33, 79–86.

Tajbakhsh, M., Heravi, M. M., & Habibzadeh, S. (2003). Phosphorus, Sulfur, Silicon and Related Elements, 178, 361–364.

Walia A., Kang, & Silverman, R. B. (2013). Journal of Organic Chemistry, 78, 10931–10937.

Wettergren, J., & Minidis, A. B. E. (2003). Tetrahedron Letters, 44, 7611–7612.

Witschi, M. A., & Gervay-Hague, J. G. (2010). Organic Letters, 12, 4312–4315.

Zhang, X., Luo, K., Chen, W., & Wang, L. (2011). Chinese Journal of Chemistry, 29, 2209–2212.

CHAPTER 17

INDUSTRIAL APPLICATIONS

DIPTI SONI, JITENDRA VARDIA, and RAKSHIT AMETA

CONTENTS

Industrial manufacturers always put there best efforts to produce new products, build new plants or automate the industries to fulfill the various demands of the society. Microwave heating is a quick and efficient method of heating materials compared to convectional or infra-red (I.R.) methods. Microwave-assisted synthesis normally increases the production rate as well as the quality of product in shorter time. Now a days microwave heating is used in different industries like pharmaceutical, chemical, polymer, food and others.

Fast heating, small space requirements and electronic control make use of microwave attractive in these industries and side wise, automation also increases through microwaves. Microwave is mainly used for drying in processes technology. Microwave vacuum drying is used for thermosensitive products with low thermal conductivity as drying process works fast and directly to dry material from the inside out.

17.1   ADVANTAGES

There are some of the salient features, which make the use of microwave more interesting. These are:
- Microwave generates higher power densities, enabling increased speeds of production and decreased production costs.
- Microwave systems are more compact, requiring smaller equipments and limited space.
- Microwave is a noncontact technology for drying.
- Microwave energy is precisely controllable and can be turned on and off instantly; thus, eliminating the need for warm-up and cool-down. Lack of high temperature heating surfaces greatly reduces the amount of product that is burned or overheated otherwise.
- Microwave energy provides uniform energy distribution. This results in more uniform temperature and moisture profiles, improved yields and enhanced product performance. This makes it possible to eliminate such disadvantages of convective drying as case-hardening, surface cracking, and local overheating.
- The use of industrial microwave systems avoids combustible gaseous by products; thus, eliminating the need for any environmental permits and improving working conditions.
- Microwave heated food products tend to retain a higher percentage of flavors and nutrients compared to conventional heating.
- Microwaves reduce production run times and reduce both; cleaning times as well as chemical costs.

Microwaves are used for processes of drying, cooking, sterilization, pasteurization, tempering (thawing) and blanching.

17.2   SCALE-UP

After many applications of the microwave heating technology, it is possible to successfully perform batch-wise chemical synthesis at multigram scales. Moseley et al.
(2008) designed seven different types of commercial microwave reactors for limited scale-up. Bowman et al. (2008) also scale-up microwave-promoted reactions to
the multigram level using a sealed-vessel microwave apparatus, while Amore et al.
(2007) scale-up the esterification reactions up to 3 mol.

The target of reaching kilogram scales still remains a dream in many of cases.
This is especially due to the penetration depth limitations of microwaves. The idea
of scaling up batch procedures of very small volumes (5 mL) by increasing reaction volumes (Liters) does require modifications in instrumentation as well as its
process. The first reason for its limitation is selective nature of microwave heating
and this is not understood explicitly for most of the case studies before actually
approaching the scale-up. Secondly, almost in all the case studies, the process was
developed in a mono-mode type microwave cavities and then shifted to multimode
microwave cavities for scaling.

However, there are a few possibility of designing a microwave setup, which
satisfies the requirements of large batch processes, assuring homogeneous heating.
One of the good options for scale-up is to switch from batch operation to continuous
operation at early stages in the process development. Scale-up studies on continuous
operation reveal evaluations of commercially available continuous flow microwave
reactors (Moseley and Lawton, 2007; Bergamelli et al., 2010; Bagley et al., 2010)
of microreactors and microwave integrated reactor setups specifically desired for a
process. Microwave-assisted flow synthesis at an industrial scale requires a proper
design of multitubular reactors integrated with microwave heating, which should
primarily overcome the limitation of the penetration depth of microwaves (~0.013
m). Moreover, operation under microwave heating should also allow accurate temperature control by precise tuning and quantification of the microwave energy distribution. So depending on a specific case, efforts should be made to design desired
microwave system, the reactor configuration and the catalytic system.

This is the main issue in the industries. Two main approaches have evolved for
scaling up microwave reactors. The first approach scales up single-mode reactors
through a flow-through reactor, and the second scales up multimode reactors to a
batch reactor.

17.2.1   FLOW-THROUGH MICROWAVE REACTOR

A flow-through microwave reactor is capable of scaling up single-mode yield from
grams to kilograms. An advantage of the flow-through reactor is its ability to perform dual-mode operations between liquid and solid phase reactions and it allows
scale-up from grams to kilograms with full temperature and pressure control. It

provides precise monitoring and control of the process parameters. Though this flow-through single-mode microwave reactor has been a breakthrough product in fulfilling requirements of a chemist pertaining to liquid phase reactions, the process has its own limitations in regard to handling solids and mixture-based reactions.

17.2.2 BATCH REACTOR

Multi-mode reactors are another breakthrough related to successful scaling up. Batch reactors are capable of accommodating larger volumes of reactants in a multimode operation in one time, and therefore, they can conduct reactions to produce higher yields. Batch reactors have been developed to scale-up the yields of reactions. However, the second and third generation equipments have been used successfully in industry at present.

Wharton (2011) developed a large-scale continuous flow microwave reactor that is capable of producing 10 kg of product per working day. The reactor can be used in a wide range of reactions, applicable to a variety of industries with a significant reduction in reaction times and increase in reaction yields. Reduction in reaction time of a Suzuki coupling from 2 h to 1 min with no loss of yield, and synthesis of dihydropyrimidine with a 2-fold increase in yield and reduction in reaction time from 8 h to 4 min are good examples. In an ionic liquid synthesis, reaction time was also reduced from 4 h to 2 min. Use of a continuous flow system, lowers the inventory of hazardous or unstable products and intermediates and makes scale-up process safer one than using a batch process.

Morschhäuser et al. (2012) performed microwave-assisted continuous flow synthesis on industrial scale. This continuous flow microwave system was based on a transmission line short-circuited waveguide reactor concept. The continuous flow reactor is capable of operating in a genuine high-temperature/high-pressure process window (310 °C/60 bar) by using microwave transparent and chemical resistant cylindrical γ-Al_2O_3 tube as the reaction zone. The system can be operated in an extremely energy efficient manner, using 0.6–6 kW microwave power (2.45 GHz). The reactor has been applied for processing four chemical transformations in a high-temperature/high-pressure region with 3.5–6.0 L/h^{-1}.

17.3 INDUSTRIAL APPLICATIONS

17.3.1 PHARMACEUTICAL INDUSTRIES

The main applications of microwave in pharmaceutical industries are drying, sterilization, thawing, production of ointments, sustained release dosage forms, drug synthesis and drug extraction.

17.3.1.1   DRYING

Microwave drying is a new drying technology that shortens drying times, reduce drying defects and increase the potential for product innovation. It is mainly used to remove water and/or polar organic solvents from brittle and heat sensitive powders, granules, bulk drugs, pastes, slurries, etc. (Anon, 1991).

Microwave vacuum dryers are more efficient as compared to sole microwave dryers. This is because of the fact that when a product is to be dried, it is to be kept under vacuum. The liquid evaporates at low temperature and thus, drying is completed at low heat exposure. It saves time, money and heat labile substance (Waldron, 1988). A rotary vacuum microwave dryer and sterilizer is a combination of vacuum and microwave heating. It provides high quality drying and sterilization of any substances and materials at relatively low temperatures avoiding conventional drying problems.

17.3.1.2   BENEFITS OF MICROWAVE DRYING

It can be used to dry very fine structures as well as particles of large dimensions.
  • It reduces drying time.
  • It can be easily integrated into automated systems.
  • It provides the development of new products.
  • It increases productivity and flexibility.
  • It reduces handling errors and production costs.

Bremecker (1983) used microwave to measure the moisture content in the pharmaceuticals. Moisture can be detected online for microwave vacuum dryer using near Infrared Spectroscopy (NIS); of course, if the moisture content of product is 6% or less. Thus, drying can be made a continuous process with precise control of moisture (White, 1994).

Application of microwave for drying of pharmaceuticals led to its utility in single pot processes, that is, incorporating mixer granulator and dryer (Pearlswing et al., 1994a; 1994b). In this method, the process is continuous and faster; thus, reducing loss of product during change of processes. Therefore, it helps in cost reduction and increase in annual yield.

Poska (1991) and Mandal (1995) have shown the comparability of physico-chemical characteristics of granules dried in microwave vacuum processors as compared to tray dryers as well as fluid bed dryers.

17.3.1.3   STERILIZATION

Microwave sterilization is a new method and it plays a significant role in synthesis (Honda et al., 1998; Sasaki et al., 1998a). The sterilization is brought about by microwave dielectric heating effect. The efficiency of a microwave sterilizer was tested for two heat labile drugs, ascorbic acid and pyridoamine phosphate, both in solution form. The results showed that reduction of bio-burden was equal to that of autoclaving, but autoclaved drugs showed certain deteriorations in quality, which is not observed in microwave sterilized drugs. Therefore, microwave sterilization holds an upper hand over autoclave sterilization (Sasaki et al., 1998b).

Sasaki et al. (1996) developed continuous microwave sterilization (CMWS) of some injection ampoules. They observed that sterilization effect of microwave was mainly due to production of heat and there is no other nonthermal mechanism. Thus, they used spores of heat resistant species, that is, *Bacillus stearothermophilus* for validation of microwave in sterilization. The high temperature and short time sterilization by microwave heating were evaluated. *Bacillus stearothermophilus* spores were used as biological indicator. The lethal effect of microwave sterilizer was equal to that of autoclave. The reliability of microwave sterilization in CMWS was confirmed using more than 25,000 test ampoules containing biological indicators (Sasaki et al., 1998c).

Microwave sterilization is now used for sterilization of heat labile drugs, where a high temperature is generated for sterilization in a shorter period of time and thus, it creates a possibility of making this process continuous. Such a process can give benefit to the industry due to its economy and lesser production time. Groning and Janski (1985) have carried out treatment in a simple microwave oven for 20 to 60 seconds showing a reduction in germ count. 60% preparations were sterile and remaining 40% preparations showed 60–90% reduction in germ count.

Tensmeyer et al. (1981) developed a sterilization method for empty glass containers, using a laser and a microwave oven. Sterilization procedures were evaluated at power levels upto 2000 Watts. It was observed that spore destruction was logarithmically related to the exposure time. *B. subtilis* spores were used to check the efficiency. All the spores got destroyed with no particulate contamination being introduced into the glass containers.

17.3.1.4 THAWING

Thawing is the process, in which freeze stored drugs are brought to normal physiological temperature before administration especially, in case of an injection. Thawing can be done using microwaves. The stability of many drugs, both; physical and chemical were not affected after microwave thawing. Microwave thawing reduced process cost as well as preparation time.

Tidy et al. (1988) showed the effect of infusion volume, its load size and microwave power on rate of thawing. Frozen infusions of 100–500 mL volume were thawed evenly and reproducibly without overheating. Microwave thawing also caused about 10% reduction in microbial count. Cloxacillin sodium, Flucloxacilline sodium and Ticarcillin disodium were reconstituted in 0.9% sodium chloride and in 5% dextrose solutions, stored frozen for upto 9 months, and then the stability of these antibiotics were assayed following microwave thawing (Vigneron et al., 1992).

The stability of intravenous Augmentin (Amoxycillin sodium and Clavulanate potassium) in a range of vehicles was investigated. It was observed that aqueous solutions frozen at –20°C and thawed by microwave radiation lost more activity than those stored at 25°C (Ashwin et al., 1987). The stability of 6 antibiotics in intravenous (I. V.) fluids in polyvinyl chloride (PVC) containers after freezing and microwave thawing were established. All antibiotics (except Ampicillin) retained 90% or more potency on microwave thawing after storage at –20°C for 30 days, and after subsequent storage at room temperature for 24 h (Holmes et al., 1982).

17.3.1.5   OINTMENT PRODUCTION

Various ointments were prepared according to official monograph of German Pharmacopoeia 9 (DAB 9). During this, heating ability of microwave was also evaluated. The products were prepared at various uncontrolled temperatures including the one, adjusted to official preparation method. Ointments prepared by this method (both controlled and uncontrolled) were evaluated for organoleptic, microbiological and rheological qualities of ointments. The ointments prepared with microwave and traditional methods did not show any significant difference (Moll and Maue, 1988).

17.3.1.6   PHARMACEUTICAL DOSAGE FORM DEVELOPMENT

Microwaves have been used successfully to prepare pharmaceutical dosage forms such as agglomerates, gel beads, microspheres, nanomatrix, solid dispersion, controlled release tablets formulation and tablet film coating (Joshi et al., 1989; Bergese et al., 2003; Vandalize et al., 2004; Moneghini et al., 2008). The use of microwave opens a new approach to control the physicochemical properties and drug delivery profiles of pharmaceutical dosage forms without the need for excessive heat, lengthy process or toxic reactants. Alternatively, the microwave can be used to process excipients prior to their use in the formulation of drug delivery systems (Wong, 2008).

17.3.1.7   DRUG EXTRACTION

Conventional techniques, that is, Soxhlet extraction, for the extraction of active constituents are time and solvent consuming, thermally unsafe and the analysis of numerous constituents in plant material is limited by extraction step. High and fast extraction ability with less solvent consumption and protection offered to thermolabile constituents are the attractive features of this new and promising microwave-assisted extraction (MAE) technique (Mandal et al., 2007).

17.3.1.8 ADVANCED MICROWAVE EXTRACTION SYSTEM

The highest performance and best safety features are the qualities of an advanced microwave extraction system in the industry. It offers fast heating of vessel together with homogeneous microwave distribution throughout the cavity.

17.3.1.9 DRUG SYNTHESIS

Microwave synthesis has the potential to influence the field of medicinal chemistry in three major phases of the drug discovery process. These are:
- Generation of a discovery library.
- Hit-to-lead efforts.
- Lead optimization.

A common theme point of interest in all these processes is the rate (speed). This greater speed provides a competitive advantage, and allows for more efficient use of expensive and limited resources, faster exploration of structure–activity relationships (SARs), enhanced delineation of intellectual property and more timely delivery of crucially needed medicines (Brittany, 2004).

Magano and Dunetz (2012) have reported methods for carbonyl reductions on large scale ($\geq$100 mmol) applied to the synthesis of some drugs in the pharmaceutical industry. They have discussed the most common and reliable methods for the reduction of aldehydes, ketones, carboxylic acids, esters, amides, imides, and acid chlorides, with representative examples, detailed reaction and workup conditions. They have also highlighted the advantages and limitations of each reducing agent with special emphasis on safety, cost, and amenability to scale-up.

where X = H; Alkyl, Aryl; OH, OR, NR$_2$, Cl.

Lehmann and Vecchia (2010) investigated a range of pharmaceutically relevant reactions for scale-up in a kilo-lab environment using a commercial batch microwave reactor. The scale-up issues were discussed, taking into account the specific limitations of microwave heating in large-scale experiments. Examples of scale-up from 15 mL to 1 L were presented. They demonstrated that the synthesis of compounds on >100 g scale is feasible in one batch. This new technology reduced reaction times significantly and the productivity of scale-up laboratory has been enhanced. It was observed that production rates of several hundred grams per day were achieved using microwave technology.

Girardin et al. (2013) reported the first kilogram-scale synthesis of MK-6096, which is an orexin receptor antagonist in clinical trials for the treatment of insomnia.

They introduced chirality on the α-methylpiperidine core in a biocatalytic transamination using a three-enzyme system with excellent enantioselectivity (>99% ee). Low diastereoselectivity of the lactam reduction was overcome by the development of a camphor sulfonic acid salt formation. A chemoselective O-alkylation with 5-fluoro-2-hydroxypyridine was optimized and developed. Overall, 1.2 kg of MK-6069 was prepared in nine steps and 13% overall yield.

MK-6096

Appell et al. (2013) reported the synthesis of (S)-N-Boc-bis(4-fluorophenyl) alanine, an intermediate in the synthesis of denagliptin from the synthesis of a 12 g principle sample to a >900 kg cGMP manufacturing campaign. The chiral center was established by the asymmetric hydrogenation of the sterically crowded precursor, ethyl 2-acetamido-3,3-bis(4-fluorophenyl)acrylate. The ability to isolate various intermediates in a physical form that would readily allow filtration, washing, and ultimately purification underpinned the successful manufacturing campaign.

Bergamelli et al. (2010) have investigated six pharmaceutically relevant reactions covering a range of physical parameters in a commercially available microwave flow reactor. The reaction conditions were scaled-up from tube or large batch scale microwave conditions, largely without change. Energy consumption measurements were also considered. This microwave flow reactor provided potentially successful manufacture for five out of six reactions investigated, where homogeneous reactions solutions could be obtained. Production rates between 0.5 and 3.0 mol^{-1}/h (1−6 L^{-1}/h) have been achieved. Here, 24−144 L of the product have been obtained that amounts to 12-22 mol/day.

17.3.2　CHEMICAL AND FINE CHEMICAL INDUSTRIES

Chemical and fine chemical industries are also benefited by the microwave-assisted organic synthesis and a number of organic compounds have been prepared under microwave exposure in limited time. It has proved to be a cleaner, time saving process and that too with good to excellent yields.

Benoit et al. (2008) studied the large-scale preparation of 2-methyloxazole-4-carboxaldehyde. They have described a method for the preparation of 10 Kg batches of 2-methyloxazole-4-carboxaldehyde. Here, reduction of the corresponding N-methoxy-N-methyl amide to the corresponding aldehyde using lithium aluminum hydride has been reported followed by workup and isolation by crystallization.

Reichart et al. (2013) carried out the preparation of n-alkyl chlorides in high yields and selectivity by direct uncatalyzed chlorodehydroxylation of the corresponding n-alcohols with 30% aqueous hydrochloric acid applying continuous flow processing in a high-temperature/high-pressure regime. Optimum conditions for the preparation of n-butyl and n-hexyl chloride involve the use of a glass microreactor chip, a reaction temperature of 160–180 °C (20 bar backpressure) and a residence time of 15 min.

Stadler et al. (2003) reported the direct scalability of microwave-assisted organic synthesis in a prototype laboratory-scale multimode microwave batch reactor. They have successfully scaled up several different organic reactions including the transformations involving multicomponent chemistry (Biginelli dihydropyrimidine and Kindler thioamide synthesis), transition metal-catalyzed carbon-carbon cross-coupling protocols (Heck and Negishi reactions), solid-phase organic synthesis, and Diels-Alder cycloaddition reactions using gaseous reagents in prepressurized reaction vessels typically from 1 mmol to 100-mmol scale. In all these cases, it was possible to achieve similar isolated product yields on going from a small scale (~ 5 mL processing volume) to a larger scale (maximum 500 mL volume) without changing the previously optimized reaction conditions. These are some good examples of direct scalability.

A series of synthetic transformations were successfully and safely scaled up to multigram quantities using focused microwave irradiation with a continuous flow reaction cell by Wilson et al. (2004). It was developed in-house and can be easily adapted to commercially available instrumentation. They have studied aromatic nucleophilic substitution (S_NAr), esterification, and the Suzuki cross-coupling reaction. It was observed that the product yields were equivalent to or greater than the yields under conventional thermal heating conditions.

Arvela et al. (2005) have scaled-up representative Suzuki and Heck couplings in water using ultra-low catalyst concentrations in an automated batch stop-flow microwave apparatus. They reported that this scale-up methodology showed proof of concept and it was easy, fast and low cost to run.

Loones et al. (2005) have reported batch wise scale-up of Buchwald-Hartwig aminations under microwave irradiation. Multi-mode (where several vessels were irradiated in parallel per batch) as well as single-mode (where one vessel was irradiated per batch) platforms were successfully used for this purpose with trifluoromethylbenzene (benzotrifluoride, BTF) as amination solvent. The obtained yields indicated a direct scalability in BTF for all these aminations. They have used a most convenient system, which allows an automatic continuous batch wise production without the necessity to load and unload the reaction vessels manually.

Microwaves have not only been used successfully in pharmaceutical and fine chemicals industries but also its use has also been explored in various other industries.

17.3.3 POLYMER INDUSTRIES

The microwave-assisted vulcanization of rubber compounds is the most important application of microwave heating to polymeric materials in terms of number of installed plants (Krieger, 1992). The processing capacity is about 500 kg/h. Control of the process and especially the extent of installed microwave output heat is typically 1 kW per 30 kg (approximately) of the product per hour. Continuous microwave belt furnaces are also available, where microwave generators are arranged in a spiral around the longitudinal axis of the cylindrical chamber to achieve a more uniform energy distribution.

A new industrial microwave system for curing of carbon fiber reinforced plastics has also been developed. This system is especially optimized for processes like injection molding or curing of that matrix and a modular system technology in connection with autoclave fabrication processes (Feher et al., 2006).

A microwave work station equipped with four single-mode microwave reactors, capping and decapping stations, robotic arm, transport and rack storage system, pipetting robot with DMSO, and acetone stations, drying and gassing stations has been designed. The microwave instruments are directly driven by a software via an Ethernet connection, while the scheduler maximizes the throughput of the system by allowing parallel multitasking (Chamoin, 2006).

Paulus et al. (2007) reported for the first time the use of different continuous-flow microwave reactors for polymerizations, using the cationic ring-opening polymerizations of 2-ethyl-2-oxazoline as a model system. In addition, they correlated the broader molecular-weight distributions with residence time distributions of the continuous-flow reactors using methyl orange as the flow marker.

17.3.4 CERAMIC INDUSTRIES

Microwave processing of ceramic materials has reached a high degree of maturity. The removal of solvent or moisture is a critical step in the generation of ceramic products in the ceramic production industry. The use of microwaves in this industry was limited to the effective removal of solvents from solid samples. It is estimated that for materials with water content below 5%, microwave drying is more energy efficient than conventional drying methods.

Ceramics are widely used in electrical components, sanitary-ware industries, and in many other industries. Microwaves have also found application in the sintering process. Sintering is the process of welding together the powdered particles of a substance or mixture by heating it to a temperature below the melting point of the components. The particles stick together and form a sinter. Initial studies in microwave-aided sintering were carried out by using a 400 W microwave-tuned waveguide applicator, to affect the sintering of alumina and silica rods at >1700 °C. Since then, a wide range of materials has been processed with microwaves. A frequency range of 28 GHz was used to facilitate the generation of homogeneous profiles of the materials sintered. However, microwave sintering has not become an economically viable replacement for conventional methods. In the near future, improved products and process simplification may lead to an economically viable microwave-sintering operation.

17.3.5   NANOPARTICLES

Many researchers have identified the difference in the presence of hot spots (which locally enhance or promote some selected reactions or transformations). The narrow temperature distribution obtained by simulation can justify the formation of nanoparticles (having a narrower particle size distribution) with respect to conventionally heated synthetic routes in case of nucleation and growth of nanoparticles (microwave hydrothermal synthesis). The large-scale production of nanoparticles requires the development of microwave reactors, which can reflect the laboratory temperature profile homogeneity. It will provide a new dedicated continuous-flow reactor, made of two twin prismatic applicators for a microwave-assisted process in aqueous solution. The reactor can produce upto 1000 L/day of nanoparticles colloidal suspension at ambient pressure and relatively low temperature and hence, it can be considered a green chemistry approach.

17.3.6   FOOD PROCESSING

The food related industry not only uses microwaves for processing but also developed products and product properties especially for microwave heating as very large number of microwave ovens are used in households. This way of product enhancement is called product engineering or formulation.

17.3.7   BAKING AND COOKING

Microwave heating inactivates enzyme fast enough to prevent the starch from extensive breakdown, and develops sufficient CO_2 and steam to produce a highly porous material as compared to conventional baking (Decareau, 1986). One difficulty to be overcome was a microwavable baking pan, sufficiently heat resistant and not too expensive for commercial use.

The main use of microwaves in the baking industry now a days is the microwave finishing, when the low heat conductivity leads to considerable higher baking times in the conventional process. A different process that can also be accelerated by application of microwave heating is (pre)cooking. It has been established technique for (pre)cooking of poultry, meat patties and bacon. It has been observed that microwave processing of chicken, beef, bacon, trout, and peanut oil does not change the fatty acid composition of these products, nor it produces transisomers (Helmar and Marc, 2007). Microwave is the main energy source, to render the fat and coagulate the proteins by an increased temperature. In the same time, the surface water is also removed by a convective airflow. Another advantage of this technique is the valuable by product namely rendered fat of high quality, which is used as food flavoring (Schiffmann, 1986).

17.3.8   THAWING AND TEMPERING

Thawing and tempering have received much less attention in the literature than most other food processing operations. Thawing is usually regarded as complete, when all the material has reached 0°C and no free ice is present. This is the minimum temperature at which the meat can be boned or other products cut or separated by hand. Lower temperatures (e.g., –5 to –2°C) are acceptable for product that is destined for mechanical chopping, but such material is `tempered' rather than thawed. These should be no confusion in these two processes because tempering only constitutes the initial phase of a complete thawing process. Thawing is often considered as simply the reversal of the freezing process.

Appearance, bacteriological condition and weight loss are important, if the material is to be sold is in the thawed condition but are less so, if it is for processing. The main detrimental effect of freezing and thawing meat is the large increase in the amount of proteinaceous fluid (drip) released on final cutting; yet the influence of thawing rate on drip production is not very clear.

James and James (2002) reported that there was no significant effect of thawing rate on the volume of drip in beef or pork. Several authors concluded that fast thawing rates would produce increased drip, while others showed it to be opposite. The results are conflicting and provide no useful design data for optimizing a thawing system. With fish, fruit and vegetables, ice formation during freezing breaks up cell structure and fluids are reduced during thawing. In microwave tempering processes, the heating uniformity and the control of the end temperature are very important, since a localized melting would be coupled to a thermal runaway effect.

17.3.8.1   DRYING

A typical drying curve of a foodstuff can be subdivided into phases. The first period is one of constant drying rate per unit of surface area. During this period, the surface is kept wet by the constant capillary-driven flow of water from within the particle. The factors that determine and limit the rate of drying in the so-called constant rate period. It describes the state of the air, temperature and relative humidity as well as air velocity.

In drying, the main cause for the application of microwaves is the acceleration of the processes, which are limited by low thermal conductivities without using microwaves, especially in products of low moisture contents. Correspondingly, sensorial and nutritional damage caused by long drying times or high surface temperatures can be prevented. The possible avoidance of case hardening, due to more homogeneous drying without large moisture gradients is another added advantage. Two cases of microwave drying are possible (i) Drying at atmospheric pressure; and (ii) that with applied vacuum conditions.

Combined microwave air dryers are more widespread in the food industry, and can be classified into a serial or a parallel combination of the both methods. Applied examples for a serial hot air and microwave dehydration are pasta drying and the production of dried onions (Metaxas et al., 1996) whereas only intermittently successful in the 1960s and 1970s was the finish drying of potato chips. The combination of microwave and vacuum drying also has a certain potential. Microwave-assisted freeze drying is well studied, but no commercial industrial application can be found, due to high costs and a small market for freeze dried food products (Knutson et al., 1987). Microwave vacuum drying with pressures above the triple point of water has more commercial potential.

Microwave energy overcomes the problem of very high heat transfer and conduction resistances, leading to higher drying rates. These high drying rates correspond also to lower shrinkage and the retention of water. In parsley, for example, most of essential oils are present as a separate phase with high boiling temperature. In fast drying conditions (high microwave energy input) only the small amount of volatile essential oils (that is dissolved) is lost, whereas there is not enough time to resolve the remaining oil in the separated phase (Erle, 2000).

In contrast, the retention of water-soluble aromas, as in apples, is not as advantageous, since the microwave energy generates many vapor bubbles, so that the volatile aromas have a large surface to evaporate. Nevertheless, the low pressures limit the product temperatures to lower values, as long as a certain amount of free water is present and this helps to retain temperature sensitive substances like vitamins, colors, etc. So in some cases, the high quality of the products could make this relative expensive process, economical also.

17.3.8.2 QUALITY

In general, the quality of microwave dried food products is somewhere between air-dried and freeze-dried products. The reduction of drying times can be quite beneficial for the color and the aroma. Venkatesh and Raghavan (2004) dried rosemary in a household microwave oven with good aroma retention while Krokida and Maroulls (1999) measured color and porosity of microwave-dried apples, bananas, and carrots. Khraisheh et al. (2004) compared air-dried and microwave dried potatoes and found a reduction of shrinkage and improved rehydration for the latter.

Yongsawatdigul and Gunasekaran (1996) showed that color and texture of microwave-vacuum-dried cranberries were better than those of air-dried samples. If we look specifically at the retention of aroma, it becomes necessary to distinguish between two basic cases. In most foods the aroma molecules are present in very small amounts, so that they are likely to be dissolved in the water phase. In this situation, the volatility of the aroma molecule in water is essential.

Schiffmann (2001) has listed a number of formerly successful applications that have been discontinued. Among these are the finish drying of potato chips, pasta drying, snack drying, biscuits and crackers. It is apparently not always the microwave process itself but rather changes in the circumstances of production that makes competing technologies more successful.

17.3.8.3 PASTEURIZATION AND STERILIZATION

Studies of microwave-assisted pasteurization and sterilization have been motivated by the fast and effective microwave heating of many foods containing water or salts. Rosenberg and Bog (1987) reviewed it in detail.

17.3.8.4 BLANCHING

Blanching is an important step in the industrial processing of fruits and vegetables. The most common method is a thermal process that can be performed by immersing vegetables in hot water (88–99°C,), hot and boiling solutions containing acids and/or salts, steam, or microwaves. Blanching is carried out before freezing, frying, drying and canning. The main purpose of this process is to inactivate the enzyme systems that may cause color, flavor and textural changes, such as peroxidase, polyphenoloxidase, lipoxygenase and pectin enzymes. The efficiency of the blanching process is usually based on the inactivation of one of the heat resistant enzymes: peroxidase or polyphenoloxidase.

Blanching has additional benefits, such as the cleansing of the product, the decreasing initial microbial load, exhausting gas from the plant tissue, and the preheating before processing. A moderate heating process such as blanching may also release carotenoids and make them more extractable and bioavailable (Arroqui et al., 2002).

The use of microwaves for food processing has increased through the last decades. Some of the advantages compared with conventional heating methods include speed of operation, energy savings, precise process controls and faster start-up and shutdown times (Kidmose and Martens, 1999). Microwave blanching of fruits and vegetables is still limited. Some of the advantages compared with conventional heating methods include speed of operation and no requirement of additional water. Hence, there is a lower leaching of vitamins and other soluble nutrients, and the generation of wastewater is eliminated or greatly reduced.

Microwave heating involves conversion of electromagnetic energy into heat by selective absorption and dissipation. When heating rapidly, the quality of fruits and vegetables such as flavor, texture, color and vitamin content are better kept (Dorantes-Alvarez et al., 2000). However, rapid heating can also lead to problems of nonuniform heating, when excessively high-energy transfer rates are used (Ohlsson, 2000).

The food processed by this novel technology is safe for consumption, because the microwave energy is changed to heat as soon as it is absorbed by the food, it cannot make the food radioactive or contaminated (Occupational Safety and Health Administration, OSHA). When the microwave energy is turned off and the food is removed from the oven, there is no residual radiation in the food.

17.4 MISCELLANEOUS

Porch et al. (2012) discussed the potential of using microwave techniques in the refinement of heavy fraction of petroleums such as bunker oil. Measurements of the dielectric properties of heavy oils at 2.45 GHz using a highly sensitive resonant cavity method, and also over a broader frequency range (100 MHz to 8 GHz) using a coaxial probe technique has also been reported. It was found that the dielectric loss is very small even in these heavy oils, but still it may be sufficiently large to provide efficient conversion of microwave energy into heat on untreated samples, and could be massively enhanced by means of a microwave-absorbing additive (e.g., carbon black).

Apart from the applications of microwaves in drying, baking, cooking, thawing, pasteurization, sterilization, blanching, etc., it has now being used in many industries like pharmaceuticals, fine chemicals, polymers, paper & wood, food, etc. But it has its own limitations in scaling up of the process to some extent. There is a great hope that the microwave process will find its own place in industries because of its greenness and the time is not far off to replace many conventional technologies of synthesis by this emerging technology.

KEYWORDS

- **Fine Chemicals**
- **Food Processing**
- **Nanoparticles**
- **Pharmaceuticals**
- **Polymers**
- **Scale-up**

REFERENCES

Amore, K. M., & Leadbeater, N. E. (2007). Macromolecular Rapid Communication, 28, 473–477.

Anon (1991). Manufacturing Chemist, 62, 36–37.

Appell, R. B., Boulton L. T., Daugs, E. D., Hansen, M., Hanson, C. H., Heinrich, J., Kronig, C., Lloyd, R. C., Louks, D., Nitz, M., Praquin, C., Ramsden, J. A., Samuel, H., Smit, M., & Willets, M. (2013). Organic Process Research and Development, 17, 69–76.

Arroqui, C., Rumsey, T. R., Lopez, A., & Virseda, P. (2002). Journal of Food Engineering, 52, 25–30.

Arvela, R. K., Leadbeater N. E., & Collins, Jr. M. J. (2005) Tetrahedron, 61, 9349–9355.

Ashraf, A., Aboul, F., Mourad, E., Hassan, H., & Eman, A. (2007). Journal of Organic Chemistry, 3(11), doi: 10.1186/1860–5397–3–11.

Ashwin, J., Lynn, B., & Taskis, C. (1987). Pharmaceutical Journal, 238, 116–118.

Bagley, M. C., Fusillo V., Jenkins R. L., Lubinu M. C., Mason, C. (2010). Organic & Biomolecular Chemistry, 8, 2245–2251.

Banerjee, R., Liu, J., Beatty, W., Pelosof, L., Klemba, M., & Goldberg, D. E. (2002). Proceedings of the National Academy of Sciences of the United States of America, 99, 990–995.

Baxendale, I. R., & Ley, S. V. (2000). Bioorganic & Medicinal Chemistry Letters, 10, 1883–1986.

Benoit, G. E., Carey, J. S., Chapman, A. M., Chima, R., Hussain, N., Popkin, M. E., Roux, G. R., Tavassoli, B., Vaxelaire, C., Webb, M. R., & Whatrup, D. (2008). Organic Process Research and Development, 12, 88–95.

Bergamelli, F., Iannelli, M., Marafie, J. A., Moseley, J. D. (2010). Organic Process Research and Development, 14, 926–930.

Bergese, P., Colombo, I., Gervasoni, D., Depero, L. E. (2003). Material Science and Engineering: C, 23, 791–795.

Bierbaum, R., Nuchter, M., Ondruschka, B. (2005). Chemical Engineering & Technology, 28, 427–431.

Bowman, M. D., Schmink, J. R., McGowan, C. M., Kormos, C. M., & Leadbeater, N. E. (2008). *Organic Process Research and Development, 12*, 1078–1088.

Bremecker, K. (1983). Indian Journal of Pharmaceutical Sciences, 45(1), 78–81.

Brittany, L. H. (2004). Aldrichimica Acta, 37, 66–76.

Chamoin, S. (2006). High-Throughput Microwave Synthesis at Novartis, in, Advances in Microwave-assisted Organic Synthesis. MAOS Conference and Exhibition. Budapest.

Decareau, R. V. (1986). Food Technology, 99–105

Dorantes-Alvarez, L., Barbosa-Caanovas, G., & Gutiearrez-Loa Pez, G. (2000). Blanching of fruits and vegetables using microwaves, in Barbosa-Caanovas G and Gould G, Innovations of Food Processing, Lancaster: Technomic, 149–162.

Erle, U. (2000). Untersuchungen zur Mikrowellen-Vakuumtrocknung von Lebensmitteln, Ph.D. Thesis, Universität Karlsruhe.

Gardner, M. J., Hall, N., Fung, E., White, O., Berriman, M., Hyman, R. W., Carlton, J. M., Pain, A., Nelson, K. E., & Bowman, S. (2002). Nature, 419, 498–511.

Girardin, M., Ouellet, S. G., Gauvreau, D., Moore, J. C., Hughes, G., Devine, P. N., O'Shea, P. D., & Campeau, L. C. (2013). Organic Process Research and Development, 17, 61–68.

Groning, R., & Janski, U. (1985). Pharmazeutische-Zeitung, 130, 2621–2625.

Groning, R., & Janski, U. (1985). Pharmazeutische-Zeitung, 130, 2621–2625.

Gupta, V. D., & Stewart, K. R. (1986). Journal of Clinical and Hospital Pharmacy, 11, 47–54.

Helmar, S., & Marc, R. (2007). Cambridge: Woodhead, 20–312.

Holmes, C. J., Ausman, R. K., Kundsin, R. B., & Walter, C. W. (1982). American Journal of. Hospital Pharmacy, 39, 104–108.

Honda, K., Ebara, V., Iijima, K., Honda, K., Shimizu, K., & Miyake, Y. (1998). Europeon Journal of Parenteral and Pharmaceutical Sciences, 3, 39–47.

James, S. J., & James, C. (2002). Cambridge: Woodhead Publishing, 159–190.

Joshi, H. N., Kral, M. A., & Topp, E. M. (1989). International Journal of Pharmaceutics, 51, 19–25.

Khraisheh, M. A. M., McMinn, W. A. M., & Magee, T. R. A. (2004). Food Research International, 37, 497–503.

Kidmose, U., & Martens, H. J. (1999). Journal of the Science of Food and Agriculture, 79, 1747–1753.

Kidwai, M., Bhushan, K., Sapra, P., Saxena, R., & Gupta, R. (2000). Bioorganic & Medicinal Chemistry, 8, 69–72.

Kidwai, M., Kumar, K., & Kumar, P. (1998). Journal of Indian Chemical Society, 75, 102–103.

Kidwai, M., Misra, P., Bhusan, K. R, Saxsena, R. K., & Singh, M. (2000). Monatshefte für Chemie, 131, 937–943.

Knutson, K. M., Marth, E. H., & Wagner, M. K. (1987). Lebensmittel-Wissenschaft und -Technologie, 20, 101–110.

Krieger, B. (1992). Polymer Material Science & Engineering, 66, 339–340.

Krokida, M. K., & Maroulls, Z. B. (1999). Drying Technology, 17, 449–466.

Lehmann, H. & Vecchia, L. L. (2010). Organic Process Research and Development, 14, 650–656.

Loones, K. T. J., Maes, B. U. W., Rombouts, G., Hostyn, S., Diels, G. (2005). Tetrahedron, 61, 10338–10348.

Magano, J., & Dunetz, J. R. (2012). Organic Process Research and Development, 16, 1156–1184.

Mandal, T. K. (1995). Drug Development and Industrial Pharmacy, 21, 1683–1688.

Mandal, V., Mohan, Y., Hemalatha, S. (2007). Pharmacogonosy Reviews, 1, 7–18.

Metaxas, A. C. (1996). Foundations of Electroheat- A Unified Approach, Chichester: John Wiley.

Moll, F., & Maue, R. (1988). Deutsche Apotheker Zeitung, 128, 1871–1873.

Moneghini, M., Bellich, B., Baxa, P., Princivalle, F. (2008). International Journal of Pharmaceutics, 361, 125–130.

Morschhäuser, R., Krull, M., Kayzer, C., Boberski, C., Bierbaum, R., Püschner, P. A., Glasnov, T. N., & Kappe C. O. (2012). Green Processing and Synthesis, 281–290.

Moseley, J. D., & Lawton, S. J. (2007). Chemistry Today, 25, 16–19.

Moseley, J. D., Lenden, P., Lockwood, M., Ruda, K., Sherlock, J. P., Thomson, A. D., & Gilday, J. P. (2008). Organic Process Research and Development, 12, 30.

Ohlsson, T. (2000). Minimal processing of foods with thermal methods,' in Barbosa- Caanovas G and Gould G (Eds.), Innovations of Food Processing, Publishing, Lancaster: Technomic.

Paulus, R. M., Erdmenger, T., Becer, C. R., Hoogenboom, R., & Schubert, U. S. (2007). Macromolecular Rapid Communication, 28(4), 484–491.

Pearlswing, D. M., Robin, P., & Lucisano, L. (1994). Pharmaceutical Technology, 28–36.

Pearlswing, D. M., Robin, P., & Lucisano, L. (1994). Pharmaceutical Technology, 44–60.

Poska, R. (1991). Pharmaceutical Engineering, 11, 9–13.

Reichart, B., Tekautz, G., & Kappe, C. O. (2013). *Organic Process Research and Development*, 17(1), 152–157.

Rosenberg, U., & Bog, W. (1987). Food Technology, 92–121.

Sasaki, K., Honda, W., & Miyake, Y. (1998). PDA Journal of Pharmaceutical Science and Technology, 52, 5–12.

Sasaki, K., Fukumura, M., & Miyake, Y. (1998). European Journal of Parenteral and Pharmaceutical Sciences, 3, 73–84.

Sasaki, K., Honda, W., Iijima, K., Ehara, T., Okuzawa, K., & Miyake, Y. (1996). PDA Journal of Pharmaceutical Science and Technology, 50, 172–179.

Sasaki, K., Honda, W., Ohsawa, S., Miyake, Y., & Kawashima, Y. (1998). Archives of Practical Pharmacy (Japan), 58, 125–135.

Schiffmann, R. F., (1986). Food Technology, 94–98.

Schiffmann, R. F. (2001). Microwave processes for the food industry, in Datta, A. K., & Anantheswaran, R. C. (Eds.), Handbook of Microwave Technology for Food Applications. New York: Marcel Dekker.

Stadler, A., Yousefi, B. H., Dallinger, D., Walla, P., Eycken, E. V., Kaval, N., & Kappe, C. O. (2003) *Organic Process Research and Development*, 7, 707–716.

Tabor, E., & Norton, R. (1985). American Journal of Hospital Pharmacy, 42, 1507–1508.

Tensmeyer, L. G, Wright, P. E., Fegenbush, D. O., & Snapp, S. W. (1981). Bulletin of the Parenter Drug Association, 35, 93–97.

Tidy, P. J., Sewell, G. J., & Jefferies, T. M. (1988). Pharmaceutical Journal, 241, R22–R23.

Vandalize, M. A., Romagnoli, M., & Monti, A. (2004). Journal of Controlled Release, 96, 67–84.

Venkatesh, M. S., & Raghavan, G. S. V. (2004). Biosystems Engineering, 88, 1–18.

Vigneron, J., Laurelli, F., Phaypradith, S., & Hoffman, M. (1992). Journal de pharmacie de Belgique, 47, 504–522.

Waldron, M. S. (1988). Pharmaceutical Engineering, 8, 9–13.

Wharton, Y., (2011). EPIC Symposium Series No. 157, 117–121.

White, J. G. (1994). Journal of Pharmacy research, 11, 728 –732.

Wilson, N. S., Christopher, R. S., & Gregory P. R. (2004). *Organic Process Research and Development*, 8, 535–538.

Wong, T. W. (2008). Current Drug Delivery, 5, 77–84.

Yongsawatdigul, J., & Gunasekaran, S. (1996). Journal of Food Processing and Preservation, 20, 145–156.

CHAPTER 18

FUTURE PROSPECTS

RAKSHIT AMETA

CONTENTS

In last three decades or so, there has been a growing interest to use microwave irradiation in organic synthesis. Microwave heating has proved to be a valuable tool for synthetic chemists not only because it can improve yields of the products and enhance the reaction rate but also as it is a safe and convenient method for heating reaction mixtures to elevated temperatures. Domestic microwave ovens were used earlier but now-a-days these are almost replaced by improvised scientific microwave apparatus with all controls like temperature, pressure, etc. Apart from being safer, this new technology has emerged out as a green chemical pathway as it fulfills some of the major criteria of green chemistry.

Microwave-assisted organic synthesis (MAOS) has become an important tool in this rapid paced, time sensitive field. It has been developed as a promising technology for a number of applications in organic synthesis where protocols with higher yields and purified products are highly desired. There is an increasing demand of different targets of some novel drugs and active pharmaceutical ingredients (APIs) by pharmaceutical industries. It has certain problems at present like large-scale production and that too in limited time. These all can be fulfilled by improving this technology by associating it with combinatorial and computational chemistry. It has its own limitations like limited scalability and some hazards, but these can be overcome by modifications of existing microwave instruments.

The chemists are normally blamed for ever increasing environmental pollution, but one should not forget that they are providing a lot of comfort to the society by synthesizing various useful products, like drugs, polymers, textiles, detergents, insecticides, fragrances, edible materials, etc. without which our life may be miserable. As a side effect, the environment is being polluted. In this context, microwave chemistry has its own importance over traditional chemistry. Conventional chemical routes require a variety of organic solvents depending on the solubility of starting materials (reactants). Some of these are toxic and in some cases carcinogenic also. On the other hand, many chemical reactions can be carried on solid support and in solvent-free conditions under microwave irradiation and thus, avoiding the harms caused by these solvents. In brief, one can say that microwave enhanced chemical reactions are relatively safer, faster, cleaner and more economical than traditional reactions carried out by gray chemical routes. It helps in developing cleaner and greener synthetic routes.

Addition reactions are very important from atom economic point of view as all the atoms of reactants are incorporated in the final product and nothing goes as a waste. Some work has been done on nucleophilic addition with a major stress on Michael addition reaction, but little information is available on the other counter part, the electrophilic addition. Only some reactions, involving electrophilic addition to fullerenes, have been reported and a lot more is to be investigated in this direction.

Substitution, elimination, protection, deprotection and many more reactions can not be left out because these are not green chemical and produce some or the other by products. The chemistry cannot be completed without these reactions and hence, studies related to these reactions involving some green catalyst and microwave radiations may be coupled with solid support or solvent-free conditions should also be carried out to get the desired products.

Condensation, coupling and such reactions are helpful in synthesizing relatively larger molecules by the combination of two or more than two smaller fragments, but with the formation of some by products. Such reactions can be carried out under microwave irradiation and these should also be investigated with more efforts towards one pot, so that the time can be saved.

There are a number of rearrangement reactions established so far. These rearrangements are beneficial from the angle of green chemistry as here also, every atom of the starting material is available in the product. Therefore, these reactions are considered to follow 100% atom economy. Some rearrangements have been studied in details like Beckmann, Claisen, Curtius, Wolff, Baker-Venkataraman, Pinacol-Pinacolone rearrangements, etc. but many such rearrangements such as Benzil-Benzilic, Overmann, Meyer-Schuster, etc. have not been properly investigated under microwave exposure in spite of advantages.

Most of the natural products of medicinal importance, synthetic pharmaceutical drugs and formulations, veterinary products, agrochemicals, etc. contain heterocyclic moiety as well as other ring systems with smaller, medium and large sizes. Microwave-assisted synthesis may prove beneficial in giving these compounds in an ecofriendly route and that too in limited time, which is a need of the day.

Nanochemistry is also spreading its arena because the nanoparticles have larger surface area as compared to macro and microparticles. It may be a point of debate at present, about the utility or the problems created by these particles, which will decide their future, but one thing is certain that these nanoparticles of metal, nonmetals, their compounds like oxides, sulfides, etc., can be easily synthesized by microwave technique in different shapes like nanoflower, nanorods, nanotubes, etc. At present, there are some tedious methods available for preparing these nanoparticles, but this can be simplified by using microwaves.

Carbon nanotubes (single walled or multi walled) have a number of applications like potential use in solar panels, thermal management of electronic circuits, lithium ion batteries, desalination, oscillators, etc. A variety of functional groups can be substituted on these CNTs and such functionalized carbon nanotubes are likely to have some fascinating properties. Microwave irradiation can help in achieving this goal.

In this rapidly developing world, most of the materials like wood, stone, metal, etc. are being substituted by some or the other kind of polymer. Therefore, there is no second opinion that the present period has been termed as 'Polymer Era.' But this replacement is quite costly from the angle of the disposal of this polymeric material. Many countries are suffering from this problem and searching for a proper solution. Microwave can assist in finding a solution to this burning problem by depolymerization of a variety of polymers.

There has been a dramatic increase in the use of microwaves as an energy source to promote various synthetic transformations. The reaction time has been reduced from hours or days to minutes or even seconds in some cases apart from increased productivity and ultimately enhanced efficiency.

Microwave-assisted organic synthesis has proved its utility as a valuable tool to ease some of the bottlenecks in drug discovery process. In last few years, there has been a dramatic increase in the use of microwave heating within the pharmaceutical industry to facilitate the chemical synthesis of new chemical entities, because of the fact that reaction rates could be enhanced, in some cases even to the order of thousand folds. The modified microwave instruments coupled with some controls have fuelled the introduction of this technique into pharmaceutical R&D laboratories of organic synthesis. Future prospects for this nascent technology should be explored so that this technique moves from the laboratory bench tops to scale up in the industries. There is a pressing demand on the pharmaceutical industry to increase their output. In this context, MAOS is quite appealing as it can increase the rate with which new chemical entities can be synthesized in limited time. This synthesis should also be equally supported by purification, isolation and characterization processes and that too timely.

The dynamic range of temperatures afforded by microwave instrumentation available today enables synthetic chemists to carry out synthetic transformations not achievable through conventional heating. Because of the strong collaborative efforts between academia and R&D laboratories of industries and developers of microwave instruments, the regular improvements in instrumentation have resulted in its refinement. These instruments should provide solutions to the problems of reproducibility, controllability and safety, so commonly experienced with the domestic microwave ovens. This technology is still underused in the research laboratories in general and particularly in industries.

MAOS has also found applications in areas like peptide synthesis, proteomics, etc. This is all possible by using microwave irradiation. Low temperature reactions (gentler reaction conditions) via microwave energy have only recently been recognized particularly in biochemical applications.

Supercritical form of water is less polar in nature and therefore, it is more effective in dissolving different organic substrates. In addition, the use of water is preferred in terms of green chemical approach, as water is environmentally more benign solvent than any other traditional organic solvents.

Safety is very important for any chemical based industry, especially when the sample is irradiated with some electromagnetic radiations as in this case of microwaves. Microwaves travel at the speed of light and therefore, they can be turned off immediately upon reaching the desired temperature of the reaction. The acceleration of reaction rates, rapid heating and reactions in sealed vessels needs additional safety measures. The industries using MAOS should take full care like explosion proof reactors, shut down mechanism, particularly in case of over heating or over pressurization and also some mechanism for proper ventilations for closed vessel reactions. A homogeneous microwave field, magnetic stirring, pressure sensors (for closed vessel reactions helps in avoiding excessive pressure build-up), temperature sensors (for temperature control), the rate and power of microwave irradiation are some of the necessary regulatory measures. Multi-mode instruments with these and some other safety features are now well suited for larger reaction volumes and parallel synthesis, too.

Two approaches, which have received the attention for process development are (i) stop flow or batch processing and (ii) continuous flow. In stop flow or batch processing, the process has to be repeated a number of times to obtain the desired amount of material while continuous flow systems are successful with varying degrees of success, because of the difficulty in handling heterogeneous reaction mixtures and viscous liquids. A continued focus in this area will find some practical solutions in future.

Single-mode microwave reactors have been very successful in the past few years in the field of process development and optimization. The use of microwave technology for the development of completely newer routes for organic synthesis and for large-scale microwave production of chemical substances can make it a fully accepted industrial technology in future. It is a welcome addition to the existing technologies. There is a need to develop techniques that can ultimately provide products on a multikilogram scale. Large reactors based on continuous flow and stop-flow techniques are required to enable production of multikilogram desired compounds.

Microwave dielectric heating has come to the forefront of chemical research, because the use of microwave instrumentation to heat reactions is a great and important shift for nearly all chemists involved in organic synthesis. Therefore, the use of microwaves as an energy source requires a mindset change of these chemists.

This timescale of these microwave-assisted reactions is in minutes and it enables a facile and rapid scoping of reaction conditions, for example, time, temperature, reagents and solvents. This rapid optimization can be used to rapidly identify routes for the synthesis of novel chemical entities. Microwave-assisted organic synthesis is no longer a curiosity now, it is a rapidly growing technology, but it has not been used with full potential.

MAOS has not been used to its full advantages in reactions of synthetic importance. This limitation has been overcome by the development of some newer and modified microwave reactors to some extent but still, there are many more miles to go before it is properly accepted by chemical based industries. It will solely depend upon the approach of the chemists using this technology, whether its potential will be duly recognized or not?

Apart from some existing problems of using microwaves on large scale, one can hope that microwave-assisted organic synthesis has a bright future in years to come and it will provide many more reactions, the green chemical pathways.

KEYWORDS

- Active Pharmaceutical Ingredients
- Microwave-assisted Organic Synthesis
- Microwaves
- Scientific Microwave Apparatus

INDEX